T0298945

Future Internet Services and Service Architectures

RIVER PUBLISHERS SERIES IN COMMUNICATIONS

Volume 15

This series focuses on communications science and technology. This includes the theory and use of systems involving all terminals, computers, and information processors; wired and wireless networks; and network layouts, architectures, and implementations.

Furthermore, developments toward new market demands in systems, products, and technologies such as personal communications services, multimedia systems, enterprise networks, and optical communications systems.

- Wireless Communications
- Networks
- Security
- Antennas & Propagation
- Microwaves
- Software Defined Radio

For a list of other books in this series, see final page.

Future Internet Services and Service Architectures

Editors

Anand R. Prasad

NEC Corporation, Japan

John F. Buford

Avaya Labs Research, USA

Vijay K. Gurbani

Bell Laboratories, Alcatel-Lucent, USA

Aalborg

ISBN 978-87-92329-59-2 (hardback)

Published, sold and distributed by:
River Publishers
P.O. Box 1657
Algade 42
9000 Aalborg
Denmark

Tel.: +45369953197
www.riverpublishers.com

For my parents Jyoti and Ramjee Prasad – A.R.P.

For my wife Gina and our daughter Jacqueline – J.F.B.

For my family – V.K.G.

Table of Contents

Part 4: Event Distribution

Part 5: VANETs

Preface

The Future Internet is a term that has been widely adopted to represent a research vision and agenda of how the Internet architecture might evolve to address limitations of the existing design. Currently there is no agreed upon model of the Future Internet, and proposals vary from incremental changes to major re-design. A number of architectural issues in the current Internet are targeted, possibly through changes to existing protocols, by addition of new layers, and use of new networking technologies. Improvements in various areas are sought, such as addressing, performance, scalability, security, and quality of service. In addition, support for emerging network technologies and new applications is needed.

Due to the wide range of topics related to Future Internet, this book focuses on Future Internet Services. The purpose is to present state-of-the-art results in services and service architectures in the Future Internet, and to identify challenges including business models, technology issues, service management, and security, and to describe important trends and directions.

This material is intended to be accessible to a wide technical audience, and is written for researchers, professionals, and computer science and engineering students at the advanced undergraduate level and higher who are familiar with networking and network protocol concepts and basic ideas about algorithms.

Future Internet Services and Service Architectures is intended to provide readers with a comprehensive reference for the most current developments in the field. It offers a broad coverage of important topics with 20 chapters covering both technology and applications written by international experts, which are organized into the following five parts:

- Future Internet Services – This part provides four chapters which present recent proposals for a new architecture for the Internet including end-to-end networking covering wireless networks and mobile devices. Service delivery in the future Internet is a key focus of these chapters.

- Peer-to-Peer Services – The P2P network overlay can be viewed as an virtual network and this paradigm is being considered as a new layer in several Future Internet designs. The peer-to-peer architecture provides a highly scalable platform for delivering both application and network services. In this part, five chapters explore the P2P architecture and its use for streaming services, communication services, and service discovery.
- Virtualization – Virtualization is an important paradigm that is seeing increased interest in recent years for its potential for resource management, supporting heterogeneity, and isolation. The five chapters in this part describe computational mobility, the convergence of virtualization and peer-to-peer, and virtualization at the endpoint, in the cloud, and in the network.
- Event-Distribution – Publish/subscribe mechanisms have long been used by distributed computing services, in particular for applications which require time-sensitive delivery of notifications. The two chapters in this part present recent developments in load balancing and in enabling publish/subscribe services in sensor networks.
- VANETs – Among the emerging networking technologies that are expected to be important to the evolution of the Internet are VANETs. Vehicular Ad Hoc Networks (VANETs) are a type of broadband mobile network technology which are designed for vehicle-to-vehicle and vehicle-to-infrastructure connectivity for moving vehicles. The four chapters in this part provide an introduction to VANETs, routing, services and system architecture.

Future Internet Services and Service Architectures is complemented by a separate volume, *Advances in Next Generation Services and Service Architectures*, which covers emerging services and service architectures, IPTV, context awareness, and security.

Acknowledgements

Thanks are due to the staff at River Publishers.

Finally, we thank our families for their support and understanding while we worked on this book.

Anand R. Prasad
John F. Buford
Vijay K. Gurbani
December 2010

PART 1

FUTURE INTERNET SERVICES

1

Next Generation Services in Future Internet

Manuel Palacín[1], Francesc I. Rillo[2], Antoni Oller[3], Juan López[4], Daniel Rodríguez[5], David Rivas[5] and Jesus Alcober[3]

[1]*Research Group on Networking Technology and Strategies, Universitat Pompeu Fabra, Barcelona 08108, Spain; e-mail: manuelpalacin@gmail.com*
[2]*Department of Electronic Engineering, Universitat Politècnica de Catalunya, Barcelona 08034, Spain*
[3]*Department of Telematic Engineering, Universitat Politècnica de Catalunya, Barcelona 08034, Spain*
[4]*Department of Computer Architecture, Universitat Politècnica de Catalunya, Barcelona 08034, Spain*
[5]*I2CAT Foundation, Barcelona 08034, Spain*

Abstract

This chapter shows a historical view of the Internet and its evolution by studying three aspects: Computation, Information and Media. It also describes the paradigms on the current Internet within the structure of advanced and multimedia services. The deficiencies and limitations of the current Internet are presented, concluding that its limit is near.

Subsequently, two different strategies are proposed. First, an evolving strategy exists that consists of applying patches when new necessities appear and a new revolutionary or clean-slate strategy that proposes redesigning the Internet from the beginning by studying all the emerging requirements.

Keywords: Internet, future Internet, SOA, Web 2.0, Internet of things.

Anand R. Prasad et al. (Eds.), Future Internet Services and Service Architectures, 3–22.

1.1 Introduction

The Internet has been in constant evolution and during all this time it has been consolidated in different stages: genesis, current Internet and future Internet.

The origin of the Internet goes back to the sixties as a research project in communication networks inside the military field. This research produced an experimental network of four nodes (ARPANET) that transmitted basic and reliable information in network failure conditions.

As the equipment and connections evolved, so did the services offered by ARPANET. Remarkable is the introduction of electronic mail (1972), the TCP/IP protocol (1974) and services like FTP, news and World Wide Web (WWW), which was the service that finally popularized the Internet to the masses.

In addition, the network offers a new computation paradigm that allows applications to be distributed on a client-server scheme in comparison to the common ones until they became centralized applications. The WWW allowed working with basic but structured contents, the well-known HTML language.

The emergence of Web 2.0 started the consolidation stage of the Internet. In Web 1.0, applications were limited to a server that offered services and contents and a client with a passive role of service consumer. In Web 2.0 clients acted as content generators using dynamic applications becoming active participants of the Web. With the boom of applications related to the development of Web 2.0, the implementation of applications based on services has grown as media services advanced. On the one hand, they are based on Service-Oriented Architecture (SOA), i.e. applications are built from a cloud of cooperating services. Therefore interfaces and the possibility of reusing the services of third parties are offered. On the other hand, currently there are more continuous media services (e.g. audio, video) with applications such as conferencing, video streaming, gaming and high definition media (see Media Axis in Figure 1.1). This fact and the continuous media integration (CMI) [1] are also appreciable in the IP Multimedia Subsystem (IMS).

The Internet revolution from Web 2.0 modifies the current Internet architecture introducing different dimensions or axes: information, computation and media.

On the information axis, the large amount of information needs a new system of classification called Folksonomy, which is based on collaborative creation and management of tags to annotate and categorize content. The idea of adding semantic metadata to the Web leads to the Semantic Web, where the

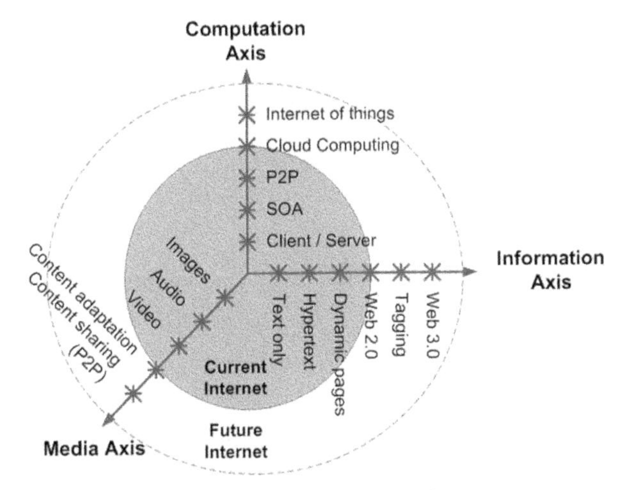

Figure 1.1 3D Internet axis.

use of descriptive languages of ontologies and the new concept of the future Web 3.0 appear (see Information Axis in Figure 1.1).

The existing content on the Internet has also evolved during the last years along the content axis. At the beginning only plain text was used as content, however the increasing network capacity allowed one to work with new multimedia content (audio and video) that is used in new applications (media streaming, video on demand, peer-to-peer content sharing). Finally, the addition of intelligence to the network and the context-awareness lead to offer the user personalized and adapted content everywhere.

In the last dimension, the Computation Axis in Figure 1.1, new computation paradigms like cloud computing that uses the infrastructure virtualization, or the Internet of Things, call into question the current Internet architecture.

Now it is believed that the current Internet architecture will soon reach its limit (routing capacity, QoS requirements and application service provisioning, and new features), and its future depends on how to deal with different aspects such as scalability, ubiquity, security, robustness, mobility and context-awareness.

Over the last years, a discussion about how to overcome upcoming Internet limitations has been initiated. From an academic point of view, one could incline for a revolutionary or clean-slate way of solving the problem by fully redesigning the overall Internet architecture. From a more practical point of view, one tends to incline for an evolution of the current Internet by patching its deficiencies, when novel services with constricting requirements appear. These two approaches of future Internet (evolutionary and revolutionary) will be analyzed in order to be validated in several scenarios with new services and computing paradigms.

1.2 Current Internet

In the case of applications deployed over the current Internet we see the tendency towards an evolution, from the first Web 1.0 to the future Web.

1.2.1 Evolution of Internet

Web 1.0 was the genesis of the state of the World Wide Web and any website design style, which started with the interconnection of the United States Department of Defense's main computers in 1962 (the Internet creation), and its subsequent development of a working Web at CERN at the end of 1990. The term was definitely used with the release of the WWW to the public in 1991. In Web 1.0 the application client-server paradigm was limited to a server that offers services and contents and a client with a passive role as service consumer (i.e. the client was a dummy machine and the server processed the whole computational load). It is commonly thought that one can only define Web 1.0 when used in relation to its successor Web 2.0.

Web 2.0 applications appeared offering new possibilities and pattern designs. They change the philosophy of application deployments and client consumption. Clients (as shown in Figure 1.2) act as content generators using dynamic applications becoming active participants of the Web. This new attitude combined with a set of new technologies over enterprise platforms results in a boom of new applications that promote the use of the Web.

The essential attributes of Web 2.0 are rich user experience, user participation, dynamic content, scalability, openness, freedom and collective intelligence by way of user participation [2].

One of the biggest steps in realizing Web 2.0 is the transition to semantic markup, or markup that accurately describes the content applied (XML). The extensibility of XML allows understanding a document created with XML; a third party can easily understand its structure and process it. Plenty of XML-based Web 2.0 applications have appeared (RSS, Atom, SOAP, XHTML and for most office-productivity tools).

The rich user experience in Web 2.0 are acquired through the Rich Internet Applications (RIA), which is a mix between the Web applications

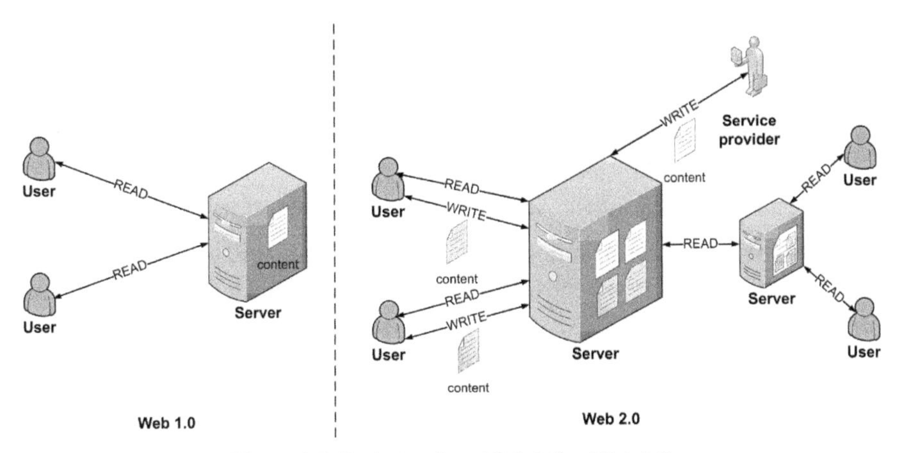

Figure 1.2 Evolution from Web 1.0 to Web 2.0.

and the classical desktop applications. The objective of RIA is introducing into the Web desktop features like dynamism, editors, menus, etc. There are technologies for RIA Web applications like FLEX that interact with a Flash screen, instead of a Web page, and interact with the server when necessary. In a similar way, HTML5 establishes some new elements and attributes that reflect the typical use of modern Web sites [3].

The user participation and collective intelligence in Web 2.0 bring together several revolutionary applications in terms of information dissemination. In this context applications like Wikis appeared in 1995, free Web applications where new entries could be added by any user of the Web and could be corrected or updated by anyone. Other applications like Blogs (1993) and Syndication are based on any user/application being a content generator (plain text, images or multimedia files) and anyone can subscribe to this feed acting as a follower. This revolution affected the news world by opening an easy and direct way of news dissemination. Moreover, Podcast joined these applications including updated multimedia content. Additionally, another form of blogging appeared in 2005, namely microblogging. The content in microblogs is much smaller than in his big brothers, and it can consist of a short sentence, an image or a video (e.g. Twitter). Microblogging has the potential to become a new informal communication medium, especially for collaborative work within organizations, used for marketing and in PR.

Last, but not least, and in a horizontal way, a new social movement called Social Networks has appeared, in which users are connected in groups and share contents about their lives and feelings (messages, pictures, videos, etc.).

Individuals on Social Networks are connected by one or more specific types of interdependency. The effect of these developments has been magnified when applying this context to Web 2.0; interconnections between users have been multiplied in a way that individuals become groups and societies operate incredibly easier and faster.

We can enumerate other Web 2.0 applications as follows:

- Mashups are applications that create contents by using or combining third-party contents (e.g. Amazon, eBay, Flickr, Yahoo, YouTube, Google maps, etc.).
- The Content Management System (CMS) is an application that permits the creation of a framework for content creation and management by participants.
- Tags are keywords or terms assigned to a piece of information by a user or an author.

In parallel, other collaborative applications are Peer-to-Peer (P2P) systems where files are shared by users and shared data bases are created. This kind of network breaks the client-server architecture. Nodes are not separated depending on their functionalities; they have some data and lack other data (peers); they are final user machines. P2P networks are scalable since as long as more peers join the networks, more processing capacity is needed, but, at the same time, more processing capacity is added, and the same inversely when a peers leaves the network. P2P Networks are basically classified by topology: Centralized and Not Centralized (structured, not structured).

1.2.2 Semantic Web

This large amount of information, previously described, made that new applications need a new system of classification called Folksonomy, based on collaborative creation and management of key words or tags to annotate and categorize content.

If the etymological formation of the term is taken into account, the definition of Folksonomy can be seen as "classification managed by people". Therefore, Folksonomies are linked to social environments where users collaborate in the description of a same piece of information.

In fact, there are two types of Folksonomies according to Vander Wal [4]:

- Wide Folksonomy: The creator of the content does not have any influence on the tags that are given to the content. In this case, the users are the ones who set tags to the content.

- Narrow Folksonomy: Only the creator of the content or a small group of people is allowed to set tags to the content.

Nevertheless, the lack of terminology control in Folksonomy has led to serious problems of Synonymy (different tags with the same meaning), Homonymy (same tag with different meanings) and Polysemy (same tag with different meanings related between them) [5].

Following the idea of adding semantic metadata to the World Wide Web it leads to the Semantic Web, where the use of descriptive semantic languages and the new concept of the future Web appear.

With this purpose, the World Wide Web Consortium (W3C) has supported the development of the "Web Ontology Language", also known as OWL, which consists of a family of knowledge representation languages for authoring ontologies [6]. In computer and information sciences, ontologies correspond to formal representations of the knowledge by a set of concepts within a domain. Ontologies provide a shared vocabulary which in turn can be used to describe the domain and the relationships within it.

OWL is characterized by the use of formal semantics and RDF/XML-based serializations for the Semantic Web [7]. In this case, RDF stands for "Resource Description Language" and is a W3C data model used as a method for conceptual description or modeling of information for web resources. In order to do that, RDF is based on making statements about resources. Therefore, "triplets", as they are known in RDF terminology, are used. Triplets correspond to subject-predicate-object expressions.

The Semantic Web will provide a new revolution of applications that will understand human behavior and preferences using these semantics. The Semantic Web will manage applications and services as an indexed semantic data base using artificial intelligence.

Therefore, future services will be applications that can reason based on logic descriptions and intelligent agents. This data processing is performed using rules that express logical relationships between concepts and data network. Thus, the Semantic Web will be able to offer to the client a new experience in which he will be the star of the movie, i.e. the managing of applications and services will be realized with the objective of satisfying all the client necessities.

1.3 Advanced Media Services

Web 2.0 has leveraged the creation of advanced media services, based on SOA, which decouple the interface of requesting a service from its implementation. Additionally, new services introduce continuous media [1], adding new features and experiences to the user.

IMS (IP Multimedia Subsystem) has been a standard way to provide this kind of services in a telco environment.

Therefore, advanced media services can be simplified in development terms thanks to the use of Service-Oriented Architectures and the IP Multimedia Subsystem (IMS), by reusing existing simple services. Besides, the combination of existing services allows creating added-value media services (e.g. videoconferencing can be integrated into a distributed agenda or into machine-to-machine networks). In the following section, the possibilities offered by these platforms are described.

1.3.1 Service-Oriented Architectures SOA

In this section, Service-Oriented Architectures and some key terms related to this development system paradigm will be explained. First, the SOA concept will be explained, together with some elements that take part in a Service-Oriented Architecture, like the Enterprise Service Bus, Web Service technologies and Service Orchestration/Choreography. Finally, a new concept where services and infrastructures can be virtualized (the Cloud Computing paradigm architecture) will be introduced.

1.3.1.1 SOA Concept

Service-Oriented Architecture (SOA) is a concept of software architecture that defines the usage of services to provide solutions to the business requirements. The OASIS group [8] defines SOA as:

> A paradigm for organizing and utilizing distributed capabilities that may be under the control of different ownership domains. It provides a uniform means to offer, discover, interact with and use capabilities to produce desired effects consistent with measurable preconditions and expectations.

SOA creates a flexible architecture, which allows reconfiguring over time. In fact, "agility" has been identified as the largest single feature for SOA. This attribute has more value when the target is a larger system that may change.

SOA enables the implementation of highly scalable architectures that reflect the business of the organization. Furthermore, it provides a well-defined layout of exposure and invocation of services which enables the interaction between different proprietary systems or third-party services.

To understand a SOA, it is necessary to identify some terms that form part of the system [9]:

- Service: function without a state, self-contained, which accepts calls and returns responses through a well-defined interface. Services do not depend on the state of other functions or processes. The specific technology used to provide the service is not part of this definition. It allows the use of asynchronous services.
- Stateless concept: a stateless service does not maintain or depend on any pre-existing condition. In SOA, services are not dependent on any other service. Services get all the information needed to respond in the call request. Since services are "stateless", they can be sequenced (orchestrated) in many sequences (sometimes called buses or pipelines) to perform the business logic.
- Provider: function that provides a service in response to a request from a consumer.
- Consumer: function that consumes the result of the service provided by a supplier.
- Orchestration: sequences, coordinates services and provides additional logic to process data. It does not include data presentation.

SOA modeling methodology for applications is called service-oriented analysis and design. The service-oriented architecture is both a framework for software development and a framework for implementation. When most people talk about service-oriented architecture they refer to a set of resident services on the Internet or an intranet using web services. There are several standards related to Web services like XML, HTTP, SOAP, WSDL, UDDI or REST. However, we must consider that SOA does not necessarily need to use these standards to be "service oriented" but its use is strongly recommended. In a SOA environment, nodes of the network make its resources available to other participants in the network as independent services which are accessible in a standardized way. Most definitions of SOA identify the use of Web Services (using SOAP and WSDL) in its implementation.

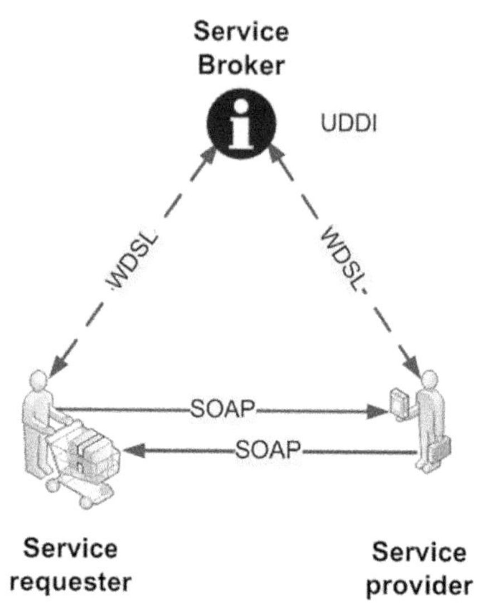

Figure 1.3 Web Services architecture.

1.3.1.2 Web Services

Web Services can be seen as small pieces that form a SOA. As we have explained previously, Web Service technology is the typical one used for designing applications in SOA, but not exclusively. However, it is interesting to explain them since this technology is the most extended. The World Wide Web Consortium defines Web Services [8] as follows:

> A Web service (as shown in Figure 1.3) is a software system designed to support interoperable machine-to-machine interaction over a network. It has an interface described in a machine-processable format (specifically WSDL). Other systems interact with the Web service in a manner prescribed by its description using SOAP messages, typically conveyed using HTTP with an XML serialization in conjunction with other Web-related standards.

Web Services are typically application programming interfaces (API) or web APIs that can be accessed over a network, such as the Internet, and executed on a remote system hosting the requested services. In common usage, the term refers to clients and servers that communicate over the Hypertext Transfer Protocol (HTTP) protocol used on the web. Such services tend to fall into one of two types: Classical Web Services (Heavy WS) and RESTful Web Services (Light WS).

Classical Web Services use Extensible Markup Language (XML) messages that follow the Simple Object Access Protocol (SOAP) standard and have been popular with traditional enterprise. In such systems, there is often a machine-readable description of the operations offered by the service written in the Web Services Description Language (WSDL). WSDL is not a requirement of a SOAP endpoint, but it is a prerequisite for automated client-side code generation in many Java and .NET SOAP frameworks (frameworks such as Spring, Apache Axis2 and Apache CXF being notable exceptions). Some industry organizations, such as the WS-I, include both SOAP and WSDL in their definition of a web service.

More recently, REpresentational State Transfer (REST) or RESTful Web Services have regained popularity, particularly with the Internet companies. The use of the PUT, GET and DELETE HTTP methods, alongside POST are often better integrated with HTTP and web browsers than SOAP-based services. They do not require SOAP XML messages or WSDL service-API definitions and so are lighter.

In a Web 2.0 context, a development in web services called Web API can be found, where emphasis has been moving away from Simple Object Access Protocol (SOAP)-based services towards more direct Representational State Transfer (REST) style communications. Web APIs allow the combination of multiple web services into new applications known as mashups.

1.3.1.3 Service Composition: Orchestration and Choreography

We have seen that Service-Oriented Architectures are formed by small entities called Services. But these services must be coordinated to develop bigger entities or complex services. In this context the Service Composition concept is introduced. In service composition two methods that describe how to compose a service appear: the Service Orchestration and the Service Choreography.

The orchestration of a service defines how the overall functionality of the service is achieved by the cooperation of other service providers. It describes how the service works from the provider perspective. It uses standard languages like WS-BPEL promoted by OASIS [8].

Choreography describes the behavior of the service from a user point of view. For this task standard languages like WS-CDL, promoted by W3C Web Services Choreography Working Group, exist [8].

The orchestration and the choreography differences are based on analogies: orchestration refers to the central control (by the conductor) of the behavior of a distributed system (the orchestra consisting of many players),

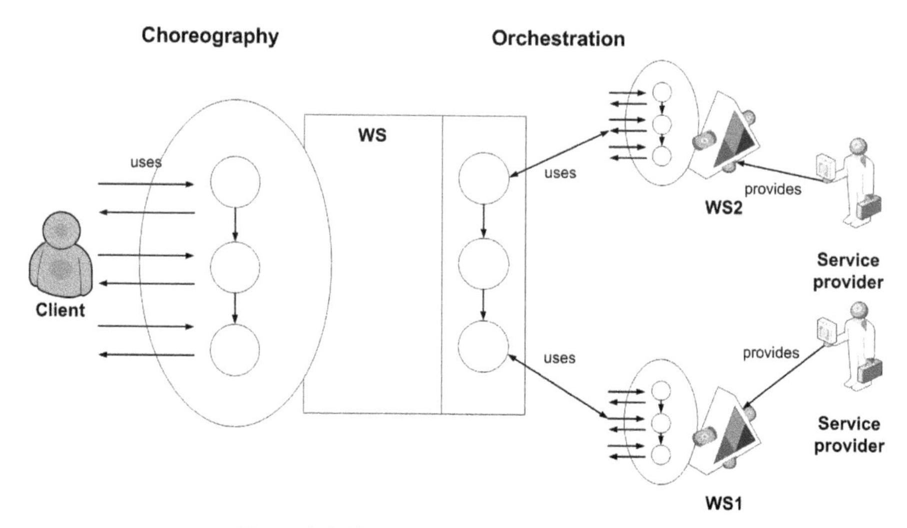

Figure 1.4 Choreography versus orchestration.

while choreography refers to a distributed system (the dancing team) which operates according to rules but without centralized control. Therefore, it is important to highlight that Orchestration is controlled in a centralized way, while Choreography is controlled in a distributed way (Figure 1.4).

In other words, orchestration differs from choreography in that it describes a process flow between services, controlled by a single party. More collaborative in nature, choreography tracks the sequence of messages involving multiple parties, where no party truly "owns" the conversation [10], although they both can be used indistinctly in several cases.

There are many service composition languages, but no definitive solution has appeared. In the following paragraphs the more promising languages are described and finally compared.

BPEL: Business Process Execution Language (BPEL) has been designed specifically as a language for definition of business processes. BPEL supports two different types of business processes:

- Executable processes allow us to specify the exact details of business processes. They can be executed by an orchestration engine. In most cases BPEL is used for executable processes.
- Abstract business protocols allow us to specify the public message exchange between parties only. They do not include the internal details of process flows and are not executable.

BPEL builds on top of XML and Web Services. It is an XML-based language which supports the Web Services technology stack, including SOAP, WSDL, UDDI, WS-Reliable Messaging, WS-Addressing, WS-Coordination and WS-Transaction. A BPEL process specifies the exact order in which participating Web Services should be invoked. This can be done sequentially or in parallel. With BPEL, we can express conditional behavior, for example, a web service invocation can depend on the value of a previous invocation. We can also construct loops, declare variables, copy and assign values, define fault handlers, and so on. By combining all these constructs, we can define complex business processes in an algorithmic manner. BPEL is thus comparable to general purpose programming language such as Java, but it is not as powerful as Java. On the other hand, it is simpler and better suited for business process definition. Therefore, BPEL is not a replacement, but rather a supplement to modern languages such as Java.

WSCI: The Web Service Choreography Interface (WSCI) is an XML-based interface description language that describes the flow of messages exchanged by a Web Service interacting with other Web Services [11]. WSCI describes the observable behavior of a Web Service. This is expressed in terms of temporal and logical dependencies among the exchanged messages, featuring sequencing rules, correlation, exception handling, and transactions. WSCI also describes the collective message exchange among interacting Web Services, thus providing a global, message-oriented view of the interactions. WSCI does not address the definition and the implementation of the internal processes that actually drive the message exchange. Rather, the goal of WSCI is to describe the observable behavior of a Web Service by means of a message flow oriented interface. This description enables developers, architects and tools to describe and compose a global view of the dynamic of the message exchange by understanding the interactions with the web service [12].

In terms of orchestration, a combination of both would be the more complete solution.

- WSCI describes how a service interacts with others, in other words, its observable behavior. A kind of sequence diagram where the interchanged messages and the possible errors/exceptions appearing during the execution can be seen.
- BPEL is more process-oriented. A state machine where the behavior of the system (all the WebServices) is globally formalized.

1.3.2 Continuous Media Services: Services in IMS

The IP Multimedia Subsystem (IMS) is a global, access-independent and standard-based IP connectivity and service control architecture that enables various types of multimedia services to end-users using common Internet-based protocols.

IMS uses the Session Initiation Protocol (SIP) for session setup and teardown, while Diameter is used as an AAA (Authorization, Authentication and Accounting) protocol. IMS is currently actively used, and (as it will be explained in the next subsection) as time goes by more functionalities are being added to this architecture.

IMS decomposes the networking infrastructure into separate functions with standardized interfaces between them. Each interface is specified as a reference point, which defines both the protocol over the interface and the functions between which it operates [13].

The connectivity layer comprises devices (routers, switches, etc.) both for the backbone and the access network (multiple type of network access permitted).

The control layer comprises network control servers to manage call or session set-up, modification and release. Several roles of SIP servers, collectively called CSCF (Call Session Control Function) are used to process SIP signaling packets in the IMS. In the security and user subscription functions, they are the AuC (Authentication Center) and the HSS (Home Subscriber Server) that communicate with CSCF in order to provide security management for subscribers. The interconnections with other operator networks and/or other types of networks are handled by border gateways (BGCF).

The service layer comprises application and content servers to execute value added services for the user. Generic service enablers as defined in the IMS standard are implemented as services in a SIP application server, which is the native server in IMS.

At an IMS terminal level the devices must behave like a SIP User Agent (UA) to support SIP signaling and additionally support RTP flows for multimedia services.

1.3.2.1 IMS Enablers and Third-Party Enablers

IMS is a horizontal architecture, i.e. it has sets of procedures that can be used by services, in a way that service composition becomes much easier and faster. Those sets of procedures are the so-called IMS Enablers, which provide Internet and communication services with enhancing features such as the default enablers offered by IMS [14]:

- Accounting enabler: this provides an application with a Diameter messaging interface, to enable communication with the accounting server through Rf accounting Web service using either a session or event offline charging method.
- Call Control enabler: this provides an application for a number of different call control scenarios across multiple networks.
- Instant messaging enabler: this provides a set of applications that a service can use to send and receive messages.
- Media control enabler: this provides a set of standard applications for controlling media server functionality across different control technologies, hiding the differences in semantics between these control languages.
- Presence enabler: this aggregates presence information from a number of different sources in a persistent scalable manner. It also provides the ability to publish the presence information to a number of different sources.
- Signaling enabler: this is used to build SIP specific services, whereby the application developer does not need to understand the complete signaling details of the SIP protocol and can instead program at a higher level of abstraction.

As a major advantage, those enablers can be woven into existing Internet services, becoming value-added services and expanding the capabilities of the service implementation.

Another advantage is the fact that the list of enablers does not finish with the IMS offer. Third parties are allowed to create new enablers in order to broaden the possibilities of service composition with IMS, and a more complex IMS service creation appears: the composition of complex services by combining simple services, IMS enablers or third-party enablers. Connecting services is a complex process and to ease it two terms appeared: Services Orchestration and Service Choreography, as mentioned in previous sections.

The Orchestration of services defines a process flow in which services interact at a message level, including the execution order and the related business logic. The Choreography of services describes the part that each of them plays in the interaction.

In Figure 1.5, we can see the architecture for service composition in IMS. The IMS Core manages all the requests from the user; whenever it is not able

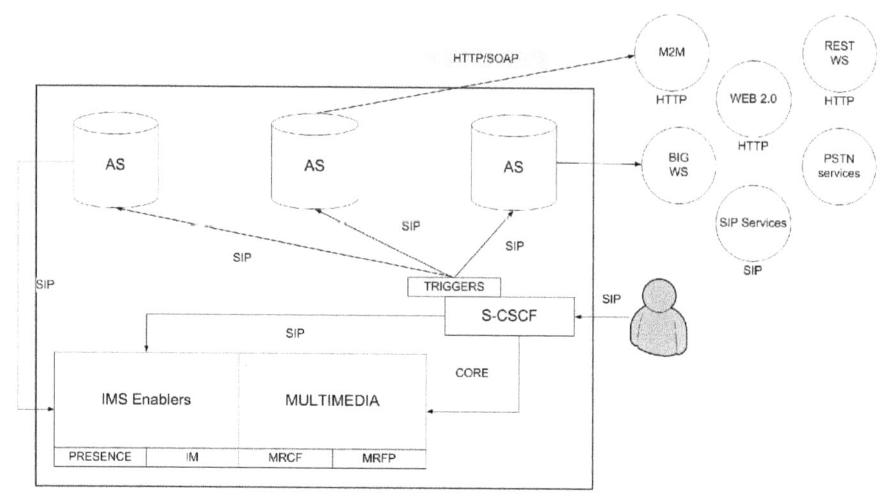

Figure 1.5 IMS service management.

to satisfy a request, the Trigger delegates this duty to another component like a SIP Application Server (AS).

1.4 Future Internet

The new Internet applications and the user increment have increased the complexity and the requirements of the Net to the point that the limit of charge and stress is getting closer. The Internet was designed with the initial purpose of interconnecting wired networks. Features like quality of service (QoS), security and new computing paradigms were not taken into account, but nowadays they have become of critical importance.

Wireless communications have introduced new requirements like ubiquity, mobility, identification, energy constrains, location, nomadism, context-awareness and other device-capability issues. The current Internet architecture is too "ossified", complex and rigid. During its more than 40 years of existence, the Internet has received a number of patches to solve problems and adding new capabilities, thus applying an evolutionary maintenance. The moment has arrived when it is imperative to add new features adapted to the Future Internet. Millions of users (and devices) can cause scalability problems in the routing and resource discovery. New applications like high quality media streaming will consume more resources which, along requirements like QoS and security, will be critical.

In the coming years [19], a revolution in the use of the Internet will happen. The concept of "The Internet of Things" has recently come up, which allows linking in an intelligent way, objects, services and things with the network, together with ubiquity.

In the near future, physical objects will be digitally tagged and will operate as a collective consciousness. Thus, any object will be able to be identified with relevant information and functions on the cyberspace. For example, when approaching an empty box of cereals to a reader integrated in the freezer, the user will be given the option to buy a new one online or to find recipes using this product. The huge quantity of identifiers exposes several problems very difficult to solve with the technologies involved in the current Internet, i.e. URLs and DNS. Furthermore, new semantic identification schemes and addressing of nodes, resources and services are needed.

Regarding network capabilities, the future Internet will grow not only in number of devices but also in the different requirements of communication between nodes. The future Internet architecture will need to allow the local interconnection of hundreds of devices with price-performance ranging from very low cost to very high bandwidth. For example, in the same local network of a house, dozens of chip sized sensors will broadcast sensor data at few bytes per second speeds while the video conference device will require a high capacity low latency point to point connection with a videoconferencing terminal in another country. This will require that Quality of Service is natively integrated into the future protocol stack at all levels and that service discovery mechanisms allow devices to locate partner services.

At the same time, when we are considering a huge number of nodes in the network, it is necessary to facilitate mechanisms for the automatic configuration of both, devices and the network elements. Depending on the QoS requirements of the application (e.g. bandwidth, latency, security level) the network will reconfigure itself to accommodate to these requirements. Autonomic networks will reduce the need of manual interventions and provide a linkage between application intentions and network behavior.

There exists a new approach to solve these problems, a disruptive solution that designs a new network architecture from scratch [18]. This methodology bases its strategy in two design principles: Role-based architecture (RBA) and service-oriented architecture (SOA).

Role-based architectures [16] are based on a modular model without layers. The roles represent blocks that offer specific functionalities (e.g. packet forwarding, packet fragmentation, etc.). The protocol stack is replaced by a protocol heap and each node executes the roles that it supports.

Service-oriented architectures offer a flexible computation paradigm that allows organizing and using distributed capabilities among different domains. Furthermore, they offer a natural environment for building a dynamic architecture by using service composition [17].

Communications are composed of basic network functions (as services) orchestrated by means of a work-flow that models the desired functionality and behavior. The network is defined as a set of nodes and services: atomic services (i.e. functions used by the TCP/IP and OSI layers) or composed services.

On the other side, new network characteristics such as context-awareness, allow the services to act and react depending on the context. This provides adapted and personalized services within a new framework of added-value and ubiquitous services.

1.5 Conclusions

This work has drawn a review of the Internet evolution, based on three points of view (computation, information and media) during the last 40 years and, at the same time, it has shown the challenges that will have to be assumed by the Internet in the future. The Future Internet architecture must be defined or redefined using rules that offer a flexible and dynamic framework. This framework should provide ubiquity, by offering services anywhere, anytime and anyhow, oriented to services and resources interconnection, with a low overhead and, also, be an open architecture, allowing the interconnection of a number of different and heterogeneous devices, as the Internet of Things.

It should also provide mechanisms for the diagnosis and the configuration that improve the usability of the Internet by reducing manual interventions. These mechanisms give the network an active role that allows it to act depending on the current context. In this sense, the evolutionary strategies that are based on maintaining the backwards compatibility are reaching a point in which changes and updates are too complex and time-consuming, so a disruptive or clean-slate solution is justified. Most authors suggest the most flexible design principles for this new architecture, e.g. role-based architectures and service-oriented architectures. This proposal breaks with the current layered model based on the OSI stack and overcomes its rigidity by avoiding the problem of redundancy and the unnecessary use of functionalities. For example, on a machine-to-machine network, the complete use of all the TCP/IP stack is not efficient, in terms of performance or energy consumption, because many elements of the stack are not necessary. In a Role-based architecture, a network node will only implement the roles it needs in its protocol heap. On the other hand, a service-oriented architecture allows organizing and using distributed capabilities on a flexible and normalized way by applying service composition.

References

[1] A. Rios, M. Hurtado, S. Sallent, J. Alcober, and A. Oller. SIP: Advanced media integration. In *SIP Handbook: Services, Technologies, and Security of Session Initiation Protocol*, pages 15–42, CRC Press, 2009.

[2] T. O'Reilly, What is Web 2.0. Design patterns and business models for the next generation of software. MPRA, Paper No. 4580, posted 7 November 2007.

[3] I. Hickson. HTML5: A vocabulary and associated APIs for HTML and XHTML. W3C Working Draft 24, June 2010.

[4] T. Vander Wal. Social Design for the Enterprise Workshop in Washington, DC Area, June 27th 2009. Available at: http://www.vanderwal.net

[5] Harry Halpin, Valentin Robu, and Hana Shepherd. The complex dynamics of collaborative tagging. In *Proceedings of the 16th International Conference on the World Wide Web (WWW'07)*, Banff, Canada, pages 211–220, ACM Press, 2007.

[6] Michael K. Smith, Chris Welty, and Deborah L. McGuinness (2004-02-10). OWL Web ontology language guide. W3C, Retrieved 15 July 2008.

[7] Patrick Hayes (10 February 2004). RDF semantics. Resource description framework. World Wide Web Consortium, Retrieved 18 April 2010.

[8] OASIS. OASIS Committees by Category: SOA. Consulted on June 2010. Available at: http://www.oasis-open.org/committees/tc_cat.php?cat=soa

[9] J. Green (Ed.). *An Implementor's Guide to Service Oriented Architecture: Getting It Right*. Westminster Promotions, 2008.

[10] C. Peltz. Web services orchestration and choreography. *SOA World Magazine*, June 2003.

[11] D. Van Than. Web Service Orchestration: An open and standardized approach for creating advanced services. Eurescom, 2011.

[12] A. Arkin, S. Askary, S. Fordin, W. Jekeli, K. Kawaguchi, D. Orchard, S. Pogliani, K. Riemer, S. Struble, P. Takacsi-Nagy, I. Trickovic, and S. Zimek. Web Service Choreography Interface (WSCI) 1.0. W3C Note 8, August 2002.

[13] D. Rodríguez, J. López, A. Oller, A. Rios, J. Alcober, and F. Minerva. Hybrid fixed-mobile P2P super-distribution. NEM Summit, 2008.

[14] Oracle Fusion Middleware. Oracle SDP – Enablers and future proofing your investment. An Oracle White Paper, April 2006.

[15] M. McHugh. The need for IMS enabler innovation. *IMS Magazine*, 2(5), October 2007.

[16] R. Braden, T. Faber, M. Handley, and N.J. Princeton. From protocol stack to protocol heap – Role-based architecture. HotNets-I, October 2002.

[17] B. Reuther, and P. Müller. Future internet architecture – A Service Oriented Approach. *Information Technology*, 6:383–389, 2008.

[18] David D. Clark and David L. Tennenhouse. Architectural considerations for a new generation of protocols. In *Proceedings of the ACM Symposium on Communications*

Architectures & Protocols (SIGCOMM'90), Philadelphia, PA, USA, September 24–27, pages 200–208. ACM, 1990.

[19] D. Clark, C. Partridge, R.T. Braden, B. Davie, S. Floyd, Van Jacobson, D. Katabi, G. Minshall, K.K Ramakrishnan, T. Roscoe, I.Stoica, J. Wroclawski, and L. Zhang. Making the world a different place. *ACM SIGCOMM Computer Communications Review*, 35(2):91–96, July 2005.

2

SpoVNet: An Architecture for Easy Creation and Deployment of Service Overlays

Roland Bless, Christian Hübsch,
Christoph P. Mayer and Oliver P. Waldhorst

Institute of Telematics, Karlsruhe Institute of Technology (KIT), 76128 Karlsruhe, Germany; e-mail: bless@kit.edu

Abstract

This chapter describes the *Spontaneous Virtual Networks* architecture (SpoVNet), which eases creation and deployment of service overlays. First, we describe the requirements for supporting service overlays in today's Internet. For instance, service overlays must work across heterogeneous networks and should be able to incorporate mobile and/or multi-homed users as well. Next, we describe the components of SpoVNet and how they relieve service overlay programmers from several complex tasks. Finally, we illustrate the application of the provided SpoVNet service interface by giving an example for creating a service overlay for streaming videos to a group of users.

Keywords: SpoVNet, overlay, network substrate, service overlays.

2.1 Introduction

The Internet has become an indispensable part of daily personal and professional life. It offers a growing variety of communication possibilities, e.g., e-mail, instant messaging, web, file sharing, video and TV streaming, social networks, online communities, or Internet telephony. With the growing popularity of such applications, user demands increase, for example regarding

Anand R. Prasad et al. (Eds.), Future Internet Services and Service Architectures, 23–47.

availability and robustness of communication, various aspects of security (e.g., data authenticity), or support of ubiquitous mobile access. Therefore, the Internet is exposed to constant change to which it must respond with appropriate new supporting protocols and services.

Overlay networks (or *overlays* for short) have become a successful way to create new services and applications above an existing physical network infrastructure (referred to as *underlay*). They establish their own overlay topology, consisting of logical connections between participating devices. This enables use of custom addressing and routing schemes inside the overlay, thereby overcoming various limitations of the underlay. In recent years a certain class of overlays, so-called *Peer-to-Peer* (P2P) systems, has become extremely popular [9]. Many well-known examples exist, e.g., BitTorrent or Skype. P2P systems enable implementation of new services and applications solely on end-systems, i.e., they do not require changes to the established infrastructure. A main feature of such P2P systems is that each end-system contributes own resources (e.g., storage, processing power, or forwarding resources) in an organized manner to allow for efficient use of these distributed resources. Important characteristics of P2P systems are fault tolerance, self-organization, and massive scalability. Supporting these properties, P2P overlays have established themselves as driving force for innovation of new services and applications [26].

Currently, P2P systems tightly couple application and delivery service. For instance, some early IP-TV solutions, like Zattoo, established P2P networks for providing global multicast distribution, which is still not natively available in the Internet infrastructure. This service is integrated into the Zattoo player, making reuse in other applications infeasible. However, developers of both applications and services would benefit from a separation of overlays and applications. In fact, proposals for future Internet architectures include a layer of *service overlays*, which will provide such novel services to applications [10].

As a major challenge, service overlays comprise distributed service elements, "raising the barrier to deployment" [10]. Since logical connections between devices constitute a building block of an overlay, it is crucial that connections between all pairs of devices within a service overlay can be established. This is difficult, since devices (1) may be scattered across different access networks that may be located behind middleboxes such as *Network Address Translation* (NAT) gateways or firewalls, (2) potentially use different network layer protocols like IPv4 and IPv6, or (3) use no network layer protocol at all, for example, if connected via Bluetooth RFCOMM in the link layer. Furthermore, devices may be mobile and change between different access networks, or may be located in more than one access network in parallel. All of these obstacles complicate the development and deployment of service overlays. To this end, innovation can be significantly fostered by providing a framework for dealing with these obstacles transparently, allowing the service overlay developer to focus on implementation of the functionality of the service itself. In fact, Clark [10] argues that virtualized platforms for service overlay deployment can "lower the barrier to innovation and competition at this level".

The *Spontaneous Virtual Networks* (SpoVNet) project develops an architecture for easy implementation and simple deployment of service overlays in current and future networks [25]. The architecture primarily enables applications and services to *spontaneously* set up and sustain a *virtual network* among all participants, using its own identifier-based addressing scheme. As its major goal, the architecture avoids manual configuration and dedicated management by employing self-organization, with the overall objective to regain controllability of today's more and more complex networks. To this end, it uses overlays at various levels: In addition to the consistent implementation of services by overlays, the SpoVNet architecture itself employs overlays to provide basic connectivity. Using this concept, heterogeneity of devices and underlying networks as well as mobility, multi-homing, and security issues are transparently handled. We have implemented this part of SpoVNet's functionality in the open-source software package *ariba* [1] and successfully demonstrated its features and capabilities [16, 17].

The remainder of this chapter is structured as follows: We identify the major features a framework for service overlay creation should provide and discuss related work in this context in Section 2.2. The architecture developed by the SpoVNet project is described in Section 2.3. An exemplary case study in Section 2.4 provides insight for building overlay-based services with SpoVNet. Finally, Section 2.5 provides conclusions and future research directions.

2.2 Creating Overlay-Based Services

In this section we will provide an overview of important requirements that must be addressed when developing a framework for the creation of overlay-based services with respect to the problems discussed in Section 2.1. We will begin by outlining the specific features that such a framework should offer.

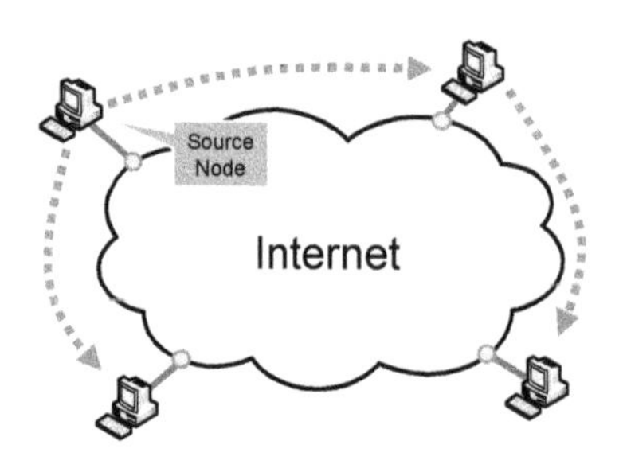

(a) Common network view from Overlays Developers' perspective.

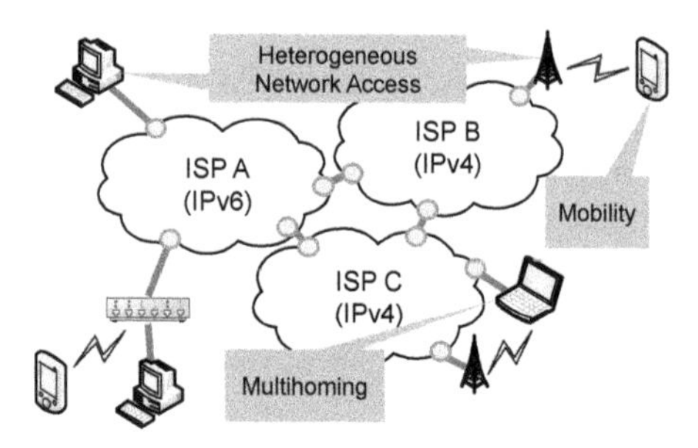

(b) Network view with consideration of mobility, heterogeneity and multihoming aspects.

Figure 2.1 Common simplistic network view versus network heterogeneity consideration.

We will then present existing related work and discuss the extent to which these approaches fulfill the requirements.

2.2.1 Required Features

We use the example of an overlay-based multicast service to describe the features that a framework for deployment of service overlays should support. Implementing multicast functionality using overlays is often referred to as *Application Layer Multicast* (ALM) and is quite well understood widely [15]. In general, an ALM approach constructs a distribution overlay among all participating nodes. Figure 2.1(a) shows such a distribution overlay in the form of a tree. Multicast messages are forwarded along this tree from node to node. Every message is generated by one source node and is received by all other nodes. Since forwarding capabilities are only accomplished by the nodes themselves, ALM implements multicast delivery without any support from lower layers, thus increasing flexibility and lowering deployment cost.

One challenge in the scenarios depicted in Figure 2.1(a) is to construct an efficient distribution tree with respect to some optimization criteria (e.g., latency determined by the depth of the tree) given some constraints (e.g., a limited out-degree of all nodes due to constrained uplink bandwidth). However, there is a large set of additional challenges with which the developer of a service overlay must cope. Most algorithms for the construction of a distribution overlay assume that it is possible to establish a direct connection between each pair of nodes in the underlay. This implicitly presumes a very idealized picture of the Internet as it is shown in Figure 2.1(a). Unfortunately, the evolution of the Internet led to a quite different picture, as shown in Figure 2.1(b).

Here, devices may be mobile and move during an ongoing overlay communication session. Since IP by itself does not support mobility of devices between networks, overlay connections must be re-established in such cases. Moreover, overlay connection establishment may be hindered by middleboxes such as NAT gateways or firewalls. Also, devices participating in a service overlay may be located in access networks that use different protocols, e.g., IPv4 vs. IPv6 on the network layer, or no network layer protocol at all, e.g., if connected using a Bluetooth RFCOMM connection. Such devices can only establish connections by using other devices as proxies. Finally, there may be devices with multiple connections to the Internet. In such cases, it might not be clear which connection should be used by the service overlay. This holds true, in particular, if each connection has different characteristics, e.g., in terms of bandwidth or latency.

In summary, Figures 2.1(a) and 2.1(b) show that challenges for service overlay construction in a current network can be separated into two groups: (1) implementing core service functionality, e.g., construction of an efficient distribution tree in the ALM example, and (2) dealing with network obstacles brought up by an evolving Internet. In order to "lower the barrier to innovation and competition" [10] the major focus of a framework for deployment of service overlays should be on transparently handling the second group of challenges, thereby allowing a service overlay developer to focus on the first. To this end, a framework for the creation of overlay-based services should provide a set of core functionalities:

- *Provisioning of end-to-end connectivity*: The framework should support connection establishment between any two devices participating in the services overlay regardless of their location, access networks, and protocols.
- *Mobility support*: The framework should transparently maintain connections when devices move between different networks.
- *Support for multi-homing*: If a device is located in multiple access networks, the framework should select the most appropriate network based on requirements specified by the service overlay. Furthermore, the framework can increase robustness by remapping a connection to a different access network if the connection to the formerly selected network fails.

- *Support heterogeneous characteristics of access networks*: Although the framework should handle most issues related to connection establishment and maintenance transparently, it might be beneficial for the developer to be aware of the characteristics of an access network, e.g., available bandwidth or latency. To this end, the framework should provide mechanisms to obtain these kinds of information generically.

Parts of the aforementioned functionalities can be provided by other mechanisms, e.g., by manually configuring middleboxes for allowing end-to-end connectivity and using *MobileIP* [19] for mobility support. Similarly, *Identifier/Locator Network Protocol* (ILNP) [5] provides an ID/Locator split mechanism realized within IPv6 addresses and by using DNS. However, such mechanisms require manual configuration or support by dedicated infrastructure (e.g., home and foreign agents or DNS servers) and thus contradict the basic principle of services overlays.

Besides the functional features described above, a framework for the creation of overlay-based services should provide a set of *non-functional features*:

- *Scalability*: Since scalability with respect to the number of nodes is an essential principle of a service overlay, this scalability should not be limited by the framework.
- *Self-configuration and self-maintenance*: Since a service overlay is self-organizing, the framework for creating it should not require any manual configuration or maintenance.
- *Extensibility*: The applicability of the framework should not be limited to a specific type of delivery service (e.g., multicast only), but should be extensible to other types of delivery services (e.g., publish/subscribe, content-based routing, etc.).
- *Usability*: Last but not least, the framework should be easy to use by providing a set of well-defined and powerful interfaces for the developer.

In the following section, we will review existing frameworks for service overlay development and deployment with these features in mind.

2.2.2 Existing Frameworks for Creating Overlay-Based Services

A larger number of frameworks exist that support overlay-based service creation. Their approaches and resulting properties differ greatly, because some originated from research projects, while others came from a software development perspective. Some prominent examples of the first type include

SATO [24], Hypercast [20], FreePastry [2], UIA [13], MACEDON [22] and Overlay Weaver [23]. JXTA [3] is a prominent example of the second type.

All of these approaches provide most of the non-functional features described in Section 2.2.1. For example, all approaches aim (to some extent) at scalability and try to avoid manual configuration. Most of the projects provide some kind of abstractions for easy access to overlay features in order to increase usability. For instance, Hypercast offers *overlay sockets* that can be used by the application and hide all overlay-related operations, such as neighbor discovery and maintenance. SATO provides the *Ambient Service Interface* for hiding underlying transport complexity. Overlay Weaver provides an interface similar to the common API for key-based routing (KBR) [11], that enables the addressing of nodes by IDs. JXTA organizes devices in *peer groups* and connects them by *pipes*. Extensibility is possible with different granularity. For example, Hypercast permits the addition of overlay sockets for constructing new overlay structures. FreePastry enables the creation of services like Scribe-based multicast on top of the basic connectivity it provides. MACEDON enables the specification of overlay protocols by finite state machines, then automatically translates those protocols to running code. Overlay Weaver allows for the easy exchange of the routing algorithm with the KBR interface, while re-using basic features such as connection management and iterative/recursive routing procedures.

While all of the projects support non-functional features to some extent, the picture is quite different with respect to functional features. At first glance, all projects provide end-to-end connectivity between all devices in an overlay. However, the underlying assumptions of these devices are different. For example, FreePastry provides a NAT traversal mechanism, but does not support heterogeneous protocols on the network layer, i.e., it assumes all devices to be located in the (IPv4-)Internet. The same is true for both MACEDON and Overlay Weaver, which establish connections using TCP or UDP sockets. They also both support generating code for simulation/emulation environments, which also use a homogeneous address format. SATO supports heterogeneous protocols on the network layer if it runs on top of the NodeID architecture [4]. Hypercast, UIA, and JXTA provide support for heterogeneous protocols. However, in such scenarios messages are always forwarded along the overlay, even if shorter underlay connections exist. Mobility support is implemented in all projects by maintaining a fixed overlay identifier that is mapped to (eventually changing) network addresses. However, in many cases, (overlay-)connections are not re-established transparently after a device has moved due to mobility. With respect to support for multi-homing, most protocols can handle interfaces to multiple access networks as well as multiple interfaces to a single access network. Nevertheless, none of the approaches supports transparent selection of the access network best suited for the demands of the service overlay. SATO is the only project that explicitly incorporates support for heterogeneous access links by using cross-layer information.

We conclude from this discussion that none of the protocols supports all functional features identified in Section 2.2.1. In particular, direct communication between devices together with fully transparent support for mobility and multi-homing is not supported by any of the projects. In the next section, we will present an architecture that closes these gaps.

2.3 The SpoVNet Architecture

The *Spontaneous Virtual Networks* (SpoVNet) project developed an architecture for the seamless realization and deployment of overlay-based services and applications in current and future networks [25]. Its main objective is to provide an enabling platform for the spontaneous and flexible creation of overlay-based services, based on the concept of spontaneous virtual networks. Such virtual networks are set up on a *per-application context*, comprising all participants that run the same application instance. The same application can run, however, in different contexts, e.g., a local group of students on a university campus, or as a real global group comprising all participants worldwide. These two different application contexts would usually make up two different SpoVNets.

In the following section, we will refer to a system running SpoVNet software – e.g., PCs, notebooks, PDAs, or smartphones – as a *SpoVNet Device*. We will refer to SpoVNet Devices that take part in a common application instance as the SpoVNet Instance. Finally, a communication endpoint within a SpoVNet Instance will be referred to as a *SpoVNet Node*. Generally, multiple SpoVNet Nodes joined to the same or different SpoVNet instances can be operated on a single SpoVNet Device.

Figure 2.2 shows an architectural view of SpoVNet, roughly divided into (top down): *Developer and Legacy Interface*, *Base Overlay*, *Base Communication*, *Security Component*, and *Cross-Layer Component*. We will first give a short overview of the overall architecture in Section 2.3.1, before we provide detail on the architectural components in a bottom-up style in Sections 2.3.2–2.3.6.

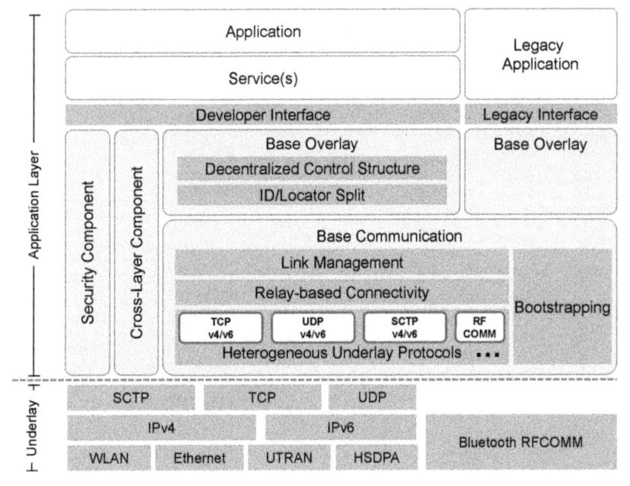

Figure 2.2 Architecture of SpoVNet with main components and sub-components.

2.3.1 Architectural Overview

Figure 2.3 and Table 2.1 visualize the mapping of mechanisms to features in the SpoVNet architecture. SpoVNet's *Service Interface* (IF) – presented in Section 2.3.6 – provides a uniform interface for the creation and management of SpoVNet instances and communication between SpoVNet nodes. SpoVNet nodes are addressed by *application-specific node identifiers*, thus allowing applications to use a consistent and stable addressing scheme. The *Base Overlay* (BO) [7] (cf. Section 2.3.3) can route messages to nodes addressed by these identifiers. It serves as basic control structure implementing an *ID/Locator Split*, i.e., it maps identifiers to current *locators* that represent network addresses in the underlay. Thus, it provides an abstraction from the physical network attachment of the SpoVNet device. Routing all service user data along the Base Overlay's topology would be inefficient. Therefore, SpoVNet is able to transparently create direct transport connections from end-to-end across the underlay, which is an important feature for constructing efficient service overlays. The Base Overlay is used to initially exchange the required locator sets between both communication peers in order to enable them to communicate directly from end to end in the underlay, if possible.

In the SpoVNet architecture creation and maintenance of such direct transport connections is accomplished by the *Base Communication* (BC). Each SpoVNet device runs the Base Communication, while all SpoVNet nodes running on such a device use the same Base Communication instance

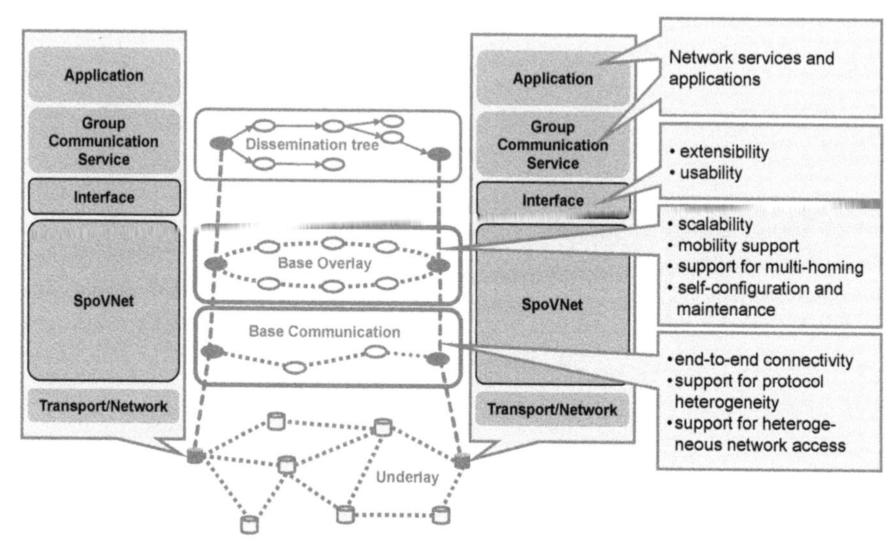

Figure 2.3 Networks built up on various layers and features in respective components.

below the respective Base Overlays. This enables performance optimization, e.g. for common utilization of physical connections by multiple logical overlay links. Finally, the *Cross Layer Component* we will present in Section 2.3.5 gathers cross-layer information and performs measurements used for protocol optimization.

2.3.2 Base Communication

The *Base Communication* (BC) enables the interconnection of heterogeneous end-systems with different network access technologies and protocols. It conceals underlay details from the Base Overlay, so that the latter does not have to deal directly with heterogeneity (e.g., different address families like IPv4 or IPv6), limited connectivity, or mobility. In order to avoid circular dependencies between Base Communication and Base Overlay, the Base Overlay handles locator-specific information transparently and simply passes it down to the Base Communication when necessary. Moreover, the Base Communication does not use identifiers as addressing information but rather employs pure underlay addresses.

Table 2.1 Overview of SpoVNet mechanisms and features implemented.

Mechanism	Feature	Mobility Support	Heterogeneous Characteristics	End-to-End Connectivity	Support for multi-homing	Scalability	Self-conf. & maintenance	Extensibility	Usability
BC	2.3.2.1 Heterogeneous Underlay Proto.	×	×	×	×				
	2.3.2.2 Link Management	×		×	×			×	×
	2.3.2.3 Relay-based Connectivity		×	×			×		
	2.3.2.4 Bootstrapping						×		×
BO	2.3.3.1 ID/Locator Split	×			×				
	2.3.3.2 Decentralized Control Structure			×		×	×		
	2.3.3.3 Distributed Hash Table						×		×
CL	2.3.5 Cross-Layer Component		×		×	×	×	×	×
IF	2.3.6.1 Developer Interface							×	×
	2.3.6.2 Legacy Interface								×

2.3.2.1 Heterogeneous Underlay Protocols

Achieving end-to-end connectivity between SpoVNet nodes in the face of heterogeneity in the network is challenging. End-to-end connectivity should be maintained even if the node is mobile and may change its point of network access. This network access may even use a different network protocol (e.g., IPv6 instead of IPv4) or possess different connectivity characteristics (e.g., when moving into an Intranet with private addresses behind a NAT gateway) than the previous point of network access.

Providing support for multiple heterogeneous underlay protocols increases the likelihood of communication between two devices and accomplishes communication between otherwise incompatible devices. While handling this multitude of protocols is complex, SpoVNet provides a uniform framework for seamless connectivity. To accomplish such an abstraction, the SpoVNet stack is implemented above the transport layer and can employ different transport protocols (e.g. TCP, SCTP, UDP), network protocols (e.g. IPv4, IPv6), and link layer protocols (e.g. Ethernet, Bluetooth). Notably, not all devices may run the same set of protocols on each of those layers, thus inhibiting direct end-to-end connections. SpoVNet therefore constructs and maintains *relay paths* that allow for indirect end-to-end connectivity between two SpoVNet devices by using other SpoVNet devices as *relay* (cf. Section 2.3.2.3 and [17, 21]). Potential *relay devices* running different protocol stacks simultaneously are found through self-organization via the overlay (more details are presented in Section 2.3.2.3).

2.3.2.2 Link Management

Today's transport connections are directly coupled to the locators of a device (cf. Section 2.3.3.1), leading to complex handling in case of mobility, middleboxes, or multihoming. For instance, detecting which transport connections are currently working or which ones are actually broken after a handover requires extra application logic.

SpoVNet implements the concept of *virtual transport links*, which are not coupled to device locators and built upon one or multiple piecewise transport connections (concatenated by relays). Virtual transport links, or short *links*, provide a transport abstraction for communication between *SpoVNet Nodes*.

Virtual transport links permit bidirectional communication and end at a SpoVNet Node. Similarly to SpoVNet Nodes, virtual transport links are addressed by a separate flat link identifier (cf. Section 2.3.3.1). How a virtual transport link maps to transport connections, or which protocols are used for implementing the link, is transparent to the service or application that requested the link. The use of requirements-oriented interfaces detailed in Section 2.3.6.1, makes it possible to influence the protocol selection to some extent.

Virtual transport links can either be established explicitly on request or automatically. We therewith differentiate between *manual links* and *auto links*. While manual links are set up explicitly, auto links are established automatically when a message is sent to a SpoVNet Node without a given specific link identifier. In this case, a link is established internally and maintained until it is closed down due to a longer idle period. In case of auto links, management of the links is highly simplified and programming errors reduced, at the cost of reduced control over how the link is created. The concept of virtual transport links makes it possible to easily build further overlays and applications upon SpoVNet, simplifying usage while providing transparent end-to-end connectivity even in case of mobility or multihoming.

2.3.2.3 Relay-Based Connectivity

As indicated earlier, establishing direct transport connectivity is often not possible due to different network protocols or middleboxes. First, direct transport connectivity in the underlay must be evaluated. In cases where such direct connectivity may not be available, potential relay devices must be detected and selected for constructing an indirect end-to-end transport path between both devices. This is done by dividing the network into several *Connectivity Domains*. A *Connectivity Domain* is a network domain in which every device can reach every other device via a direct underlay path. Nodes that are located in and connected to more than one connectivity domain are denoted as *relays*. The Base Communication uses paths consisting of one or more relays to forward messages between two connectivity domains. Relay paths are discovered using a link-state routing mechanism run by the relays. This requires that each relay have the ability to advertise the Connectivity Domains in which it is located, and thus, know a unique identifier for these domains.

The Base Communication performs self-organizing identification of connectivity domains and assigns a unique identifier, called *Connectivity Domain Identifier* (CDID) [21] to each connectivity domain. Allocation of CDIDs is a self-organizing process without a central entity or manual configuration. CDIDs are represented by 160 bit identifiers, which are selected randomly during the identification algorithm. The structure of a connectivity domain is not static, since reconfiguration of the network, protocol upgrade, or network dynamics caused by failures result in a continuous adaptation process. Such events can trigger partitioning and merging of connectivity domains. To detect and handle such events in a self-organizing manner, SpoVNet devices periodically perform checks to determine whether connectivity to relays within their domain is intact. If loss of connectivity to more than half of the relays is detected, partitioning or serious network problems are suspected. In this case the process of CDID adaptation is initiated and a new CDID proposal is generated and distributed as so-called conflict dataset. Nodes receiving such a conflict dataset adapt through a fixed scheme the new domain identifier, whereas conflict datasets they generated on their own can get included. Through this approach CDID proposals converge on all nodes within a short time period [21]. Alternatively, in order to detect a merge of connectivity domains, nodes periodically try to contact devices – they currently can only reach through relays – in the underlay to detect new communication opportunities. If such communication succeeds, a new CDID for the merged domain is calculated collectively.

2.3.2.4 Bootstrapping

In general, overlays face the problem of *bootstrapping*: joining of a new node into the system. This requires "finding" other nodes that are currently present in the overlay – in contrast to server-based systems that operate at well-known addresses. For contacting such nodes, it is necessary to find their locators (i.e., addresses in the underlay, e.g., IP addresses or link layer addresses) and to have a working communication path towards such a locator in the underlay. While we have shown that random probing mechanisms and address caches are effective in large P2P systems [12], they have problems in smaller and spontaneous networks.

To allow for simple usage and self-organization, SpoVNet provides various bootstrapping modules to help a node join the SpoVNet Instance or reestablish connectivity. Joining a SpoVNet can be performed through any node that is already connected to the SpoVNet instance. Therefore, the provided bootstrap modules hold great value in spontaneously deploying a SpoVNet. For example, the SpoVNet implementation currently provides support for bootstrapping by an IPv4/IPv6 UDP broadcasting protocol, multicast-based DNS (mDNS), the Bluetooth *Service Discovery Protocol* (SDP), as well as manual out-of-band configuration of bootstrap information. Bootstrap modules are not coupled to the actual protocol used for disseminating the bootstrap information; rather, all locators (cf. Section 2.3.3.1) are distributed through all bootstrap modules, e.g. distributing bootstrap information over the Bluetooth SDP module does not require actual joining through SDP. Rather, all locators are distributed through the module and the joining node tries to achieve connectivity through the common subset of supported protocols. Future work will consider the integration of novel bootstrap mechanisms – e.g., based on new features of IPv6 [8] – that can be employed in a larger scale while still being applicable to smaller SpoVNets.

The mechanisms described in this section are applicable for bootstrapping the SpoVNet Base Overlay. Bootstrapping higher layer service-specific overlays is supported through a directory-like service being implemented through a *Distributed Hash Table* (DHT) and described in Section 2.3.3.3.

2.3.3 Base Overlay

SpoVNet employs a service-independent *Base Overlay* (BO) as a control structure that is used for signaling, to provide basic connectivity, and to build a uniform flat identifier-based addressing scheme. Specifically, the Base Overlay provides end-to-end connectivity including mobility and multihoming support. Furthermore, it provides distributed storage functionality that can, for example, be used for service bootstrapping. The main functional components of the Base Overlay comprise the ID/Locator split and the decentralized control structure that will be described in the next sections.

Table 2.2 Example of NodeID and locator set that makes up the endpoint descriptor.

NodeID
c07e00c4535f64a549fd15cf41439127b785710cfa029384

Endpoint descriptor	
layer 4	tcp{31016};udp{32020};
layer 3	ip{192.168.178.23|129.13.182.17|2800:1450:8006::68};
layer 2	rfcomm{[00:26:5e:ab:f9:e7]:10};

2.3.3.1 ID/Locator Split

The Base Overlay provides two main concepts to foster the easy development of service overlays: *NodeIDs* and *LinkIDs*. Nodes are addressed using NodeIDs that are invariant to mobility or multi-homing. A NodeID maps to a set of network locators and CDIDs that describe physical network attachment points of a node, which is shown as an example in Table 2.2. A service that creates a communication context with another node requests establishment of a *virtual link* that is further identified through a *LinkID*. Similar to a NodeID, the LinkID is invariant to mobility and multi-homing. Each node maintains a set of local locators (i.e., layer 2–4 addresses) that are collectively called the *endpoint descriptor*. Creation of a link is transparent to the service: The two participating nodes exchange their endpoint descriptors through the Base Overlay. Using the Base Communication both nodes try to establish transport connections from *both* directions. After negotiating feasible transport connections, a successfully established virtual link is identified by its LinkID.

2.3.3.2 Decentralized Control Structure

SpoVNet uses a *Key Based Routing* (KBR) overlay as Base Overlay. It is transparently usable and built either upon direct transport connections or relay paths, respectively. Provided by the Base Communication, it is used as the basic control structure for signaling. Whenever a data connection between nodes is required, a connection is established by signaling via the overlay, i.e., by exchanging current locator sets. The locators are passed to the Base Communication, which establishes a connection as described earlier. Thus, the Base Overlay interacts with the Base Communication to establish end-to-end connectivity. Such connections are set up on demand and managed automatically, relieving developers from management of transport connections between nodes. The use of direct connections is, e.g., in contrast to *Identity Hash Routing* (IHR) [13] used by UIA, which uses the overlay to overcome heterogeneity and always routes messages by the Kademlia-based KBR functionality. Such behavior enables communication in otherwise disconnected networks, although at the cost of inefficient communication paths.

As described above, KBR mechanisms are an appropriate way to implement identifier-based addressing, implicitly providing the non-functional features of scalability, self-configuration, and self-maintenance. The Base Overlay is primarily used as KBR overlay that provides functionality for identifier-based addressing through NodeIDs. Since connections across the heterogeneous network are transparently handled by the Base Communication, SpoVNet is not restricted to one specific KBR protocol. Thus any protocol providing key-based routing functionality can be employed (e.g., CAN, Chord, Pastry, Kademlia). The current implementation uses a modified Chord variant [17].

2.3.3.3 DHT

The KBR routing that is part of the Base Overlay control structure (described in Section 2.3.3.2) is further used for higher layer mechanisms that provides *Distributed Hash Table* (DHT) functionality. This feature enables participating SpoVNet nodes to store arbitrary information in a distributed manner in the control structure. One of the main potential uses for this feature lies in the acquisition of service bootstrap information. A node joining a SpoVNet service can get the necessary service bootstrap information on how to enter (e.g. the identifier of the *Rendezvous Point*) from the Base Overlay (of which it is already a part). Therefore, service developers can easily rely on this convenient feature, being unburdened from carrying out complex service bootstrapping procedures. Note that this service bootstrapping is different from the general bootstrapping procedure required for entering a SpoVNet Instance (cf. Section 2.3.2.4).

2.3.4 Security Component

Central mechanisms in the security of SpoVNet are cryptographically generated NodeIDs, as well as cryptographically generated identifiers for addressing of a specific SpoVNet. These are identifiers created from public/private key pairs that provide proof of ownership. For, example, such NodeIDs provide security while establishing secure virtual links through authenticity and resistance against spoofing attacks. Nodes are herewith able to perform authenticated, integrity-protected, and encrypted communication. Creation of closed SpoVNets is an interesting concept for smaller groups. Here, an initiator can generate a cryptographically generated identifier that describes the SpoVNet instance. Any node that wants to join the SpoVNet can validate that this initiator really created the SpoVNet instance, as he can prove ownership of the SpoVNet instance identifier. Therewith, the SpoVNet is cryptographically bound to the initiating node that performs authorization, for example. The private key that is required for proof of ownership for the initiator can further be passed on to other nodes if required. Conceptually, the use of hidden SpoVNets, i.e., keeping bootstrapping information only visible to authorized nodes, is an interesting concept.

2.3.5 Cross-Layer Component

On one hand, the SpoVNet architecture copes with the heterogeneity and dynamics of the underlying network, providing a high degree of abstraction to the service or application developer. On the other hand, concealment of underlay properties can lead to suboptimal decisions while building up the service overlay. In order to permit coping with the network and its challenges while simultaneously providing insight into the structure of the network and its performance, SpoVNet provides cross-layer information by the *Cross-layer Information for Overlay* (CLIO) [14] component, shown as *Cross-Layer Component* (CL) in Figure 2.2. CLIO is able to provide manifold information that could be used for optimization and decision-making by the services, e.g., available bandwidth or path latency. Services can directly use this component to request information about various underlay properties and performance measures, which they can employ to optimize their overlay structures and to make further decisions concerning the service. Thus, they may register for the periodic reception of some specific information that is of interest to them over the whole service lifetime, or they may get information from CLIO a single time. The CLIO component relieves services of the need to implement their own mechanisms for obtaining the desired information, which can be shared between all services.

2.3.6 Interfaces

The SpoVNet architecture offers different *Interfaces* (IF) to access its functionality, depending on the type of service or application intended to be used. When developing dedicated services or applications for SpoVNet, the architecture offers a fully-featured developer interface, providing accessibility to its mechanisms. Besides this, legacy applications also can be used with the ar-

Table 2.3 Overview of SpoVNet developer interface for service and application development.

	Function	Description
Node	`initiate`	Create an application-instance specific SpoVNet
	`join`	Find other nodes in this SpoVNet instance and join
	`leave`	Leave the SpoVNet instance
	`onJoinCompleted`	Indicate join success
	`onJoinFailed`	Indicate join failure
	`onLeaveCompleted`	Indicate leave success
	`onLeaveFailed`	Indicate leave failure
Communication	`bind`	Bind a service to *ariba* using a specific ID for multiplexing
	`unbind`	Unbind a service from *ariba*
	`establishLink`	Establish a virtual link to another NodeID
	`dropLink`	Drop a virtual link
	`sendMessage`	Send a message over a link, if no link given built one up first
	`onLinkUp`	Indicate successful setup of virtual link
	`onLinkDown`	Indicate successful dropping of virtual link
	`onLinkChanged`	Indicate link mobility
	`onLinkFailed`	Indicate that the link has dropped
	`onLinkRequest`	Indicate incoming link request
	`onMessage`	Incoming messages on a virtual link

chitecture without the need for modifications by using a legacy interface [16]. We will shortly describe both interfaces in the following.

2.3.6.1 Developer Interface

The developer interface provides access to the SpoVNet architecture for services and applications. In this section we will describe the concrete developer interface that comes with our implementation of the SpoVNet Base, called *ariba*. The *ariba* developer interface exposes two sub-interfaces to the developer, as shown in Table 2.3: a node-specific interface, and a communication-specific interface. The node-specific interface defines functionality used for creating and joining the SpoVNet instance, while the communication-specific interface provides establishment of virtual links and message sending. Several services can be bound to the communication-specific part of the interface, each with its own service-specific *ServiceID* that allows for multiplexing several services and applications used in a single *ariba* instance.

Using *ariba* as a communication library requires no extensive setup, but rather an implementation of simple interface functions, therewith providing usability to the developer and extensibility inside the framework through abstraction. Due to abstraction of network addresses and explicit transport protocols – using identifiers and requirements-oriented interfaces – *ariba* allows for rapid prototyping of communication services and applications. *ariba* has been designed to operate in an event-driven way, preventing complex threading issues for service developers. Services exclusively operate on virtual transport links created through the communication-specific interface. By using event-functions, services are notified of communication-specific and node-specific events and can react accordingly. This paradigm heavily simplifies service development and allows developers to focus on the actual service functionality: By consistently using overlay networks the *Underlay Abstraction* provides for easy and comfortable creation of innovative services, e.g., a network-aware multicast service [18]. The set of interfaces of all services are accumulated to a service abstraction, which developers can use to implement new services and applications. The separation of service implementation and service interface permits the developer to exchange the specific implementation of a service without requiring any changes to the interfaces or the application using them. For instance, a SpoVNet service – like e.g. Multicast service – can be replaced by a native network service – like IP Multicast – as soon as it becomes available within the underlying network. Therewith, the SpoVNet architecture provides for a seamless migration between currently existing and future networks, keeping innovative services and applications stable and enabling their use even in today's networks without requiring native underlay support.

2.3.6.2 Legacy Interface

Besides dedicated applications and services, SpoVNet also supports existing legacy applications. The provided legacy interface appears to such legacy applications like a common socket interface that is connected to the SpoVNet architecture via a `TUN`-device, thus enabling transparent use [16]. Resolution of DNS names is performed by SpoVNet transparently through the Base Overlay.

2.3.7 Summary

This section introduced the SpoVNet architecture and its main components as a powerful framework for the development and deployment of spontaneous and flexible overlay-based services and applications. The SpoVNet architecture offers transparent maintenance of connectivity in heterogeneous network environments through the collaboration of the various components and mechanisms shown in Table 2.1. The use of persistent identifiers, the cross-layer

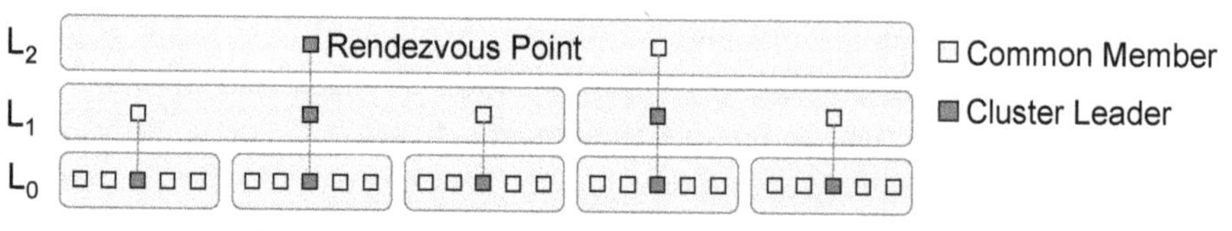

Figure 2.4 Exemplary group communication structure.

capabilities, and the easy-to-access interfaces are key features for enabling fast and to-the-point services and applications on top. We will provide an example describing the usage of this architecture in the next section.

2.4 Case Study: Building Service Overlays with SpoVNet

In this section we will provide insights into service development with the SpoVNet architecture. To demonstrate the benefits of the proposed architecture we developed a *group communication service* that builds upon the SpoVNet architecture.

For realizing group communication as a SpoVNet service we chose an unstructured overlay approach that uses a scalable hierarchy of clusters and is closely related to the well-known NICE approach [6] (cf. Figure 2.4). All participants are divided into a set of clusters with one leader being elected per cluster. Leaders themselves compose the next hierarchy layer of clusters iteratively, until only one participant, as the highest hierarchy cluster leader, is left. Nodes are clustered based on active mutual network latency measurements and a new node joining the hierarchy has to contact one dedicated node (denoted as *Rendezvous Point*) to become part of the structure. For such a service, some core mechanisms have to be implemented, namely:

1. service bootstrapping for successful joining the group communication service,
2. maintaining reachability of participating nodes in face of underlay changes,
3. actively determining mutual latencies in order to build/optimize the structure, and
4. the actual forwarding strategy inside the service structure.

While mechanism (4) is service-specific to our example, mechanisms (1)–(3) are provided by the SpoVNet architecture in a generic way such that other services can easily benefit from them.

Our sample service employs the SpoVNet architecture's developer interface to access the mechanisms (1)–(3). For addressing nodes in the service, SpoVNet identifiers are used exclusively instead of network addresses, leading to distinct node naming from the service's point of view. With these identifiers, the service builds persistent virtual links between communicating participants which are maintained transparently even in face of limited connectivity, heterogeneity, or mobility. Thus, the core mechanism (2) is solved without having to care for it inside the service itself. In order to let joining nodes find the Rendezvous Point (1), the service deploys bootstrap information in the Base Overlay directly by storing it in the provided DHT. A joining node then first becomes part of the SpoVNet instance, where it will be able to obtain this information in order to join the service overlay structure. Instead of proactively triggering periodic network measurements between participants (core mechanisms (3)), our multicast service relies on the cross-layer information provided by the corresponding cross-layer component in the SpoVNet architecture. The latency information[1] provided this way is used to determine cluster composition as well as to detect network location changes that may lead to placement adaptations in the structure. Therefore, no own measurements have to be triggered. Furthermore, the collected information in the cross-layer component could be shared by all services, thus minimizing measurement redundancies in the architecture. The service accesses the SpoVNet architectures functionality through the communication-specific interface shown in Table 2.3.

The group communication example shows how the SpoVNet architecture can be used to relieve services and applications from implementing dedicated mechanisms to handle network obstacles in order to enable easy service development. Figure 2.5 provides an overview of the architectural mechanisms that our service directly uses from the service's point of view. In our example we showed that most of these core mechanisms (except the actual service functionality (4)) are already provided by the SpoVNet architecture, allowing services in SpoVNet to be focused on their target mechanisms, making them less error-prone and less complex.

[1] While our example service only uses latency measurements, the CLIO component has a much higher flexibility in providing cross-layer information.

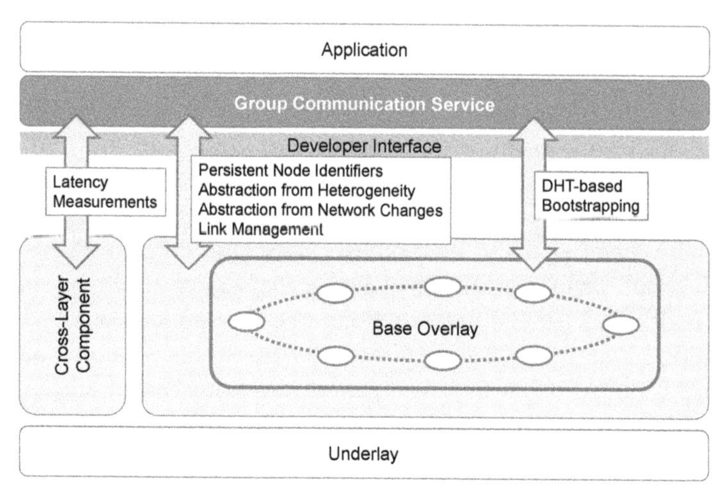

Figure 2.5 Architecture usage from a service point of view.

2.5 Conclusion and Future Directions

In this chapter, we argued that frameworks for supporting easy creation and deployment of overlay-based services can significantly foster innovation by lowering potential barriers. We identified major functional requirements for such frameworks, including the provision of end-to-end connectivity across heterogeneous networks and protocols, as well as support for mobility, multi-homing, and heterogeneous characteristics of access networks. Motivated by the observation that none of the existing frameworks meets all of these requirements, we introduced the *Spontaneous Virtual Networks* architecture. This architecture uses two closely interacting components, the Base Communication and the Base Overlay, to provide all features mentioned in this chapter. Specifically, the Base Communication supports end-to-end communication via heterogeneous underlay protocols either directly or via relay paths constructed by intermediate nodes. Furthermore, it provides transparent link management in the face of mobility and multi-homing. Integrating ID/locator split solutions like ILNP in the underlay may reduce the management effort performed by the Base Communication. The Base Overlay supports persistent addressing of overlay nodes by identifiers, which are mapped to current network addresses, implementing an ID/locator split that supports mobility and multi-homing. Furthermore, it serves as a decentralized control structure for establishing new end-to-end connections efficiently without routing data along the overlay structure. An additional cross-layer component provides information to take heterogeneous characteristics of access networks into account in order to allow for optimization.

Future research directions are three-fold: First, more sophisticated methods for establishing connections across heterogeneous networks are required. Second, novel networking paradigms like delay tolerant networking should be supported by the architecture to enhance the field of applicability. Third, sophisticated delivery services that reduce network load by exploiting multihoming must be developed.

Acknowledgements

This work is funded as part of the *Spontaneous Virtual Networks (SpoVNet)* project by the Baden-Württemberg Stiftung within the BW-FIT program and as part of the Young Investigator Group *Controlling Heterogeneous and Dynamic Mobile Grid and Peer-to-Peer Systems* (*CoMoGriP*) by the *Concept for the Future* of Karlsruhe Institute of Technology within the framework of the German Excellence Initiative.

References

[1] Ariba Website. http://www.ariba-underlay.org.

[2] FreePastry Website. http://www.freepastry.org.

[3] JXTA Website. https://jxta.dev.java.net/.

[4] B. Ahlgren, J. Arkko, L. Eggert, and J. Rajahalme. A node identity internetworking architecture. In *Proceedings of Global Internet Symposium*, Barcelona, Spain, pages 1–6, April 2006.

[5] R. Atkinson, S. Bhatti, and S. Hailes. ILNP: Mobility, multi-homing, localised addressing and security through naming. *Telecommunication Systems*, 42(3/4):273–291, December 2009.

[6] S. Banerjee, B. Bhattacharjee, and C. Kommareddy. Scalable application layer multicast. In *Proceedings of SIGCOMM'02*, Pittsburgh, Pennsylvania, pages 205–217, October 2002.

[7] R. Bless, C. Hübsch, S. Mies, and O. Waldhorst. The underlay abstraction in the Spontaneous Virtual Networks (SpoVNet) architecture. In *Proceedings of Next Generation Internet Networks (NGI)*, Krakow, Poland, pages 115–122, April 2008.

[8] R. Bless, O.P. Waldhorst, C.P. Mayer, and H. Wippel. Decentralized and autonomous bootstrapping for IPv6-based Peer-to-Peer networks. Winning Entry of the IPv6 Contest 2009 by IPv6 Council, May 2009.

[9] Cisco. Cisco visual networking index: Global mobile data traffic forecast update, 2009–2014. http://www.cisco.com/en/US/solutions/collateral/ns341/ns525/ns537/ns705/ns827/white_paper_c11-520862.html, February 2010.

[10] D.D. Clark. Toward the design of a future internet. Technical Report Version 7.0, Massachusetts Institute of Technology (MIT), 2009. http://groups.csail.mit.edu/ana/People/DDC/Future Internet 7-0.pdf.

[11] F. Dabek, B. Zhao, P. Druschel, J. Kubiatowicz, and I. Stoica. Towards a common API for structured peer-to-peer overlays. In *Proceedings of the 2nd International Workshop on Peer-to-Peer Systems (IPTPS '03)*, Vol. 2735/2003, Berkeley, CA, pages 33–44, February 2003.

[12] J. Dinger and O. Waldhorst. Decentralized bootstrapping of P2P systems: A practical view. In *Proceedings of IFIP International Conference on Networking*, Aachen, Germany, pages 703–715, May 2009.

[13] B. Ford. *UIA: A global connectivity architecture for mobile personal devices*. PhD Thesis, Massachusetts Institute of Technology, September 2008.

[14] D. Haage, R. Holz, H. Niedermayer, and P. Laskov. CLIO – A cross-layer information service for overlay network optimization. In *Proceedings of GI/ITG Fachtagung Kommunikation in Verteilten Systemen (KiVS 2009)*, Kassel, Germany, pages 279–284, March 2009.

[15] M. Hosseini, D.T. Ahmed, S. Shirmohammadi, and N.D. Georganas. A survey of application-layer multicast protocols. *IEEE Communications Surveys & Tutorials*, 9(3):58–74, July 2007.

[16] C. Hübsch, C. Mayer, S. Mies, R. Bless, O. Waldhorst, and M. Zitterbart. Demo abstract: Using legacy applications in future heterogeneous networks with ariba. In *Proceedings of IEEE INFOCOM*, San Diego, May 2010.

[17] C. Hübsch, C.P. Mayer, S. Mies, R. Bless, O.P. Waldhorst, and M. Zitterbart. Reconnecting the internet with ariba: Self-organizing provisioning of end-to-end connectivity in heterogeneous networks. *ACM SIGCOMM Computer Communication Review*, 40(1):131–132, January 2010.

[18] C. Hübsch and O.P. Waldhorst. Enhancing application-layer multicast solutions by wireless underlay support. In *Kommunikation in Verteilten Systemen (KiVS)*, Kassel, Germany, pages 267–273, Springer, 2009.

[19] D. Johnson, C. Perkins, and J. Arkko. Mobility support in IPv6. RFC 3775 (Proposed Standard), June 2004.

[20] J. Liebeherr, J. Wang, and G. Zhang. Programming overlay networks with overlay sockets. In *Proceedings of 5th Workshop on Networked Group Communications (NGC)*, Munich, Germany, Lecture Notes in Computer Science, Vol. 2816, pages 242–253, Springer, 2003.

[21] S. Mies, O. Waldhorst, and H. Wippel. Towards end-to-end connectivity for overlays across heterogeneous networks. In *Proceedings of Workshop on the Network of the Future (Future-Net), co-located with IEEE ICC*, Dresden, Germany, June 2009.

[22] A. Rodriguez, C. Killian, S. Bhat, D. Kostić, and A. Vahdat. MACEDON: Methodology for automatically creating, evaluating, and designing overlay networks. In *Proceedings of 1st USENIX Symp on Networked Systems Design and Implementation (NSDI 2004)*, San Francisco, CA, pages 20–20, 2004.

[23] K. Shudo, Y. Tanaka, and S. Sekiguchi. Overlay weaver: An overlay construction toolkit. *Computer Communications*, Special Issue on Foundation of Peer-to-Peer Computing, 31(2):402–412, 2008.

[24] M. Stiemerling (Ed.). *System Design of SATO and ASI.* Deliverable D12-F.1, Ambient Networks Project, 2006.

[25] O. Waldhorst, C. Blankenhorn, D. Haage, R. Holz, G. Koch, B. Koldehofe, F. Lampi, C. Mayer, and S. Mies. Spontaneous Virtual Networks: On the Road towards the Internet's Next Generation. *it – Information Technology*, Special Issue on Next Generation Internet, 50(6):367–375, December 2008.

[26] O.P. Waldhorst, R. Bless, and M. Zitterbart. Overlay-Netze als Innovationsmotor im Internet – Spontane virtuelle Netze: Auf dem Weg zum Internet der Zukunft. *Informatik Spektrum*, 2(288):171–185, March 2010.

3

Service Architectures for the Future Converged Internet: Specific Challenges And Possible Solutions For Mobile Broadband Traffic Management

Jochen Eisl[1], Albert Rafetseder[2] and Kurt Tutschku[2]

[1]*Nokia Siemens Networks GmbH & Co. KG, 81541 Munich, Germany; e-mail: jochen.eisl@nsn.com*
[2]*Department of Future Communication, University of Vienna, 1090 Vienna, Austria; e-mail: {albert.rafetseder, kurt.tutschku}@univie.ac.at*

Abstract

Overlay and P2P-based applications have proven to be highly successful. The issues of the resulting loads on networks are witnessed by network operators and users alike [21]. Current overlays do not synchronize with the capabilities and constraints of the underlying network. This chapter discusses traffic management mechanisms for overlay applications, and investigates how future mechanisms for the core part of mobile broadband systems can overcome today's limitations even for applications other than P2P. Steadily increasing the access bandwidth on the radio interface will not suffice to solve the issues described, but rather make congestion in the core and transport networks more likely. Thus, mechanisms in the core are needed to smartly interact with overlay applications. We show in this chapter how a *Resource Information Service* (RIS), a new functional component in the future mobile network architecture (Evolved Packet System, EPS) can be designed for deployment in a cellular operator network. The RIS coordinates traffic management on different layers. A prototype is realized using network virtualization techniques.

Anand R. Prasad et al. (Eds.), Future Internet Services and Service Architectures, 49–72.

Keywords: Evolved Packet System (EPS), overlays, Peer to Peer (P2P), traffic management.

3.1 Introduction

Application overlays such as *Peer-to-Peer* (P2P) *applications* or *Content Distribution Networks* (CDNs) have changed the use of the Internet fundamentally. A major success of overlays was their capability to enable and support the concept of a *Prosumer* [33], i.e. users are simultaneously *producer* and *consumer* of information. Overlays support this capability as they are *application-specific*, e.g. using their own naming scheme for the mediation of information [36].

This resulted in a significant rise of overlay data traffic in recent years [10]. Forecasts even indicate an acceleration of this trend, especially in public mobile networks [15]. As a result, overlays and the underlying network may experience performance degradations, e.g. from overload situations, which can already be felt by both network operators and users of today's networks [21].

Before addressing traffic management challenges for overlays and potential approaches to cope with them, we provide an overview in Section 3.2 on overlay application types which already create challenging traffic patterns and data volumes, or which likely will do so soon.

The introduction of high capacity cellular access technologies, as realized by Third-Generation Partnership Project's (3GPP's) Long Term Evolution (LTE) [5], constitutes the requirement of a highly efficient traffic management also in the core network part of cellular networks. The resources in the core networks of these systems, i.e. in the *Evolved Packet Core* (EPC), are not as expensive as on radio links, but lack support for synchronization with the overlay applications. For understanding these constraints, in Section 3.3 we provide an overview of the EPS architecture (which comprises the LTE access technology and the EPC) and of the *bearer concept*, which serves to aggregate packets into flows within cellular systems. Bearers permit the handling of packets in a flow as a whole, e.g. for efficient mobility support. The bearer concept was introduced in mobile data networks in order to handle the dynamics from mobility and to implement additional features such as support for Quality of Service (QoS). The concept of bearers is rather diametrical to the concept of *packet routing* typically used in IP networks.

Traffic management for packet-oriented and IP-based systems has been studied in research communities for a long time, so we will provide an overview of state-of-the-art concepts in Section 3.4, and then discuss in Section 3.5 how new approaches and tools can be adopted in the EPS. One of the main challenges is to integrate the concepts of traffic management at different layers, i.e. bearer-oriented vs. packet-oriented vs. application-oriented, into a unified *traffic management service.*

A fundamental deficiency in various state-of-the art solutions is the mismatch between the requirements from applications, and the capabilities provided by and constraints inherent to the underlying network. To overcome this mismatch, in Section 3.6 we introduce the *Resource Information Service* (RIS), which can resolve this lack of coordination and mediate between adjacent layers or between neighboring entities. In order to show the capabilities of this mechanism, we have established a testbed for proving the feasibility of the RIS concept based on the *Seattle* virtualization architecture [13]. This approach may not perfectly simulate behaviour in a mobile core network, but is a powerful mechanism to show the impact of resource selection at different locations in the Internet. Since this chapter describes ongoing research on traffic management in the EPS, we conclude in Section 3.7 with an overview on the implementation and evaluation tasks to be carried out in the near future.

3.2 Overview on Application Overlays

Overlays have become very popular in the early 2000s due to the success of the P2P file sharing overlay networks. Video streaming and progressive download applications provided by means of CDNs have contributed with even higher traffic volumes in recent years, and will dominate the Internet at least in the near future. In order to understand the potential impact of overlays on specific networks, an overview is provided on those application. Figure 3.1 depicts a classification of different types of overlays with prominent examples for all the mentioned categories.

P2P Applications have grown in the early 2000s to become one of the dominant applications on the Internet, at least by traffic volume. Backbone operators, Internet service providers, and access providers consistently reported P2P-type traffic volumes exceeding 50% of the total traffic load in their networks at that time [10].

P2P systems *virtualize* the shared resources, hiding the actual way they are provided by the community of peers, e.g. where data is physically stored or how a function is performed. The traffic volumes and the size and duration of traffic flows of P2P systems might impact the performance of transport

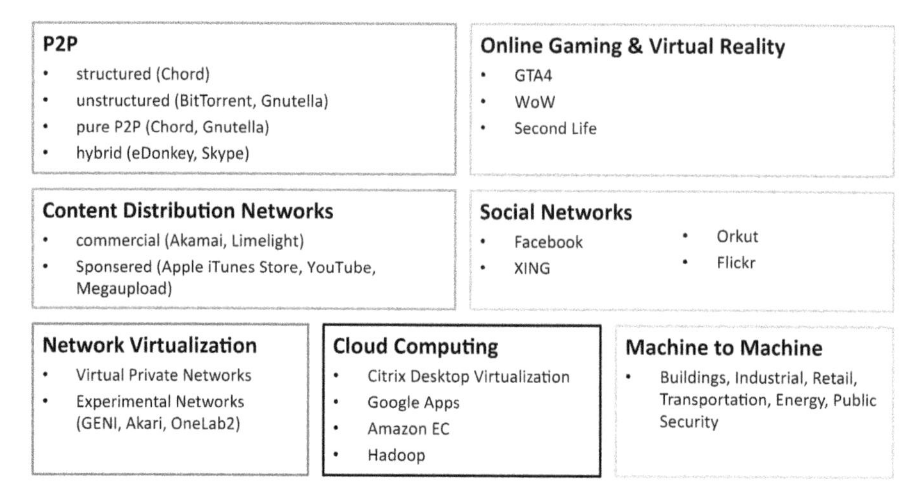

Figure 3.1 Classes of overlay applications.

networks. P2P traffic flows do not exhibit heavy tails, and are generally rather short due to multi-source downloading [34]. This also defines the time scales on which traffic management has to be performed [36].

Content Delivery Networks (CDNs) [20], especially for video distribution, are among the most popular applications on today's Internet, and are still expected to grow significantly in future. Simple CDN architectures, when implemented as an overlay on top of the current Internet architecture, may lead to increased overhead and may not perform as required. As a result, major content providers operate private WAN infrastructures to support content delivery with sufficient bandwidth. In order to distribute the content to the users, these infrastructures need gateways to the Internet, denoted as network end-points [18]. They enable the transfer of the content from the private WAN into public IP networks, so that the content becomes generally accessible to end users. Content providers typically do not announce the locations of these network endpoints, although their locations can be identified by measurements [18]. Some content providers arrange peering agreements with *Communication Service Providers* to enable a short path to the user without transiting other networks.

Virtual Reality and *Massively Multiplayer Online Games* (MMOGs), such as *Second Life* [25] and *World of Warcraft* (WoW) [12] have grown from a small base of enthusiastic users to millions of paying fans. The latest research on MMOGs from media analysts [3] still shows significant growth and innovation in the MMOG market, driven by the dominance of WoW. Typically, users connect to MMOGs using standard PC hardware, but video game consoles have also started providing access to them. Server farms apply distribution and load sharing techniques to handle the very large number of users and to reach scalability. Basically, the communication relationships of Virtual Reality and MMOGs can be classified into three categories: (a) between user clients and game servers, (b) between user clients (e.g. for direct audio communication), and (c) between servers. While traffic volumes for the communication cases (b) and (c) are hard to distinguish from other traffic, or are not accessible since it is carried mostly in private LANs, first analyses of traffic between user clients and game servers and the associated overlay topologies are available [14, 17, 31].

Online Social Networks have emerged as one of the key representatives of the so called Web 2.0 [32]. In Online Social Networks, users form an overlay network based on their social relationships. Most social networks are implemented as rather centralized web sites, e.g. Facebook or Flickr. If overlays become structured according to social network patterns, then a mismatch of the overlay and the physical network is expected in general. Traffic adaptation should leverage this mismatch, e.g. by clustering members of a certain part of the online community on the same server.

As outlined above, distributed Internet applications use overlays in new and very diverse ways. Further ways of using overlays start to get popular, e.g. *cloud computing* [38], or are being intensively discussed by researchers, e.g. *routing overlays using network virtualization.* These types of overlays aim at more general usage and higher flexibility compared to traditional VPNs [8].

We will shift our focus now from the recently evolved notion of overlay networks to current advances in the field of public mobile networks, whose design continues to support the well-known voice and text message services while at the same time acknowledging the demand for mobile high-bandwidth Internet access.

3.3 The Evolved Packet System

In this section, we will take a look at the architecture and principles of the EPS (Evolved Packet System), the technology destined to replace GSM (Global System for Mobile Telecommunication) and UMTS (Universal Mobile Telecommunications System) in public mobile networks.

GSM had initially enabled nation-wide mobile voice and text message services. The introduction of UMTS then was a major evolution in mobile communication, as it permits data services for sufficient Internet experience on mobile handhelds or PCs. Further enhancements of the UMTS radio access technology have been introduced in recent years. High speed packet access (HSPA) facilitates transmission rates up to several Megabits per second. However, the upgrade of the existing UMTS architecture to higher access speeds cannot address problems that have shown over the last few years: (a) to operate networks cost-efficiently despite the enormous growth of data traffic, and (b) to converge the diverse wireless and fixed-line access systems. The capacity of the core networks has not been an issue so far, but may have to be upgraded as well with continuous deployments of *Long Term Evolution* (LTE), the new 3GPP high bandwidth radio interface technology.

Due to the evolution of the radio access, the core network architecture of UMTS was revisited as well. As a result, 3GPP [5] initiated the *System Architecture Evolution* (SAE) project which defines a new packet-only core network, denoted as the *Evolved Packet Core* (EPC). The combination of the EPC and the LTE radio access is the *Evolved Packet System* (EPS). Although EPS is the only correct term for the overall system, the system is often referred to as LTE/SAE or simply LTE.

Let us take a look at the current architecture of the EPS. The logical architecture is characterized by a clear distinction between user and control plane entities. The core side of the user plane consists of two logical entities, the eNodeB (evolved NodeB) and the SAE-GW (Service Architecture Evolution GateWay). The eNodeB includes the radio interface to the UE (User Equipment, e.g. a mobile phone), and the SAE-GW connects to a PDN (Packet Data Network, typically the Internet). This is a great simplification over UMTS, where parallel structures for circuit-switched and packet-switched data paths exist. Traffic management is supported with features such as scheduling of packets on the radio interface or assignment of resources in the core network to specific flows (*bearers*, see below) according to network policies.

In detail, the SAE-GW combines functions of two logical sub-entities, the Serving Gateway (S-GW) and the Packet Data Network Gateway (PDN-GW or P-GW). The S-GW serves as the mobility anchor when a UE moves (causing a handover) between the same or different types of cellular networks, e.g. between UMTS and LTE. The PDN-GW enforces per-subscriber QoS and flow-based policies on data streams, allocates an IP address to the UE when it joins the network, and facilitates handover between 3GPP networks (e.g. GSM or UMTS) and non-3GPP networks (e.g. Wi-Fi or WiMAX). All entities described so far are responsible for the transport of user data between UEs and the PDN, and make up the user plane of the EPS as shown in Figure 3.2.

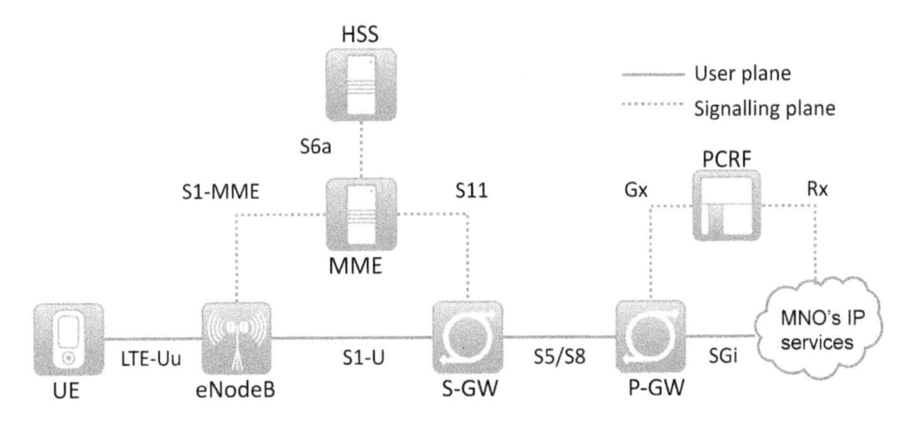

Figure 3.2 Architecture of the EPS, cf. [30].

Inside the signaling plane there is the Policy Control and charging Rules Function (PCRF) that is responsible for accounting of the transferred data volume and traffic type, and provides the per-subscriber QoS policy information to the PDN-GW as well. The HSS (Home Subscriber Server) holds user subscription information. The Mobility Management Entity (MME) is a control node that exchanges control information between eNodeBs and the core network. It performs the necessary setup and teardown, controls signaling for handover of UEs between eNodeBs, and manages the bearers accordingly.

Although designed to be an all-IP network, the different QoS requirements of different types of flows in the EPS cannot easily be met by mechanisms based on IP alone. QoS differentiation at the radio interface should, for example, determine which types of flows are admitted at a certain load on the eNodeB without initially establishing a flow and then trying to reduce its committed bandwidth afterwards. The concept introduced as a solution is the *bearer* [30], which represents a virtual connection, data path, or tunnel with unique QoS characteristics. Different types of bearers span different interfaces in the EPS, so there are radio bearers between the UE and the eNodeB, and data bearers between the entities inside the EPC. In the case of UE mobility, the associated radio bearers must be moved to any new eNodeB the UE connects to. Since the bearers act as tunnels for the IP traffic of the actual application, mobility is made to look seamless from the application's point of view, i.e. the endpoint IP address does not change.

Following this overview on the Evolved Packet System, in the next section a number of traffic management mechanisms for both fixed-line and mobile networks are discussed.

3.4 Recent Advances in Traffic and E2E-QoS Management

Traffic management mechanisms for packet-oriented and IP-based networks have been studied for a relatively long time. Despite this history, unfortunately, it has to be acknowledged that the goal of versatile and efficient *end-to-end Quality of Service* (E2E-QoS) has not yet been reached in practice. Before presenting new approaches for smart traffic steering in the mobile core, we will provide in the following an overview of major concepts which we think have contributed to, although not reached, the aim of E2E-QoS management.

3.4.1 Application-Layer Mechanisms

Application-layer mechanisms directly measure and control application metrics. Therefore, the application's performance is adjusted to deliver a good level of user experience. This comes at the expense of these mechanisms being only suitable for certain applications or requiring complex computations on application level.

Non-Transparent Caching. A well-known example of application-layer traffic management is the caching of content [24]. The web browser is configured or instructed by the web server to fetch content from a defined caching server. We denote this type of caching as *non-transparent caching* since the application is aware of this mechanism. Non-transparent caching can have multiple beneficial effects, such as reducing the load on the main server, receiving the content faster, and putting less load on the network due to preferred location of content.

Caching Peers. When caches are applied in P2P file sharing overlays, then certain peers might take over the role of caches. These peers are denoted as *caching peers* [35]. A caching peer stores selected popular content. The other peers can download this content from the cache peers as they would from ordinary peers. However, standard mechanisms and protocols for the access, discovery and management of caches that are not specific to P2P application protocols, are still missing. This lack was recently identified [29] and triggered the formation of the IETF "birds of a feather" discussion group DECADE (DECoupled Application Data Enroute) [7].

Peer Selection and Multi-Source Download. The combination of the concepts of *Multi-Source Download* and *peer selection* leads to significant performance improvements in P2P file sharing overlays. First, a peer can download in parallel parts of a file from different peers, thus exploiting parallelism. Moreover, a peer can select the providing peers on a short time scale, thus always choosing the currently optimal provider. BitTorrent [16], a P2P file sharing system, widely uses this strategy. However, care must be taken that the application-layer measurements a peer can carry out closely reflect the real properties and current status of network connections with regard to available bandwidth, delay, or congestion. Information concerning the locality of a resource inside an ISP's or a mobile operator's own network that could be used to reduce inter-domain traffic is not typically available to the application, and thus not taken into account during peer selection.

3.4.2 Network-Layer Mechanisms

Network-layer mechanisms for traffic management are more general compared to their application layer counterparts as they are usually not application-specific. On the other hand, they might lack optimal support and efficiency for certain applications for the same reason.

Packet-Based QoS Enforcement. One of the most well-known network layer mechanisms for packet-based QoS enforcement is *DiffServ* (Differentiated Services) [11,22]. DiffServ enforces QoS on a coarse-grained level. The packets are assigned to a limited number of levels, denoted as *classes*. All routers along the path of a packet stream need to implement QoS in a similar way in order to achieve consistent end-to-end behaviour. The description how a router should treat packets is denoted as the *per-hop behavior* (PHB). A typical architecture of a DiffServer router is depicted in Figure 3.3. The router comprises two consecutive stages: a *Routing Engine* and a *Queuing Engine.* The Routing Engine determines the packet's next hop as it travels through the network. The Queuing Engine handles the contention for resources among the aggregates. For example, the packets of an aggregate may be dropped arbitrarily when insufficient resources are allocated to this aggregate, and random queuing delays are introduced for the packets of the same flow.

At first view the similarities between aggregates in DiffServ and bearers in the EPS make this architecture a natural choice for QoS enforcement in the EPS. However, QoS architectures that are similar to DiffServ architecture reveal two major disadvantages [28]: They usually drop packets somewhat randomly, causing some transmissions to stall, and packets are queued be-

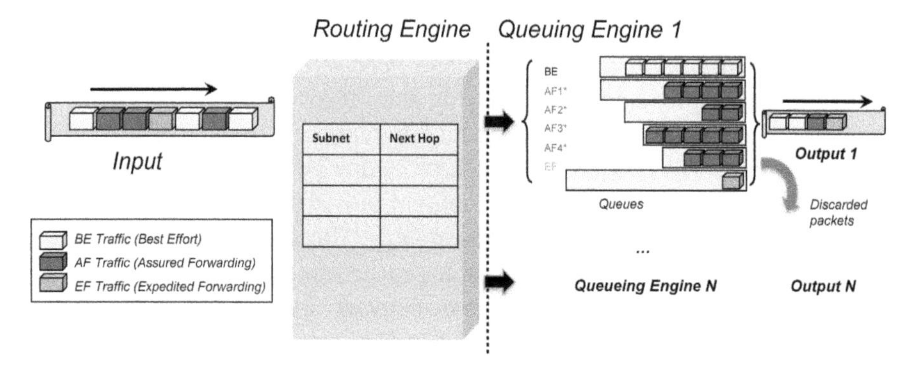

Figure 3.3 The DiffServ architecture.

cause of momentary overload, and experience substantial and non-uniform delays. The latter property reduces the throughput significantly.

Flow-Based QoS Enforcement. Hence, sophisticated flow management is suggested to overcome disadvantages of DiffServ-like QoS architectures in routers [28]. The major ideas are that each packet contains the full information to assign it to a flow it belongs to, and that flow information can be derived from the packet header or payload itself instead of being configured explicitly. The packet header includes the source and destination IP address, source and destination port, payload protocol; the payload is identifiable by Deep Packet Inspection (DPI). Sophisticated flow management identifies and imposes QoS actions before the packets of a flow are forwarded.

Figure 3.4 shows a general flow-based forwarding architecture derived from [28]. A similar hardware-oriented architecture is suggested in [37]. In Step 1, the Flow Engine computes a hash value for each packet's header. If there is a corresponding entry in the hash table, the packet either goes directly to the output port (Step 3), or it is discarded (Step 2). Using a hash table is a very efficient forwarding concept, as the Routing Engine is only consulted when the flow is seen for the first time, i.e. when there is no corresponding table entry. Then, the hash value, some header values (optionally), the port for the next hop as determined by the Routing Engine, and some flow characteristics are stored in the hash table. Afterwards, the Flow Engine keeps track of each flow's characteristics, e.g. its duration, byte count, and throughput, for identifying the flow type and for dedicated QoS enforcement. Feedback can be provided to the Flow Engine (Step 4), e.g. on queue lengths or available bandwidth.

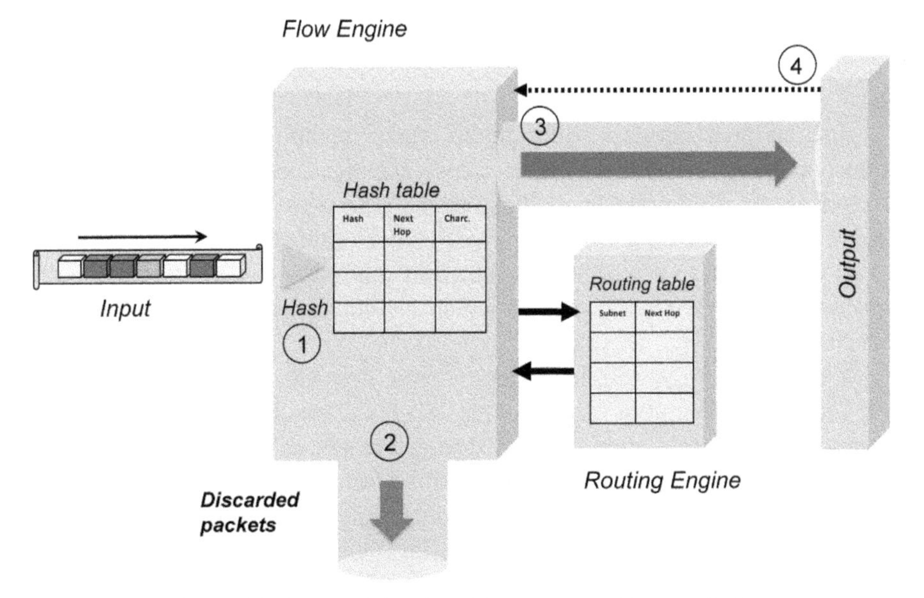

Figure 3.4 A general flow-based forwarding architecture [28].

If too many packets arrive at the router, the Flow Engine has to discard packets like in the DiffServ architecture. However, it can perform this task in a more fine-grained and intelligent way, e.g. distribute the packet loss over multiple flows, as actual flow information can be used for deciding which packets to drop from a flow, instead of relying on a synthetic, arbitrarily assigned QoS enforcement principle.

What makes the Flow-Engine-based router architecture an attractive concept for the EPS is the similarity of bearers and flows. The switching and fast forwarding capabilities of the suggested architecture can significantly reduce time for bearer setup and modification, and at the same time maintain QoS more accurately.

3.4.3 Link Layer

Traffic management on the link layer acts upon the link-layer addressing scheme such as Ethernet, and is not necessarily bound to managing IP traffic. Switches usually perform forwarding of traffic to connected hosts based on the link-layer (e.g. MAC) addresses of these hosts. Furthermore, switches may be configured to respect VLAN tags, thus creating parallel, independently manageable link-layer structures running over the same physical network. Each link-layer structure can now execute its specific switching, queuing, and prioritization schemes in order to support its specific QoS and resilience objectives.

In a cellular network, different link-layer technologies can be used to transport data between the radio access and the core part. With EPS, the introduction of Ethernet-based links in the core network may become more common, also enabling the traffic management concepts discussed here. For practical reasons, we still assume however that other transport technologies such as microwave will be used which cannot support such sophisticated mechanisms.

OpenFlow is a system for programmable flow handling on the link layer, cf. [26], which exploits the fact that most modern Ethernet switches and routers contain flow tables (tables storing the switch's hardware port identifier, VLAN ID, Ethernet addresses, IP addresses, and TCP/UDP ports of traffic seen on that hardware port) that run at line rate to implement firewalls, NAT, QoS, and collect statistics. By using OpenFlow, a network administrator can isolate network traffic, e.g. into production and research flows, thus obtaining virtual networks and *network virtualization* [9].

All OpenFlow switches must be capable to perform at least the following three basic actions on a flow: (a) to forward the flow's packets to a given hardware port (or ports), allowing packets to be routed through the network; (b) to encapsulate and forward this flow's packets to a controller via a secure channel, typically used for the first packet in a new flow, so a controller can decide if the flow should be added to the flow table or processed entirely by the controller; and (c) to drop the flow's packets, which can be used for security, to curb denial of service attacks, or to reduce spurious broadcast discovery traffic from end hosts.

Efficient flow handling capabilities make OpenFlow an appealing choice for a link-layer traffic management in the EPC. A disadvantage can be found in limited scalability. While link-layer technologies such as Ethernet are most widely used by enterprises and access network providers due to their cost effectiveness, simplicity and ease of configuration, they degrade in their performance due to their unstructured routing concepts. Systems like Seattle [23] and PortLand [27] that introduce structured link-layer architectures might point to solutions to level these limitations.

3.4.4 Mediator-Based Approaches

The traffic generated by an overlay in a physical network and the performance of the overlay depend highly on the match of the virtual structure with the physical topology. Typically, P2P overlays are either *unstructured*, i.e., random virtual connections are established between the peers, or *structured*, i.e., a certain design rule is implicitly given for the virtual topology. Neither type typically considers the capabilities of the physical resources in which they are embedded. These observations lead to the approach of *mediator-based* forming of overlay topologies. The mediator, a generally centralized entity, supports the overlay nodes in their decision how to interconnect among each others. The mediator may, for example, suggest upon request the nearest node in terms of Euclidean distance. In general, a mediator may have more knowledge about the complete networked system, e.g. about the physical topology, but it may also know the network congestion status, costs, and policies. Recently, different projects such as the IETF working group ALTO (Application Layer Transport Optimization) [2] apply the mediator concept. Even though the main purpose of the mediator is to support P2P applications, we assume that other types of applications delivered via CDN could benefit from this approach as well.

3.5 Future Traffic Management for the EPS

After presenting the traffic steering concept more generally, we turn our focus on concepts for future EPS traffic management and E2E-QoS mechanisms to the possibilities in the EPC. The main objective in the design of the mechanisms is to accommodate the requirements and the capabilities of the specific features of the EPC, its entities and the *bearer concept*. It should be stressed that most of the following suggested mechanisms were not standardized for EPS while this book was written. Some of these concepts originally were suggested for IP networks without consideration of cellular access.

Transparent caching. A possible network-layer mechanism in the EPC is *transparent caching*. Here, the data streams embedded in bearers are inspected for caching opportunities. The operation requires the function to be located at the SAE-GW, where the bearer is terminated, i.e. at PDN-GW or (in some cases) the S-GW.

Transparent caching mechanisms which support whole user objects require *Deep Packet Inspection* (DPI) for the identifications of these. Here, IP packets might be eventually modified, e.g. for the redirection of the applica-

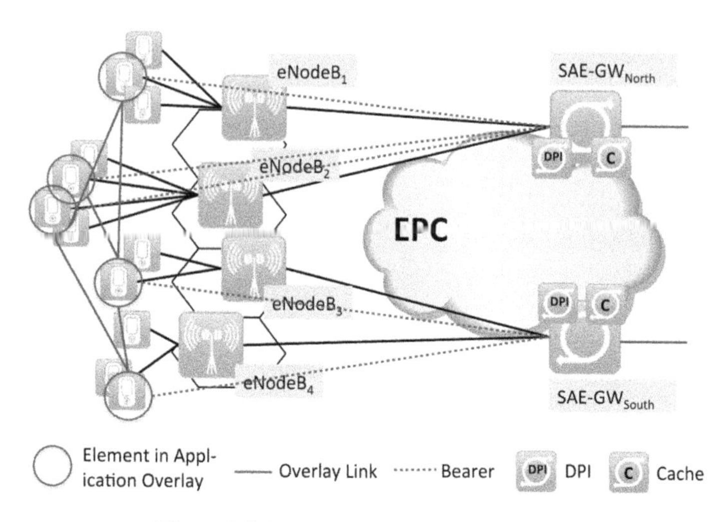

Figure 3.5 Transparent caching using DPI.

tion data flows to a certain cache entity through Network Address Translation (NAT). An initial architecture for caching using DPI is shown in Figure 3.5. The DPI function on the SAE-GW redirects the application layer flow to a cache entity. The advantage of transparent caching using DPI is the versatile applicability and fine-grain decision rules. The disadvantage of this caching mode is high computational power required for DPI, which has to be performed at line speed for a huge number of bearers in parallel.

EPC Overlay Optimization. The architecture of EPC requires a specific logical binding between EPC elements, for example that a certain eNodeB is tied to a certain SAE-GW when a bearer has been established. Physically, the interconnection of the EPC elements can be achieved by arbitrary transport connections with several hops between them. Hence, the EPC can be viewed as an overlay on top of the transport system that can be adapted and optimized. For this purpose, three optimization concepts can be identified: (a) load balancing on node level, i.e. between different nodes as a whole, (b) load balancing on functional level, i.e. between same functions on different nodes, (c) the breakout of traffic flows to the PDN. The latter concept denotes a gateway, which may be located near to a base station for the transition to the public Internet.

Load Balancing on Node Level. The most intuitive possibility for adaptive traffic management within the EPC is load balancing between similar EPC elements. This is already provided with the bearer setup to some extent, since

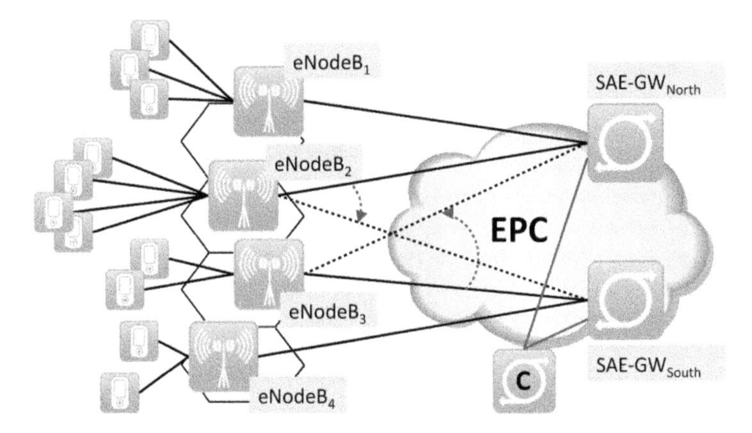

Figure 3.6 Load balancing on node level.

typically the gateway node with the lowest load is selected. Shifting bearers during their lifetime between different nodes might however be required in the future, e.g. if long-lived bearers need to be switched due to shutdown of a network node. Figure 3.6 shows the example of bearers shifted at $eNodeB_2$ from $SAE\text{-}GW_{north}$ to $SAE\text{-}GW_{south}$ and from $eNodeB_3$ from $SAE\text{-}GW_{south}$ to $SAE\text{-}GW_{north}$. This way, the number of bearers each SAW-GW has to handle can be leveled.

Load Balancing on Functional Level. The transition from a macroscopic view, i.e. how EPC components are interconnected, to a microscopic view, i.e. where EPC function are located, leads to a second class of overlay optimization possibilities. On a microscopic view, the logical functions of EPC elements are composed by a sequence of simpler functions. For example, the function of the SAE-GW can be split into a sequence of functions provided by S-GWs and PDN-GWs. The microscopic functions do not necessarily have to be executed at the same physical location. Thus, by *virtualization* and *remote execution* of these functions, the required procedures can be executed at arbitrary locations to further optimize the load within EPC. Figure 3.7 shows how the southern S-GW is used to lower the load of the northern S-GW. The virtualization aspect so far is not considered within the EPC.

Local Breakout refers to not routing application flows for roaming users via the home network of that user [6]. Traffic may alternatively be routed via a gateway node, which is closest to a considered server or peer node in the Internet. Possible strategies for local breakout depend on business models, available resources, aim of resource usage (e.g. save core network resources,

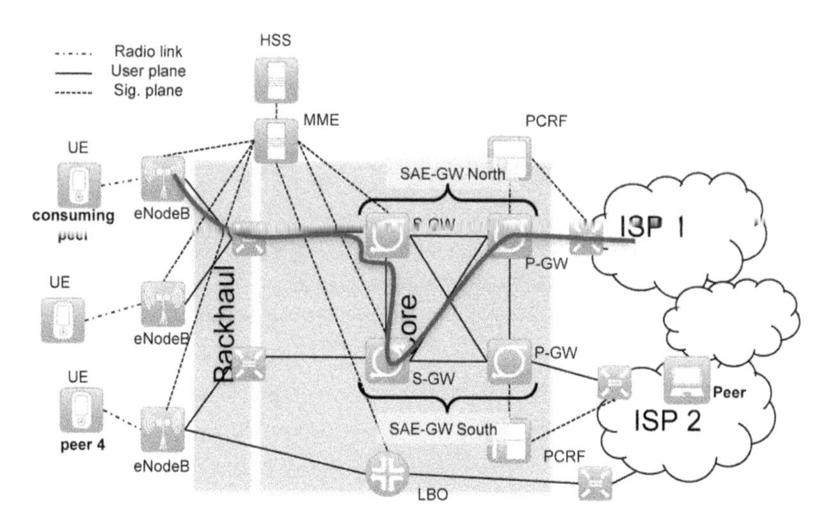

Figure 3.7 Load balancing on functional level.

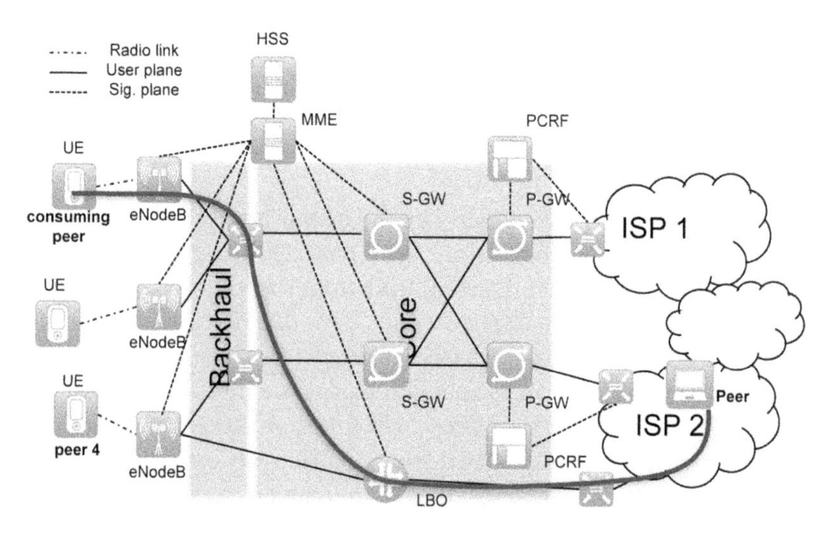

Figure 3.8 Local breakout.

but also use cheap resource as long as possible), Service-Level Agreements (SLAs) between operators, type of traffic, and policies defined in the operator domain. An example for local breakout is depicted in Figure 3.8, where the flow to a consumer peer is routed through the local breakout element to the peer located in the network of ISP 2.

3.6 Resource Information Service

The discussion above presented traffic management mechanisms for different layers in IP networks and the EPC, and also discussed cross-layer mediation of information. It was introduced how traffic management can benefit from interaction with the selection of nodes and the set-up of virtual connections. Parallels between bearers and virtual flows across mobile and fixed-line (Internet) domains may be exploited, e.g. by smart flow handling in routers and flow management in the core network. Traffic management should also consider the quality and availability of physical resources in the transport system, for example through active probing. Thus, the application overlay topology can be made to match the resources provided by the underlay in the best possible way. Through cooperation between the layers, an application overlay can optimally utilize the physical resources, cf. [19]. In order to enable cooperation between the EPS and the application overlay in a structured way, we suggest the introduction of a mediator as described previously, which is named *Resource Information Service* (RIS) subsequently.

3.6.1 The RIS Concept

The Resource Information Service offers its capabilities as a service, thus allowing for different types of cooperation between applications and the EPS. It will collect information about the EPS, the quality of application overlay nodes (or elements) in the EPS, and, if possible, also in the PDN. Finally, it may provide this information to other overlay elements for peer selection. The structure of the RIS service is depicted in Figure 3.9. The RIS might mediate information between the application overlay and traffic management mechanisms on different layers such as caching, peer selection, or flow switching, and thus influence each layer's control strategies and decisions.

A first analysis shows that RIS may be introduced in the operator's service domain, e.g. the *IMS* (IP Multimedia Subsystem), together with a new user-network interface which facilitates the exchange of content and resource information with an application client or the user. For the same purpose, a new network node interface is required to exchange similar information with an application server, which may be located within the same domain as the RIS. This interface would be needed for applications which transparently redirect a user's connection request to an alternative resource (e.g. as done by an HTTP proxy). The RIS also needs to interact with the EPC to obtain information about user profiles, possibly obtain further information about network load via a traffic monitoring entity, and to process information about policies to

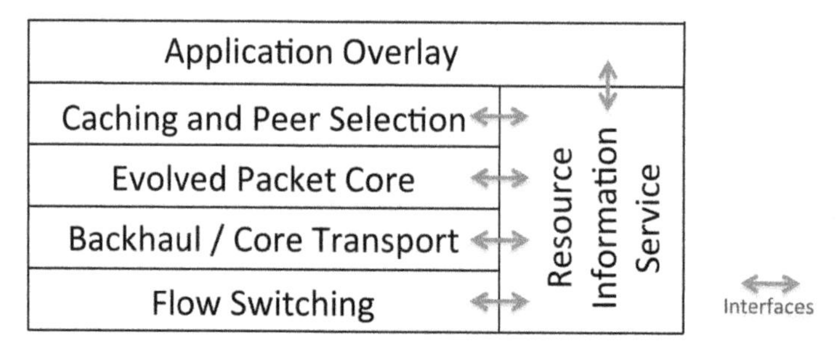

Figure 3.9 Integrated, layer-cooperative traffic management.

determine preferred resources from the network operator's perspective. The required extensions for EPC components, interfaces and interactions with bearer information require further study. In case the RIS should support only mediator functionality within the traffic management subsystem, we assume it to be a passive element in the architecture. Its task is then to provide information about the preferred location of content and to process required data related to users, policies, applications, etc. for that purpose.

3.6.2 Investigation of RIS Capabilities in a Virtual Testbed

In order to test the concept, a basic RIS prototype was developed. The prototype is able to control the switching of IP flows across different network paths whose respective properties are known. An H.264 video stream, carried over the Real-Time Protocol (RTP) in User Datagram Protocol (UDP) packets, is then used as the data source. RTP sequence numbers at the source and destination are available as an implicit measure of transmission quality.

The test network the prototype runs on employs the *Seattle* experimentation platform. Seattle [4, 13] is a distributed programmable Internet testbed system comprising donated resources, i.e. locked-down virtual machines programmable in a subset of the Python programming language. The software has been developed by the University of Washington as part of the GENI project [1]. The aim of our experiment on Seattle is to get an understanding how the throughput of flows if affected by uncoordinated actions performed by the overlay and the network infrastructure. No hardware implementations of EPS elements are involved, nor are these implementations simulated. We rather let paths through the Internet take the role of the backhaul transport infrastructure, and initiate a number of configurable virtual routers for our IP flow on remote computers, thus forming a routing overlay.

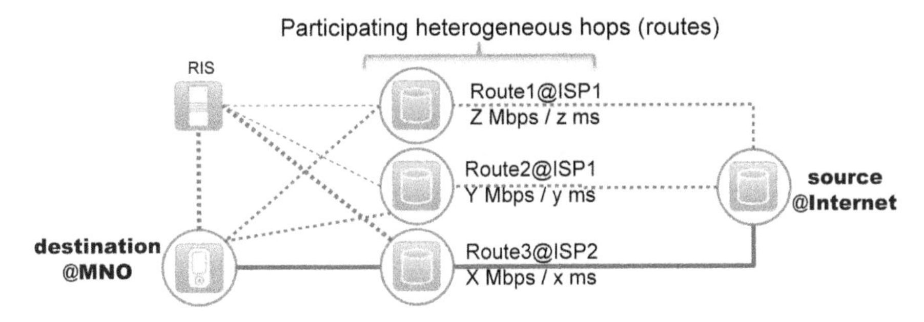

Figure 3.10 Test setup for the RIS prototype.

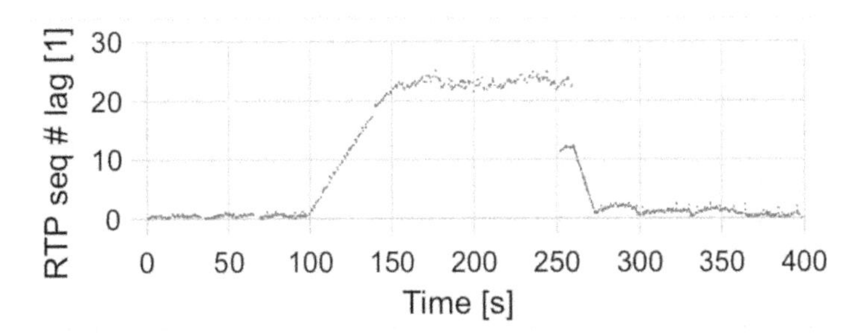

Figure 3.11 Lag of RTP sequence number between source and destination.

3.6.3 Evaluation of the RIS Test Setup

Figure 3.10 shows the considered scenario. The different routes have different properties, and switching between them is controlled by the RIS. By sending appropriate commands, the RIS can transparently reroute the stream through the overlay on the fly (similarly to how mobility of a UE is handled in the EPS). For measurement purposes, the RIS node is on the same LAN as the video source and destination in our test setup. For the evaluation, we show four distinct phases of video streaming across the routing overlay that relate to the control enforced on the route by the RIS prototype. Refer to Figure 3.11 for a trace of the lag in RTP sequence numbers between the source and the destination versus time.

Point-to-point. For reference purposes, we start with a PTP (point-to-point) route between the source and the destination. Due to both nodes being on the same LAN, the delay between sent and received data streams is very low.

Add one overlay hop. Some fifty seconds into the stream, the overlay route is modified to include the virtual router. The latency between the source and destination and the overlay node is now approximately 180 ms, still well below the noise floor in the Figure.

Overload. At 100 seconds into the stream, the network load on the overlay node is increased. Between 100 and 140 seconds, the time lag between the sent and received streams is seen to increase. When implemented fully, an integrated RIS should now trigger an automatic rerouting of the flow. In this experiment, we deliberately chose not to reroute the stream in order to analyze the mechanics of delay and loss.

It takes until 150 seconds into the experiment until the throughput of the received stream recovers, yielding two parallel traces with a distance of approximately 15 seconds. It is concluded from the measurements that during this phase, there is some loss of incoming packets whose amount suffices to let the node to keep up with the sender rate. Yet, the delay between the sent and received streams is so large that for the receiver the incoming data has no value anymore ("delay loss"). Since the loss continuously increases, we assume that for proactive rerouting of video streams, the RIS should not solely monitor the loss on a path, but also be very sensitive to variations in the overall delay.

Relaxation. Some time after 250 seconds, the overload condition on the node is relaxed. It is seen that there are now two seemingly parallel packet streams at the receiver. Some packets destined to be forwarded during the overload condition are finally played out after long delay, while fresh packets, not suffering much delay, are also forwarded. Thus, the overall receive rate is higher than the source send rate. The received stream even manages to catch up with the sent stream, so the latency again reaches 180 ms.

Final stage. For the last 50 seconds of the experiment, the PTP route is restored. As in the beginning of the experiment, the route quality is very good, that is, enough bandwidth is available, and the overall delay is low.

In the following, we discuss the effects encountered in the experiment, and some further requirements derived from that. The use of the routing overlay is transparent for the application-layer traffic as long as the required delay and bandwidth is available. The buildup of delay during the overload condition is worth discerning, and an effort was made to analyze its root causes.

The Seattle testbed consists of distributed virtual machines that try to limit their resource consumption as not to interfere with the donors' use of their hardware. In detail, the Seattle hypervisor limits resource usage by forcing the overlay application to pause when it overspends a resource. This approach is taken for every resource on a donated machine the hypervisor virtualizes, i.e. CPU time, disk space, network ports, and network bandwidth. Since the overload condition introduced in the experiment resulted in overspending the allowable bandwidth, the virtual router we initiated is forced to sleep. Therefore, it is not able to drain the socket buffer associated with the overlay connection. While the virtual router sleeps, UDP datagrams keep coming in, flooding the socket buffer and leading to packet loss. Once the hypervisor allows the virtual router to read the socket buffer again after the forced pause, outdated packets (delayed by the length of the pause) are read from the socket buffer, thus draining it. The resource limit for the overlay application probably does not suffice to completely empty the buffer, so the application is forced to pause again after reading some packets. That way, some packets already considered outdated are delayed further. The socket buffer fills up again, more packets are lost, and so forth.

This is the basic mechanism behind the buildup of delay in our experiment; the maximum delay of 15 seconds represents an equilibrium between loss, buffer size, and input and output bandwidth limits. The large amount of delay could result in an unpleasant experience for streaming; it is altogether unacceptable for any type of live video streaming. In order for overlay routing to perform well even under unfavorable conditions such as congestion, it is thus necessary that the operating system's socket buffer management and the hypervisor's resource scheduling strategies are coordinated better. Current operating system socket semantics and low-level scheduling allow for acceptable performance across applications, but it is difficult for the application to have its packets scheduled according to its own policies, as would be beneficial e.g. in our experiment. Despite seemingly correct design in either part of the experimental system, in some cases the combined performance is worse than the respective performance of the overlay and underlay alone would be, as both could for example start to queue on overload and then drop packets. We therefore consider the implementation of cross-layer coordination mechanisms as an important field for further examination and possible realization of performance gains.

3.7 Conclusion

In this chapter we detailed possible approaches for adaptive and cross-layer coordinated traffic management of overlay applications in the future Evolved Packet System (EPS). The traffic management of EPS systems may have to address several conventional layers and possibly new sub-layers such as those created through network virtualization techniques. Network virtualization may also be employed to optimize the transport layer, e.g. between radio and core part of cellular networks. Current state-of-the-art mechanisms for traffic management for overlay applications enable applications to adapt to the capabilities of an IP-based, homogeneous transport layer. The proposed mediator-based traffic management architecture – *Resource Information Service* (RIS) – exploits the multilayer nature of future EPS systems. It mediates information on resource requirements and resource quality, so that various traffic management mechanisms can be applied concurrently, in a cooperative and coordinated fashion, or exclusively. The RIS can also take into account additional network entities such as caches or local breakout elements.

Possible use cases for an RIS include policy-based resource selection (e.g. selection of peers), and the selection of the appropriate traffic management mechanism and parameters on a specific layer. More details about required extensions for functional components and interfaces within EPS and the operator's service domain are subject for further study. Investigations of the capabilities and the requirements of a prototype RIS for flow management within an Internet-based virtualized testbed is provided at the end of the chapter. The results show in a generalized way how the lack of synchronization between overlay application and network infrastructure adversely affects the delay across a routing overlay, thus harming the video stream transmitted.

References

[1] Homepage of the GENI Engineering Conference 4. Information available at http://www.geni.net/GEC4/GEC4.html.

[2] IETF Working Group ALTO (Application Layer Transport Optimization). Information available at http://www.ietf.org/html.charters/alto-charter.html.

[3] Screendigest. http://www.screendigest.com/.

[4] Seattle: The Internet as a testbed. Resources available online at https://seattle.cs.washington.edu/ Accessed 20 November 2009.

[5] Third-Generation Partnership Project. Information available at http://www.3gpp.org.

[6] 3GPP. 3GPP TS 22.278: Service requirements for the Evolved Packet System (EPS).

[7] R. Alimi, Y.R. Yang, H. Liu, D. Zhang, E. Li, and R. Zhou. Sailor: Efficient P2P design using in-network data lockers. Presented at DECADE Bar BOF at IETF 75, Stockholm, Sweden, July/August 2009.

[8] SCOPE Alliance. I/o virtualization: A nep perspective. Technical report, SCOPE Alliance, 2009.

[9] T. Anderson, L. Peterson, S. Shenker, and J. Turner. Overcoming the Internet Impasse through Virtualization. *Computer*, 38(4), 2005.

[10] N.B. Azzouna and F. Guillemin. Analysis of ADSL traffic on an IP backbone link. In *Proceedings of IEEE Conference on Global Communications (GlobeCom)*, San Francisco, CA, 2003.
[11] S. Blake, D. Black, M. Carlson, E. Davies, Z. Wang, and W. Weiss. An architecture for differentiated services. RFC 2475. Available online at http://www.ietf.org/rfc/rfc2475.txt, 1998.
[12] Blizzard Entertainment SAS. World of Warcraft. Information available at http://www.wow-europe.com/en.
[13] J. Cappos, I. Beschastnikh, A. Krishnamurthy, and T. Anderson. Seattle: A platform for educational cloud computing. *ACM SIGCSE Bulletin*, 41(1):111–115, 2009.
[14] K.T. Chen, P. Huang, C.Y. Huang, and C.L. Lei. Game traffic analysis: An MMORPG perspective. In *Proceedings of the International Workshop on Network and Operating Systems Support for Digital Audio and Video*, pages 19–24, ACM, 2005.
[15] Cisco. Cisco visual networking index: Forecast and methodology, 2009–2014. Information available at http://www.cisco.com.
[16] B. Cohen. Incentives build robustness in bittorrent. In *Proceedings of Workshop on Economics of Peer-to-Peer Systems*, Berkeley, USA, May 2003.
[17] J. Färber. Traffic modelling for fast action network games. *Multimedia Tools and Applications*, 23(1):31–46, 2004.
[18] P. Gill, M. Arlitt, Z. Li, and A. Mahanti. The flattening internet topology: Natural evolution, unsightly barnacles or contrived collapse? In *Proceedings of the 9th International Conference on Passive and Active Network Measurement*, pages 1–10, Springer-Verlag, 2008.
[19] V.K. Gurbani, V. Hilt, I. Rimac, M. Tomsu, and E. Marocco. A survey of research on the application-layer traffic optimization problem and the need for layer cooperation. *IEEE Communications Magazine*, 47(8):107–112, 2009.
[20] E. Lua J. Buford, and H. Yu. *P2P Networking and Applications.* Morgan Kaufman, Burlington, MA, 2009.
[21] J. Peterson and A. Cooper. Report from the IETF Workshop on Peer-to-Peer (P2P) Infrastructure, May 28, 2008. Available online at http://www.ietf.org/rfc/rfc5594.txt, 2009.
[22] K. Kilkki. *Differentiated Services for the Internet.* Macmillan Technical Publishing, Indianapolis, IN, 1999.
[23] C. Kim, M. Caesar, and J. Rexford. Floodless in SEATTLE: A scalable ethernet architecture for large enterprises. In *Proceedings of the ACM SIGCOMM 2008 Conference on Data Communication*, pages 3–14, ACM, 2008.
[24] James F. Kurose and Keith W. Ross. *Computer Networking: A Top-Down Approach Featuring the Internet, 4/E*, Addison Wesley, 2008.
[25] Linden Research Inc. Second Life. Information available at http://secondlife.com/.
[26] N. McKeown, T. Anderson, H. Balakrishnan, G. Parulkar, L. Peterson, J. Rexford, S. Shenker, and J. Turner. Openflow: Enabling innovation in campus networks. *ACM SIGCOMM Computer Communication Review*, 38(2):69–74, 2008.
[27] R. Niranjan Mysore, A. Pamboris, N. Farrington, N. Huang, P. Miri, S. Radhakrishnan, V. Subramanya, and A. Vahdat. PortLand: A scalable fault-tolerant layer 2 data center network fabric. *ACM SIGCOMM Computer Communication Review*, 39(4):39–50, 2009.
[28] L.G. Roberts. A radical new router. *IEEE Spectrum*, 46(7):34–39, 2009.

[29] Haibin Song, Richard Alimi, Y. Richard Yang, and Ning Zong. DECADE problem statement (slides). Presented at the 75th IETF Meeting in Stockholm, July 26–31, 2009. Available online at http://trac.tools.ietf.org/area/app/trac/attachment/wiki/BarBofs/IETF75/P2PCaching/decade-ietf75-1-ps.ppt. Accessed 29 September 2009.

[30] Matthew Baker, Stefania Sesia, Issam Toufik (Eds.). *LTE – The UMTS Long Term Evolution: from theory to practice*. John Wiley and Sons, West Sussex, UK, 2009.

[31] P. Svoboda, W. Karner, and M. Rupp. Traffic analysis and modeling for world of warcraft. In *IEEE International Conference on Communications, ICC'07*, pages 1612–1617, 2007.

[32] Tim O'Reilly. What is Web 2.0. Information available at http://www.oreillynet.com/pub/a/oreilly/tim/news/2005/09/30/what-is-web-20.html.

[33] Alvin Toffler. *The Third Wave*. Bantam Books, New York, 1990.

[34] K. Tutschku. A measurement-based traffic profile of the eDonkey filesharing service. In *Proceedings of the 5th Passive and Active Measurement Workshop (PAM2004)*, Antibes Juan-les-Pins, France, April 2004.

[35] K. Tutschku, A. Berl, T. Hossfeld, and H. de Meer. Mobile P2P in cellular networks: Architecture and performance. In *Mobile Peer-to-Peer Computing for Next Generation Distributed Environments: Advancing Conceptual and Algorithmic Applications*, chapter 16, IGI Global, Hershey, PA, 2009.

[36] Kurt Tutschku. Peer-to-peer service overlays – Capitalizing on P2P technology for the design of the future Internet. Habilitation Thesis, Institut für Informatik, Lehrstuhl für verteilte Systeme, Universität Würzburg, 2008.

[37] Q. Wu and T. Wolf. Design of a network service processing platform for data path customization. In *Proceedings of the 2nd ACM SIGCOMM Workshop on Programmable Routers for Extensible Services of Tomorrow*, pages 31–36. ACM, 2009.

[38] L. Youseff, M. Butrico, and D. Da Silva. Toward a unified ontology of cloud computing. In *Proceedings of Grid Computing Environments Workshop, GCE'08*, pages 1–10, 2008.

4

Integrating P2P with Next Generation Networks

Athanasios Christakidis[1], Nikolaos Efthymiopoulos[1], Shane Dempsey[2], Jens Fiedler[3], Konstantinos Koutsopoulos[4], Evangelos Markakis[5], Stephen Garvey[2], Nicolas de Abreu Pereira[6], Spyros Denazis[1], Spyridwn Tombros[1], Evangelos Pallis[5] and Odysseas Koufopavlou[1]

[1]*Electrical & Computer Engineering Department, University of Patras, 26500 Rion, Greece; e-mail: schristakidis@ece.upatras.gr*
[2]*Waterford Institute of Technology, Waterford, Ireland*
[3]*Fraunhofer Fokus, 10589 Berlin, Germany;*
[4]*Blue Chip Technologies S.A., 8345 Moshato, Athens, Greece*
[5]*Centre for Technological Research of Crete, Estavromenos, 71004 Heraklion, Greece*
[6]*Rundfunk Berlin Brandenburg (RBB), 14482 Potsam-Babelsberg, Germany*

Abstract

This chapter describes the major components and their interactions of a novel architecture called VITAL++ that combines the best features of two seemingly disparate worlds, Peer-to-Peer (P2P) and NGN, in particular IMS, which are then used to support multimedia applications and content distribution services. To this end, P2P is enhanced with advanced authentication, DRM mechanisms while NGN benefits from enhanced scalability, reliability and efficient distribution of service and content by exploiting P2P self organization properties. We describe novel P2P algorithms for optimizing network resources in order to efficiently distribute content among various users without resorting to laborious management operations required in NGN.

Anand R. Prasad et al. (Eds.), Future Internet Services and Service Architectures, 73–94.

Keywords: peer-to-peer, NGN, content distribution, IMS.

4.1 Introduction

The widespread adoption of the Internet technology in daily life as a major communication medium has led to the emergence of a plethora of e-applications, some of which have already become more popular than conventional telephony. Assisted by the wider roll-out of broadband communication technologies, Internet and its use has elevated digital communications to higher levels and made audiovisual communications, such as content distribution, digital TV, and video on demand affordable for everyone and mainstream among the Internet applications of today. Among them, computer based applications like the TvTube, Skype and Music City, offer rich-content to users, in real time with acceptable quality, making use of sophisticated peer-to-peer (P2P) technology algorithms for content tracking, downloading, synthesis and playback. These emerging types of applications, rich in user-created content, enabled by P2P technology, with high demands for network resources are rapidly changing the landscape of network operations and requirements creating new challenges in network and service management, configuration, deployment, protocols, etc. P2P is primarily an end-users' technology that fosters self-deployment and self organization while it achieves optimized resource utilization for the deployed applications and services. In other words P2P technology has succeeded where QoS mechanisms have failed in being deployed and operating at large scales.

On the other hand, the latest trends in telecommunication networks have led to the emergence of the first version of converged IP communication platforms known as Next Generation Networks (NGN). An initial instance of NGN is the IP Multimedia Sub-system (IMS). IMS networks constitute fully-fledged IP networks, offering, in contrast to Internet communications, quality-controllable fixed, mobile and wireless links. With IMS, users is said to be able to make ubiquitous use of operator services using 3G UMTS, WiFi and PC-based terminals. IMS is a control plane technology that primarily addresses issues of heterogeneity of access technologies, addressing schemes, AAA, security and mobility management from an operator's perspective.

So far, these two seemingly competing technologies, NGN and P2P have been deployed independent of each other, thus failing to mutually exploit their strengths towards creating a new and more powerful paradigm.

When comparing IMS and P2P [8], we compare two inherently different worlds. IMS is a technology for controlling media flows, administer sub-

	P2P	IMS
Scalability	Very good	Difficult
Single Points of Failure	No	Yes
Users as Content Providers	Yes	No
DDoS vulnerable	No	Yes
Access	Easy	Difficult
Security / AAA	Bad	Good
Topology aware	Difficult	Yes
Standardized	No	Yes
Quality of Service	No	Yes
NAT Client Problem	Difficult	No
Service Deployment	Difficult	Easy

Figure 4.1 IMS vs. P2P comparative overview.

scribers and control access to services. IMS is a very centralized architecture, which has its strengths in controlling. P2P technologies on the other side have been designed to be scalable, adaptable, and failure resilient, mainly for the distribution of media (files, streams). Figure 4.1 illustrates the complementary features of both technologies, which are discussed in the following.

As already mentioned, scalability is one of the biggest features of P2P networks, while scaling up an IMS core network can only be done by means of laborious configuration operations that increase the management overhead. P2P networks usually have no single point of failure as they are self-healing, while e.g. the HSS is a single point of failure for an IMS network. In a P2P network, users are often the only content providers, while this concept is not supported in the IMS, which clearly distinguishes between consumers and service providers. Under normal circumstances, a P2P network is not vulnerable to DDoS attacks, because an attacked node will behave as a single failure, which is subject to self-healing in the rest of the network. IMS is more vulnerable, as e.g. an I-CSCF may be flooded and put out of service for all users. Access to a P2P network can be considered easy, as there is no access control in open P2P networks. Nevertheless, password-controlled P2P solutions exist, but none of them are standardized. Access to an IMS and its services is quite complex, as cryptographic mechanisms need to be deployed for even the simplest access. Additionally, each user must be provisioned and its profile, which is stored in the HSS, must be maintained.

But in direct comparison, IMS does not only have disadvantages against P2P technologies. The complexity of access results in a much better security situation among IMS users due to the AAA management in the IMS core network for all users. Due to the fact that IMS is located in an operator network, it can also access network topology information in a standardized way from a Network Attachment Sub-System (NASS) [5]. This information must be estimated or measured in a pure P2P overlay and therefore the incorporation of information from a NASS will result in a better and more efficient overlay if presented to an overlay construction algorithm. IMS is a standardized architecture with standardized network protocols and functions, which makes development for an open market possible, while P2P overlays are usually dictated by the associated piece of software. For the same reason as for topology awareness, IMS components can influence the data paths between users and service nodes in terms of bandwidth (resource reservation), which is completely unthinkable for any pure user-driven P2P network, as user nodes cannot influence routers in any of the involved networks. P2P systems usually need to provide remarkable efforts, up to dictating the architecture of the overlay, in order to make the clients which are behind a NAT communicate with each other. In the IMS, the P-CSCF has the task to deal with NATed [6] users (proxy, holds pinhole open, all communication goes through the P-CSCF). Last but not least, probably the biggest advantage of IMS is the easy way to deploy new services, which is very expensive in a P2P system, as a P2P system is usually designed directly on the use case (e.g. file sharing).

This chapter describes the VITAL++ architecture [11] which is the result of combining and experimenting with the best features of the two worlds, namely, IMS-like control plane functionality and P2P technology. This has given rise into a combined communication paradigm that brings benefit to both users and operators and makes multimedia applications readily and securely available.

4.2 Use Cases as Motivation

The combination of P2P with NGN/IMS technologies opens numerous possibilities, a few of which will be sketched in this section. These are based on use cases implemented in the VITAL++ IST project [11].

The first use case is Remote Services Access (Geo-Blocking). Due to licensing policies AV Content on the internet is often geo-blocked, i.e. only available in certain areas. This, however, excludes users who have paid their broadcast licence fees but happen to be temporarily outside the geographic area where they live and pay their fees.

With IMS technology, viewers can be enabled to consume content they have a right to access wherever they are. A suitable area of application would be the streaming AV (IPTV) offered by national public broadcasters, which could then be made available for all rightful viewers anywhere throughout Europe. As these public broadcasters by law and regulation often cannot pay for distribution outside their business area, the use of P2P technology may be very useful to reduce costs for content distribution.

The second use case is Content Distribution in Rural Areas. In some remote rural areas, served by satellite connections or radio access, the use of P2P technologies can improve the way operators serve multimedia on-demand content. In a rural area where a number of users are connected to a broadband network using a number of satellite accesses the same content may be forwarded at different times at several satellite accesses using a high amount of bandwidth. This scenario can be improved if subscribers are connected to a local area network (wired, WiFi, etc.) and share one network access.

The network operator can minimize the use of the expensive bandwidth access using a P2P approach. In this approach a user serves contents to other users, or, alternatively, an operator whenever on-demand content requested by a user to the network is stored at a local element belonging to the operator. In both cases, when a second user in the same location asks for the same content, it is distributed from the local broadband network.

Another use case is Personalized Radio and Video Service. As a demonstrator for the usage potential of combining P2P and IMS technology, the VITAL++ consortium developed a demonstrator application which empowers a personalized radio experience way beyond broadcast radio programmes.

4.3 VITAL++ Architecture: An Overview

The VITAL++ architecture has been derived from several major design criteria, which are:

1. Minimal modification of standardized functions.
2. Easy to deploy into existing IMS networks.
3. Maximum openness for expansions.
4. Security for media and user data.

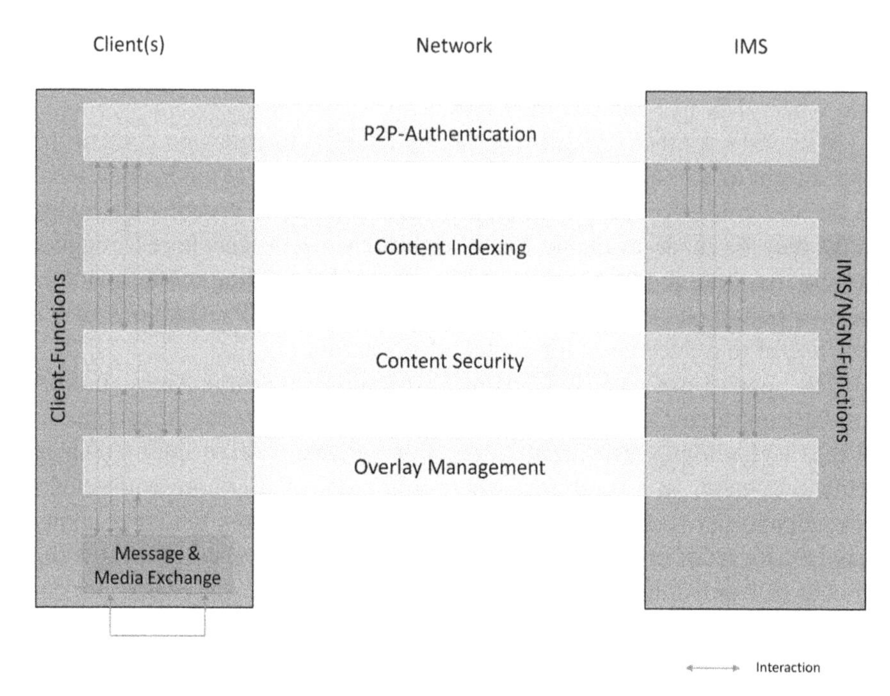

Figure 4.2 VITAL++ abstract view of the overall architecture.

5. Optimal overlays with intelligent path management.

From these design aspects, it has been decided to position IMS sided functionalities of the VITAL++ architecture in an application server; while the client sided functionalities are located directly in the client so that no additional nodes become necessary. Figure 4.2 illustrates an overview of the architecture and its functional blocks, which are explained in the remaining sections of this work. In order to address the VITAL++ challenges, a number of Sub-Architectures (SAs) that interact with each other have been defined, each one responsible to address specific design criteria. These are the P2P Authentication sub-architecture (P2PA), the Content Index sub-architecture (CI), the Overlay Managements sub-architecture that consists from Content Diffusion Overlay (CDO) and the P2P Block Scheduler and the Content Security sub-architecture (CS). Each sub-architecture spans across the client, the network and the IMS with its components. Sub-architectures may interact with each other in an arbitrary way, especially in the client, while on the NGN side there need to be well-defined interfaces. Thus the media exchange is not denoted as sub-architecture, but it interacts with these and itself in the same as well as in remote clients.

Peer-to-peer authentication (P2PA) SA is responsible for enabling clients (peers) to authenticate messages which they receive in order to ensure that clients know who has really created the message. This is a basic requirement in order to enable secure P2P messaging. Based on this, additional features can be introduced, like a secure DHT or authentic media streaming, etc.

The content index (CI) SA has three major objectives. It allows the publishing of an object from users and content providers, enables queries for objects that our system maintains and distributes, acts as a tracker and provides the initial insertion of a peer to the overlay that distributes the objects that it requests.

The OM SA is analyzed in [3]. Content Diffusion Overlay is a graph that participating peers dynamically form and maintain by each one of them selecting a small subset of peers that act as its neighbors. The purpose of the CDO is the distribution of the content, which users exchange with their neighbors in real time in the form of data (content) blocks. This graph determines the network paths that the system uses in order to distribute the content according to the user requests. The system creates and maintains one CDO for each media object that it distributes. P2P Block Scheduler ensures the on-time and stable distribution of each media block to every participating peer.

The Content Security (CS) sub-architecture has been designed to enable content providers to control the distribution of their content using a Digital Rights Management technology. VITAL++'s DRM system took its requirements from real content provider requirements and so it is related to real-world business requirements. More specifically, the requirements range from Conditional Access to streaming Content, Encryption of file-based and streamed content where appropriate, flexible rights expression, integration with Accounting, respect for privacy and consumer rights, assertion of fair-use for purposes such as backup/education, etc., and identity-based conditional access (providing a better alternative to Geo-IP blocking).

4.3.1 P2P-Authentication Sub-Architecture

The purpose of the P2P-Authentication Sub-Architecture (P2PA-SA) is to enable clients (peers) to verify the authenticity of messages which have been sent by other clients directly to them, without passing through any operator

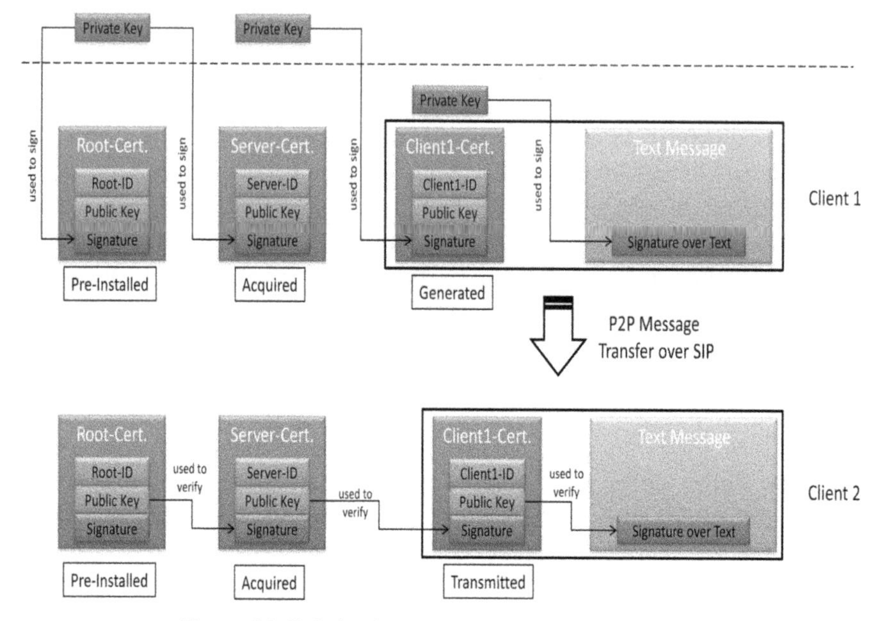

Figure 4.3 Relation between certificates and messages.

controlled entity. This envisages the security of services, which are based on pure P2P message exchange, like sharing of contacts or media, etc.

The P2P-Authentication sub-architecture works with certificates, which describe an entity and its properties. In the VITAL++ scope, three types of certificates are distinguished. The root certificate is self-signed and pre-installed in every client and P2P-authentication server module. The server certificate that is signed by the Root-CA, is pre-installed in every P2P-Authentication server module, describes the identity of the server domain and its public key and is acquired by each client during registration. The third is the client certificate that is signed by a P2P authentication server on request and describes the identity of the client and its public key.

Finally, each client is equipped with these three certificates, which allow it to perform all authenticity transactions and checks as explained in the following sections.

The relation between the certificates and their use in order to enable authentic message exchange is depicted in Figure 4.3. In every transaction there is either a certificate or signature being transported between the entities. Both are encoded as XML documents and attached as a MIME multipart message to the corresponding SIP message.

Initial certificate provision: The P2P Authentication module in the VITAL++ AS will process the registration hint from the S-CSCF and supply the newly registered user with its server certificate, signed by the common Root-CA, as illustrated in Figure 4.3.

Client certificate authorization: The client hereby generates its personal private-public key-pair and creates an unsigned certificate with its identity and public key. This is then being sent to the VITAL++ AS, which checks the identity and other fields of the certificate before he signs it with his private server key. The signed certificate is then being sent back to the client, which stores it as its own personal certificate. After performing this transaction, the client owns a valid certificate verifiable by every instance, which also knows the server certificate.

As this process is vulnerable against a man-in-the-middle attack, it is advised to encrypt this transaction. For that purpose, we suggest a Diffie–Hellman key agreement transaction before the main transaction in order to establish a common secret knowledge, which is then used to generate a symmetric key for message encryption.

Client-to-client message authentication: The sender creates a text message, which he signs with his private key, which corresponds to its own client certificate. He then sends the text message along with its own client certificate and the message signature to the receiver. The receiver can then first check the authenticity of the client certificate using its server certificate, followed by checking the message signature with the public key from the client certificate and inform the user accordingly.

4.3.2 Content Security SA

The Content Security Sub-Architecture (CS-SA) is implemented as a SIP Instant Messaging based service. The CPS is integrated within the IMS network as shown in Figure 4.5.

The User Equipment (UE) here represents a VITAL++ node. The node accesses the functions of the Content Protection Function, the logic of the Content Security Sub-Architecture) using the IMS ISc interface. The ISc is a SIP protocol connection that is used when the S-CSCF loads a trigger point corresponding to the message that has been presented to it. In our case the message is matched based on a known "service identifier" e.g. content-protection@<vital-domain> and the VITAL++ SIP header that is added to all VITAL++ messages.

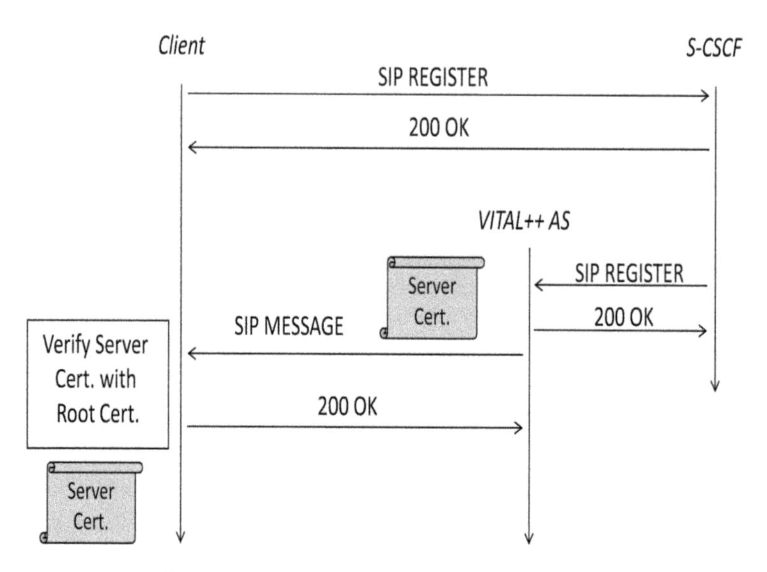

Figure 4.4 Initial server certificate acquisition.

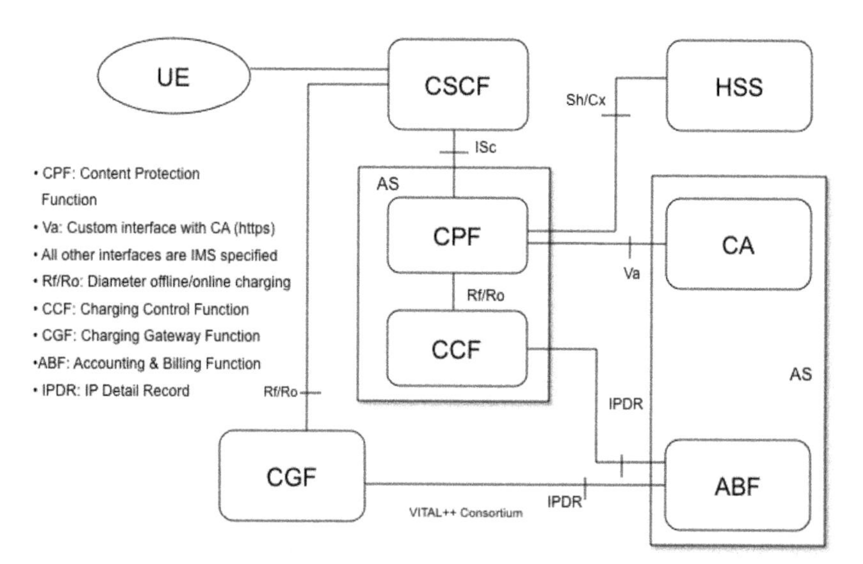

Figure 4.5 CPS integration within the IMS.

Licensing Content: The process of licensing a piece of content follows a Request/Response model and uses the SIP Instant Messaging conversation mechanism defined by the 3GPP. By re-using an existing mechanism we rely on the standard IMS authentication and message security mechanisms.

The Content Protection Function (CPF) is deployed within a standard IMS application server corresponding to the specification of the Java Community Process JSR 289 [10]. The CPF may additionally use the HSS to verify a subscriber's credentials using the Sh interface (profile information).

The content licensing process is orchestrated using a "Licensing Conductor", implemented to the design specified by the Open Media Common [1] group. In realizing this implementation, the Java Business Process Management (JBPM) open source workflow management engine was chosen to describe the licensing process. AS denotes an Application Server hosting the specified node in the CP-SA.

Identity Management: Similar to the P2P-SA, the CPF uses Public Key Infrastructure (PKI) to mutually authenticate content provider and content consumer. The CPF acts as a trusted intermediary, meaning that the content consumer and provider do not have to interact directly in the licensing process. This is necessary as the content is super-distributed among peers in the overlay. Mutual authentication means that the content consumer can be confident the licensed content is being licensed from the correct provider and has not been tampered with. The content provider similarly benefits from IMS Authentication and PKI being used to identify the consumer. A Certificate Authority (CA) is used to associate public-private key pairs with IMS identities.

Business Rules: The Drools Expert [2] rules engine is used to process business logic encoded in text-based rules. The content provider registers licensing rules with the Content Security Sub-Architecture. These rules can be parameterized and hence associated with individual users, user groups, content types, network context (e.g. user location) and billing scenarios (e.g. pre-pay, post-pay). For example: the following is true if the subscriber has a prepay account and their account balance is sufficient to afford the content item.

$$\text{Subscriber(Account\_type == “prepay”) \&\&} \\ \text{Subscriber(account\_balance)} \geqslant \text{Content(cost\_estimate)}$$

Integration with Accounting: The Accounting subsystem consists of elements including:

- A Charging Gateway Function (CGF) – An IMS charging gateway for storing usage data. The CFG exposes a diameter interface to switching and application server nodes.
- An Accounting and Billing Function (ABF) – A flexible and highly scalable accounting system based on spreadsheet worksheets. The ABF has a web service interface. It receives usage data in the IP Detail Record (IPDR) XML format and responds with an XML rating document. The rating document may be transformed into a customer bill for service and network usage.
- A Charging Control Function (CCF) – A rules-based charging decision function that evaluates whether a service can be provided to a particular user based on their charging profile and that of the service, e.g. The service may require "post-pay" and the user account may be "pre-pay" only. The CCF is implemented using JBPM workflows.

The ABF within the Accounting system associates a charging worksheet with a service or content provider and the service or content being provided. The worksheet describes additional logic for special tariffs to incentivize good behavior on the overlay e.g. relaying content. We have adopted the extensible XML-based IPDR (Internet Protocol Detail Record) Network Data Management-Usage (NDM-U) [7] scheme for charging data.

4.3.3 Overlay Management SA

The objective that we fulfill through the architecture of the OM SA is the creation and the maintenance of a scalable system, in terms of participating peers, through the distribution of its management and organization process to them. Additionally, we focus on adapting the graph to dynamic peer arrivals and departures and continuously reorganize it according to them. Special attention has been given to the adaptation of Content Diffusion Overlay (CDO) to the dynamic network conditions and exploitation of network locality in the selection of neighbors from each peer. Finally, innovative algorithms have been designed and run in CDO that deploy a P2P overlay graph structure that ensures the maximum utilization of upload bandwidth contributed by highly heterogeneous participating peers while a newly-designed P2P block scheduler exploits the properties of the P2P overlay in order to uniformly distribute the sum of the upload bandwidth resources to every participating peer.

The P2P overlay graph structure (Figure 4.6, left) consists of two interacting sub-graphs. In the first graph we insert only peers (class 1 peers) whose

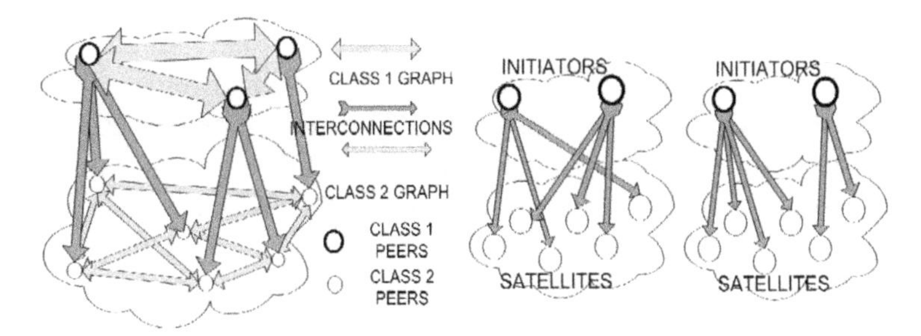

Figure 4.6 Left: The graph structure of the CDO. Right: Execution of DOMA.

upload bandwidth exceeds the bit rate of the service rate that our system has to sustain, while in the second we insert the rest (class 2 peers). These two graphs are constructed in such a way that all nodes have an equal number of connections. The interconnection between two graphs is done with connections that class 1 peers create in order to provide peers of class 2 with additional upload bandwidth resources. The number of these connections is proportional to the surplus of upload bandwidth of class 1 peers. This surplus is also assigned uniformly in peers of class 2.

In both graphs all the peers periodically execute a Distributed Optimization and Maintenance Algorithm (DOMA) that reorganizes the "neighborhoods" of CDO in order to keep the structure of the graph optimal for content delivery even during peer arrivals and departures. It also ensures high levels of bandwidth utilization. The algorithm makes use of an "energy function" that captures the impact of specific parameters, e.g. network latency, between any two nodes in the overlay. DOMA is executed between two neighbors that we note as initiators and their direct neighbors that we called satellites. Its purpose is to minimize the overall sum of the energy functions between initiators and satellites under the constraints on the number of neighbors that the aforementioned graph structure implies. In Figure 4.6 (right) the length of the arrows expresses the value of the energy function. One initiator in the left figure has twice as much surplus bandwidth as the other. We observe that the execution of DOMA minimizes the sum of energy functions while it reassigns the number of neighbors according to their upload bandwidth resources.

Every change in the underlying network, in the resources of a peer, peer arrivals and departures or execution of DOMA in neighboring nodes triggers new changes in CDO while it always converges to the desired graph structure and to a minimized sum of energies [3].

4.3.4 Content Index SA

In IMS networks, context indexing is used for distinguishing calls with respect to requested content type. Quite often in commercial P2P services content indexing is used not only for accelerating the content searching process, but as a tool for content publication, together with content description information.

In the scope of the VITAL++ network, content indexing is defined as a Sub-Architecture (CI-SA) accessed using the SIP Instant Messaging standard [4] and offering the following services:

- *Content Publication*: The content publication service can be used by IMS users interested in offering content. The service works by declaring content availability to the network that may be fed to the users through the CI-SA. Context searching and download is possible from third-parties by executing network search on the basis of content description information publicized along with the content.
- *Content Searching*: Content searching is the basic service offered, whereby users looking for particular content are browsing other users' publicized content on the basis of certain criteria. These criteria are submitted to the CI-SA and the result is fed to the requesting users as a list of descriptions of available content items. The list also contains matching criteria which are used as filters against relevance of the result for presentation to the user.
- *Overlay Bootstrapping-Maintenance*: Contrary to the regular session setup process of IMS, whereby connection parameters are negotiated during bearer set-up, the VITAL++ client has to join P2P overlays well before this process takes place, in order to be able to acquire content indexing information. For this purpose, once the user has selected a specific content item to be retrieved and reproduced locally, this has to be communicated to the CI-SA. In this case the CI-SA interacts with the Overlay Management SA (OM-SA) in order to either create a new overlay or to update an existing one. In any case the outcome of the OM-SA, which is a list of peers per overlay member, is sent either to a newly added member of an overlay or to an existing member for which the list of peers has been updated.

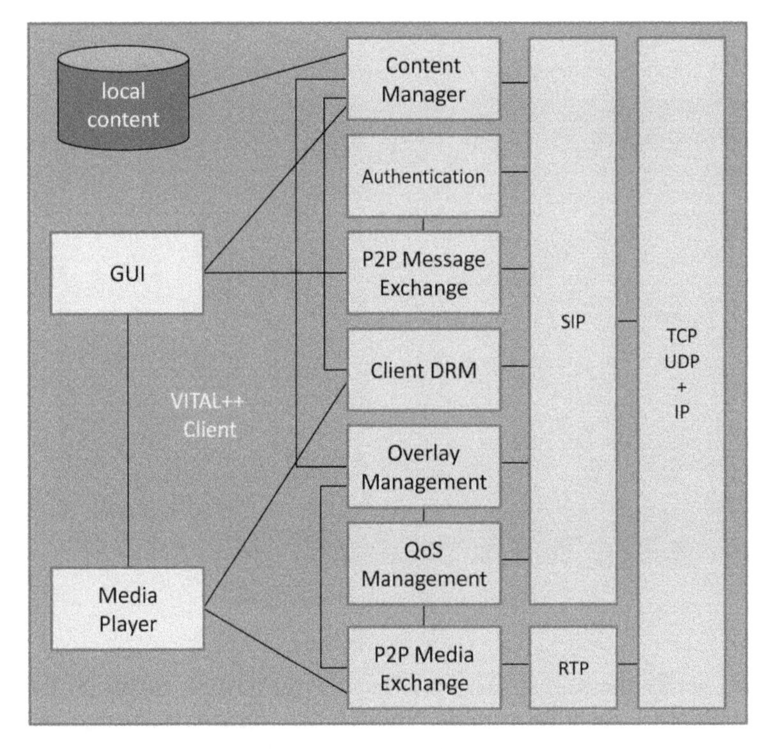

Figure 4.7 Client functional blocks.

4.3.5 VITAL++ Client Architecture: An Overview

The terms "Client" and "Peer" are used interchangeably in this document as they refer to the same thing. The VITAL++ client is a hybrid client. This means it is an IMS client and a P2P client at the same time. The IMS functionalities are used to mainly interact with an IMS core or system for exchanging control information, while the P2P part is used to exchange content with other peers.

The components of the client are directly derived from the necessity to interact with other clients and the IMS core in order to fulfill the envisaged features. Figure 4.7 illustrates the functional blocks inside the client. These are the content manager, which is responsible for publishing and discovering content as well as triggering DRM operations via the client DRM module if a license needs to be obtained. The authentication module obtains and manages certificates of VITAL++ entities (clients, application servers, root-certificate). It interacts mainly with the P2P message exchange in order to

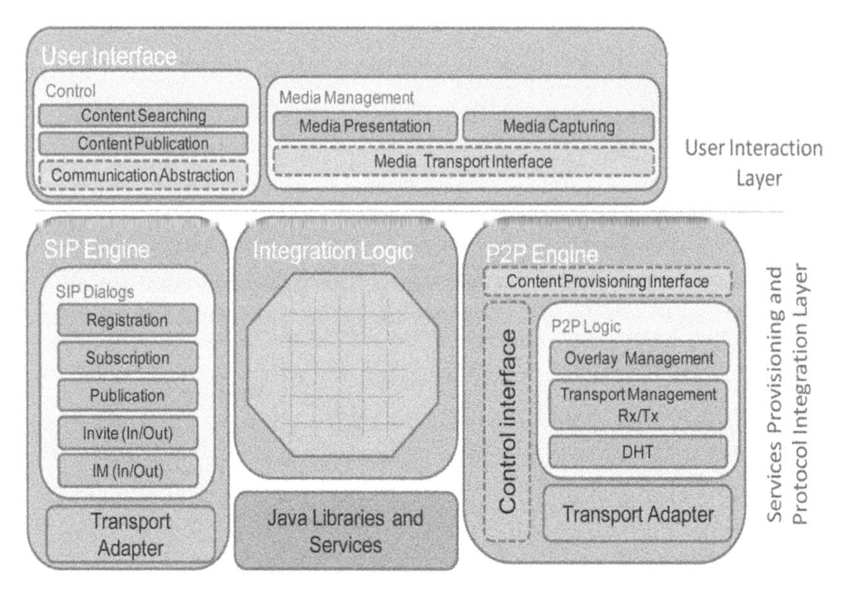

Figure 4.8 Platform components.

sign and verify messages. The latter has the purpose to exchange P2P messages with other peers for generic purposes (i.e. playlist exchange, etc.). The overlay management module obtains overlay changes from the application server and re-organizes its neighborhood accordingly, also to respect to QoS requirements, issued by the QoS management module, which can also realize QoS enforcement via NGN mechanisms. Furthermore, standard IMS client functionality is realized (not depicted) for initial IMS registration and IMS session management.

Session initiation and negotiation, which in IMS is handled by SIP Invite dialogs, has been replaced by P2P mechanisms, while media streaming is handled by the transport layer of the IMS. In this context, the VITAL++ client has been implemented following the architecture depicted in Figure 4.8.

The user interaction layer is the part of the client that interacts with the user. It provides all media playback and capturing capabilities as well as means for aiding the discovery and publication of content. This layer operates in a transport and communication agnostic manner. It produces and consumes both control information and media content. Control data are generated and processed by GUI elements that allow the use navigate in the acquired information. The control information that is produced identifies either criteria for content searching/publishing or content selection for acquisition through the underlying layers procedures. Media management contains all the required mechanisms for media representation or capturing by using of the available media libraries. Control of the media components is restricted to configuration regarding media playback or capturing leaving transport layer dependencies to be handled by the underlying layers according to UA configuration.

The services provisioning and protocol integration layer contains one SIP engine and one P2P engine. The SIP engine is built as a library that allows for the establishment of a number of SIP dialogs. The dialog objects can be configured to provide the content that is exchanged in their lifetime so that it can be processed in other application modules for the provision of a specific service.

The P2P engine provides a configuration interface through which all the control information can be applied for the proper initialization and maintenance of the engine. Additionally the P2P engine provides a content exchange interface through which media content can be transmitted to the network or retrieved and forwarded to the media handling modules.

4.4 VITAL++ P2P Functionality for Live Streaming

We have used VITAL++ architecture (client and network sides) to deliver live streaming as it has strong requirements in terms of the bandwidth that it needs, introduces high amounts of traffic in the underlying network and strict time constraints in the distribution of content as peers consume it in real time.

The multimedia stream generated by individual users and/or content providers is divided into blocks and distributed by the deployed overlay. A P2P block exchange scheduling algorithm (P2P-BESA) – also part of the VITAL++ client – ensures the distribution of each block to every user that requests the specific multimedia stream with low latency. This latency is known as setup time and it is defined as the time interval between the generation of each block from the stream producer until its delivery to every participating peer. An efficient P2P-BESA has to maximize the delivery rate of the multimedia stream with respect to the participating peers uploading capabilities while ensuring the reliable delivery of the stream in the presence of dynamic conditions such as batch peer arrivals and departures, dynamic network latencies and path bit-rates.

Neighbors in the CDO periodically exchange the set of blocks that they have. Each receiver exploits this information and proactively requests blocks from its neighbors in the CDO in order to: (a) avoid duplicate block transmis-

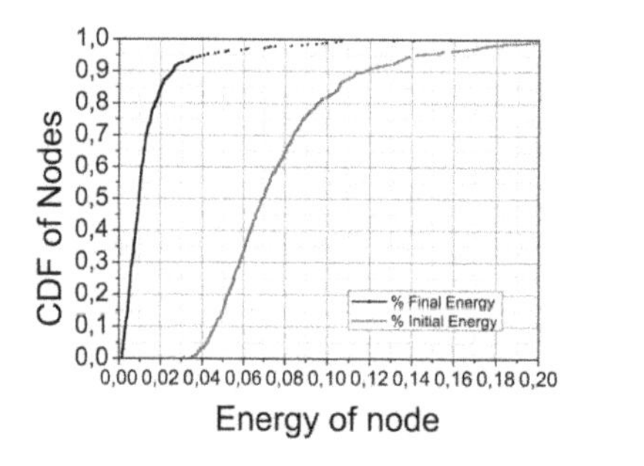

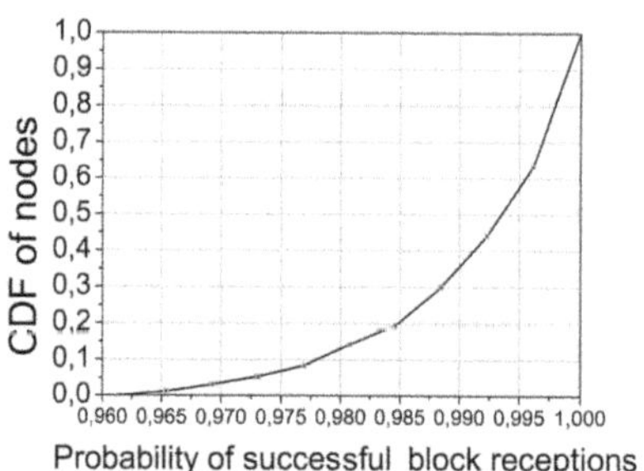

Figure 4.9 Left: CDF of the average network latency. Right: CDF of the successful block receptions.

sions from two peers, (b) eliminate starvation of blocks, and (c) guarantee the diffusion of newly produced blocks and/or rare blocks in a neighborhood.

In contrast, every time a sender is ready to transmit a new block, it examines the set of blocks that its neighbors have and using as criteria the most deprived neighbors (miss the largest number of blocks) and neighbors with high capabilities of upload bandwidth, selects one of them and transmits it to a block. Figure 4.9 depicts the performance of our system. In the left graph we demonstrate the cumulative density function of average network latency with their neighbors (energy) that peers have in a randomly formed overlay, and one built by our CDO described in [3]. We observe a reduction of energy by approximately 90%. In the right graph we depict the cumulative density function with the percentage of the successful block transmissions that each peer that participates in our system has. We mention here that the video steaming rate is 95% of the average upload bandwidth of the participating peers and the latency between the generation of a video block and its distribution in every peer in the system is 4 seconds. Through these graphs we observe the optimal and stable delivery of a video (right graph) while simultaneously our system minimizes the traffic that it introduces in the underlying network (left graph).

4.5 VITAL++ Testbed Deployment

In order to test/evaluate the proposed VITAL++ paradigm a number of heterogeneous telecommunication platforms were interconnected into a common experimental playground. This unified testbed environment includes the IMS-enabled telecommunication infrastructures from FOKUS (Fraunhofer Institut), Telekomm Austria (TA), Telefonica I+D (TID), University of Patras (UOP), and Voiceglobe, as well as an interactive DVB-T platform (at CTRC premises) acting as a Media Provider (Broadcaster) and Data injector in the services that we offer. More specifically, here we describe a scenario where live streaming TV content (IPTV) is fed from an active-user located within the DVB-T broadcasting footprint (potential Broadcaster) onto the P2P engine, via a VPN connection. The received IPTV stream is processed by the P2P engine and distributed over the entire VITAL++ infrastructure via a number of specially configured VPN tunnels, established with OpenVPN software, enabling both IMS and P2P connectivity.

To this end, a VPN server was installed in Telefonica I+D's premises and VPN clients in the remaining testbeds, in order to establish a VPN tunnel between Telefonica I+D and each partner's testbed. A DNS server was also installed to resolve the domain names of each IMS core, thus enabling the placement of calls between users registered at different IMS cores. Once basic IP and IMS connectivity was achieved, it was the turn of adapting the cores to VITAL++ needs, deploying the elements and entities required by the architecture.

First, the PCs hosting the IMS clients would also host the VITAL++ clients, both Monster's and BlueChip Technology's. It was decided that VITAL++'s users were going to be created only at Franunhofer's IMS Core, which would be considered as the home network for those users. While testing from another testbed, those users would act as roaming users accessing to their home network from a visited network, via the local P-CSCF. The Sub-Architectures defined in the project would be distributed among the testbed, not centralized, and would be located at the following testbeds. Content Indexing (CI-SA) and Overlay Management (OM-SA) at UOP. P2P Authentication (P2PA-SA) at Fraunhofer's and Content Security (CS-SA) at Waterford Institute of Technology (WIT). In this case, two VPN tunnels were required, one for the CS itself and the other for the Certificate Authority (CA) that the CS relies on. These SAs are considered as IMS Application Servers (SA), and are registered as such in the Fraunhofer IMS core, in order to route IMS traffic among them and the clients. However, besides adapting the testbeds, VITAL++ requires a number of external elements to carry out its functionalities properly. They are listed here briefly.

In the case of Voiceglobe, a commercial provider of SIP telephony, a direct connection to the testbeds' network cannot be established since its subscribers are assigned real IP addresses. In order to circumvent this obstacle, a VITAL++ Proxy has been implemented, that would appear to the VITAL++ system as a standard VITAL++ client and will enable Voiceglobe clients to download the contents made available by VITAL++.

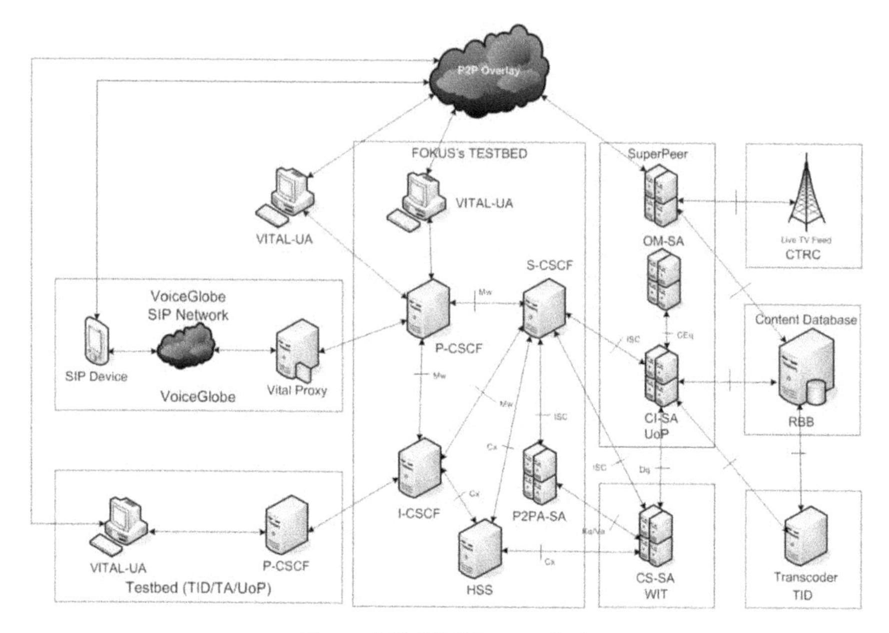

Figure 4.10 VITAL++ testbed.

A database containing the data and metadata of the contents distributed by VITAL++ is located at the premises of Rundfunk Berlin Brandenburg (RBB). This database will not be connected directly to the network comprising the VPN tunnels or visible by the VITAL++ clients, but indirectly accessed via the testbed of the University of Patras.

In order to provide the content format that best fits the terminal capabilities (i.e. bandwidth, spatial resolution) a transcoding utility has been made available at the premises of Telefonica I+D, which will automatically convert any uploaded contents into a set of predefined formats, served by respective overlays, enabling VITAL++ clients to join the most appropriate for their needs.

In the case of the IPTV injection scenario, the Centre for Technological Research of Crete (CTRC) had injected a live streaming TV content onto the UoP test-bed from an active-user (potential Broadcaster) located within the regenerative DVB-T platform at CTRC premises, via a VPN connection [9]. This stream is processed by the P2P engine and distributed over the entire VITAL++ infrastructure.

From a logical point of view, the relationships and interfaces among the different entities involved in VITAL++ are depicted in Figure 4.10.

VITAL++ client registration will be carried out by the Fraunhofer's IMS core inside the Fokus testbed or via a VPN connection from the partners in the case of roaming users from other testbeds (visited networks). In the case of VoiceGlobe subscribers, the use of VITAL++ Proxy is mandatory. Once registered, the VITAL++ Clients will be able to interact with the VITAL++ SA by sending and receiving IMS messages, in order to carry out the following tasks:

- To upload a content to RBB's systems and create as many overlays as required for transcoding purposes plus a superpeer to serve them, in case the original uploader leaves the overlay.
- To search for contents, according to users' preferences stored in a personal profile, in order to retrieve a list of recommendations.
- To play that list of recommendations, joining the overlays serving each object. Notice that the play will be sequential and the user will be given the option of skipping tracks (though not of direct selection).
- To compose a playlist with contents already available in the system or stored in its HD and publish it.
- To join the playlists of the users in broadcast mode, who will be identified as such in the user interface.
- To join the overlays offering Live TV and audio broadcast.

4.6 Conclusions

In this work we analyzed how IMS, with its centralized management and features for AAA can exploit a P2P networking paradigm to offer content distribution services that are scalable, adaptable, secure, and reliable while simultaneously offering new functionalities to the users. Through the development of the VITAL++ platform we learned that P2P authentication is a promising solution for scalable and secure content distribution services. Content security and accounting may be successfully combined with P2P overlays in order to meet business requirements. Centralized content indexing reveals P2P capabilities and does not hurt system scalability. Finally, the optimization and the dynamic adaptation of the content distribution overlay is a critical factor for the successful operation of P2P content distribution services.

Acknowledgements

This work is funded by the European project VITAL++ under contract No. INFSO-ICT-224287. We also want to thank Dimitrios Dechouniotis for his assistance.

References

[1] Open Media Commons. http://www.openmediacommons.org.
[2] Drools Expert Jboss Community. http://www.jboss.org/drools/drools-expert.html.
[3] N. Efthymiopoulos, A. Christakidis, S. Denazis, and O. Koufopavlou. Liquidstream – Network dependent dynamic P2P live streaming. *Peer-to-Peer Networking and Applications*, 4:50–62, 2011.
[4] B. Campbell et al. RFC 3428 – Session initiation protocol (SIP) extension for instant messaging, IETF, 2002.
[5] ETSI. Tispan release 1 architecture, 2005.
[6] IETF. RFC 2663, IP network address translation (NAT) terminology and considerations. http://tools.ietf.org/html/rfc2663, 1999.
[7] ipdr.org. Network data management usage (NDM-U) for IP-based services, version 3.1.1 edition, 2002.
[8] J. Fiedler T. Magedanz and J. Mueller. Extending an IMS client with peer-to-peer content delivery. In *Proceedings of the Second International Conference on MOBILe Wireless MiddleWARE, Operating Systems, Applications – ICST MOBILWARE*, 1978.
[9] E. Markakis E. Pallis, and H. Skianis. Exploiting peer-to-peer technology for network and resource management in interactive broadcasting environments. In *Proceedings of IEEE Globecom*, 2010.
[10] JSR SIP Servlet v1.1. http://jcp.org/en/jsr/detail?id=289.
[11] VITAL++. http://www.ict-vitalpp.upatras.gr/.

PART 2

PEER-TO-PEER SERVICES

5

Challenge and Chance: A P2P Distributed Services Network for Mobile Internet

Yunfei Zhang[1,*], Hui Zhang[2], Yang Li[1], Jin Peng[1], Guangyu Shi[3] and Min Zhang[4]

[1]*China Mobile Research Institute, Unit 2, 28 Xuanwumenxi Ave., Xuanwu District, 100053 Beijing, China; e-mail: zhangyunfei@chinamobile.com*
[2]*NEC Laboratories America, 4 Independence Way, Princeton, NJ 08540, USA*
[3]*Huawei, 2330 Central Expressway, Santa Clara, CA 95050, USA*
[4]*Beijing Jiaotong University, No. 3 Shang Yuan Cun, Hai Dian District, 100044 Beijing, China*

Abstract

With more types of terminals, networks, and services, Mobile and wireless Internet challenges the existing development and deployment of P2P applications and distributed services.

In this chapter, we firstly survey the prior practices of P2P and distributed services architectures; and then, we propose a unified carrier-level P2P based distributed services network (DSN), as a common overlay architecture supporting the development of distributed services like VoIP, multimedia streaming, file downloading and other popular services in heterogeneous and mobile Internet networks with more focus on management and control than existing works from the operator's point of view. Furthermore, we present a novel large-scale distributed simulator HiFiP2P, and a real test-bed called Nebula, with the purpose of verifying the effectiveness and efficiency of our DSN architecture.

*Author for correspondence.

Anand R. Prasad et al. (Eds.), Future Internet Services and Service Architectures, 97–117.

Keywords: P2P, distributed service, architecture, directions, simulator, test-bed.

5.1 Introduction

Peer-to-peer computing (P2P for short) is a disruptive technology which brings great change and chance for service architecture. Service provision is no longer constrained by server side. With more types of terminals, networks, and services, mobile and wireless Internet challenges the existing development and deployment of P2P applications and distributed services. Firstly, in a heterogeneous network environment with a mixture of fixed, mobile and wireless access, pure P2P architecture does not work well. New components like proxy and more complex network conditions introduce more functionality in the corresponding software architecture. Secondly, the lack of common overlay architecture, supporting various applications in a heterogeneous environment, results in repetitious work in application development. Thirdly, different overlay architectures may interact with the underlay network in possibly conflicting ways, which deteriorates the performance. Fourthly, different from the fixed Internet, more management functionalities are required in the architecture to create a more controllable, operational, and healthy eco-system. Lastly, no suitable test-bed is available for verifying the performance of P2P and distributed services in the heterogeneous networks.

JXTA [15] is a pioneer P2P development architecture specification begun by Sun Microsystems in 2001. Although JXTA considers mobile scenarios, it has some evident problems. Firstly, although JXTA's use of XML specifies all aspects of P2P communication for any generic P2P applications, JXTA might not be suited to some specific standalone P2P applications [13]. XML is verbose for data transmission. The network overhead of XML messaging might be more trouble than it is worth in mobile environments. Further, JXTA does not address the problem of transfer optimization in mobile and wireless environments [23], like the efficiency to allow relays to use TCP. Thirdly, JXTA does not address the peer group automated management problem and lack of protocols to form practical peer-to-peer groups [9, 13]. JXTA only allows for flat structure and all peers in one group cannot really scale well. Lastly, JXTA looks more like a service oriented framework but it seems there is no real requirements for peer-to-peer based SOA unless we have unknown or dynamic services.

DOCOMO has also proposed a P2P service architecture for mobile environments [25], which touches on the preliminary aspects of control and management of the overlay in the architecture. The DOCOMO proposal is basically proxy based with two layers: a P2P core services layer, and an application specific P2P services layer. The P2P core services layer provides (1) fundamental functionalities such as P2P network bootstrapping and efficient resource lookup, (2) mobile adaptation for search and overlay management. This is a good starting point to discuss the trend of P2P and distributed services architecture. However, the inconsistency of functionality classification in P2P core services layer handles the development of architecture. What is more, regarding control and management, some important functionalities like QoS, mobility management beyond search, etc., are lacking.

Some P2P and distributed architectures for standalone applications are also proposed. For instance, Siemens raised P2P architecture for file sharing applications with centralized index server. It does not fit with other applications [19]. Nokia brought forward a hybrid P2P architecture [19] for file sharing applications while the premise is that the terminals have multiple interfaces. It has a relatively low sharing utilization.

There are also initiatives in IETF helping to build up distributed services architecture. P2PSIP [16] defines the protocols and distributed service architecture for SIP and XMPP applications. PPSP [17] is a counterpart protocol suite of P2PSIP in IETF, where tracker based P2P streaming services architecture is introduced.

These above attempts are for silo and isolated applications. No matter what the aim is (some of them are for simplifying the problem or narrowing down the scope), they do not consider the possible interactions of different applications. Silo architecture for isolated development of applications has potential shortcomings in power consumption. Imagine that we have three types of services – VoIP, Streaming and file sharing deployed in a cloud-star environment, i.e., there are widely distributed machines, even mobile terminals like stars spread in the network, and there are deployed super-nodes in the network core like clouds gathering together for clusters. Therefore in each cluster, three kinds of services coexist. If there are mechanisms to coordinate machine consumption for different applications even in the physical layer, more benefits will be achieved. The machine number used can be 80% reduced and the utilization ratio for each used machine is five times increased with virtualization technologies, according to IBM CRL research. Therefore there are big demands for an operator or ICP to build a generic P2P architecture meeting with the following requirements:

1. A generic architecture to adapt with heterogeneous networks and terminals, including mobile and wireless terminals.
2. A generic software architecture to deal with most of the core parts of a manageable, operational and controlled P2P system.
3. A reasonable architecture to accommodate different applications with the green-network technologies.
4. And what is more, a good test-bed and simulation tool to test and deploy the key methods and services and see the validity of the proposed architecture.

In this chapter, we present our novel carrier-level P2P based distributed services network called DSN for short, which is the next generation operable and manageable distributed services architecture, for the convergence of telecommunication services and mobile Internet. Our preliminary implementations, simulation and Internet-scale test of P2P VoIP and P2P streaming system over DSN architecture demonstrate its effectiveness and promising future [29].

5.2 Challenges of P2P Architecture in Heterogeneous Environments

Nowadays, both the telecommunications network and Internet are facing some severe challenges respectively, and at the same time, they also have some advantages that attract each other. Therefore, to realize the convergence of the telecommunications network and Internet is a trend in current industrial practice. The key point is to combine the advantage of the flexibility of distributed computing for service development in Internet and the control and management strength in telecommunication network in designing the new distributed services architecture.

Several problems arise for the introduction of P2P technologies into service architecture.

First of all, as stated earlier, the randomness of peer activities makes it hard to manage the system and monitor the activities of the node status. For a controlled and manageable architecture, it is necessary to design modules to address this challenge.

Second, traffic management and topology-awareness for P2P applications. In contrast to client/server architectures, P2P applications often must choose one or more suitable candidates from a selection of peers offering the same resource or service [12]. The application cannot always start out with an optimal arrangement of peers, thus risking at least temporary poor performance and excessive cross-domain traffic. Providing more information for use in peer selection can improve P2P performance and lower ISP costs. We have researched a T2MC method to reduce such mismatches by Traceroute and 2-Means Classification Algorithm [10] and added it to the DSN architecture.

Third, mobility support for distributed applications. As we pointed at the beginning of the chapter, terminal mobility makes a quite different picture for P2P architecture. We have studied the problem how the current popular P2P streaming applications like PPLive, PPStream and UUSee support node mobility without EXPLICIT mobility support mechanisms [3] like MIP or PMIP because the low-layer mobility management mechanisms are far from wide deployment. The primary test result shows the necessity to have such mechanisms in DSN [27, 28, 32].

The fourth is QoS management. Different from the plain Internet, DSN shall support QoS assured services with multiple layer QoS mechanisms. Besides traditional access control, traffic engineering, DiffServ, and InterServ, DSN addresses overlay QoS assurances (e.g., OverQoS [1]) for better performance. There are two cases using QoS assured services. The first is high end to end bandwidth applications like Telepresence [18] by Cisco, which is an enterprise level teleconference system with perfect video quality in dedicated networks with about 4–10 Mbps bandwidth in a dedicated private network; however, there are much more complexities in public Internet. We envision that overlay QoS can solve this problem.

The second is of mobile and wireless environment. As we know, the application performance in mobile and wireless environment is quickly exacerbated due to bad network conditions. And it is hard to improve it in the physical layer. The overlay layer is a much easier place to execute network condition detection and performance improvement. If overlay QoS helps to add application-level error detections and tolerance mechanisms, the problem will become easier to solve.

All these challenges are not the basic part that typical P2P architecture can solve, but for an operator or ICP who runs lots of different P2P applications for profit, these requirements are necessary for building controlled and manageable distributed services.

5.3 Large-Scale P2P and Distributed Services Architecture for Heterogeneous Environment: DSN Design

5.3.1 System Architecture

DSN is a wireless Internet-oriented platform with carrier-level service capabilities, which is located in the core network between the service network and the bearer network. Based on distributed technologies, especially P2P, the DSN provides core network capabilities for wireless Internet services and necessary support for the operation of services. The main points about the philosophy of the DSN design are as follows:

- deeply exploring the requirements of applications for network capabilities according to Internet services;
- introducing new services and realizing the desired network capabilities in combination with existing technologies;
- maintaining the operability and manageability of the telecommunications network;
- being oriented to mobile telecommunications customers, taking the advantage of the network and highlighting the service characteristic of wireless Internet services.

Much of the core functionality of the DSN can be realized through the application of software. With the componentization of software and the underlying layer screened by middleware, the DSN software can be deployed on hardware platforms with different functions and in different locations to realize distributed deployment.

As shown in Figure 5.1, the DSN function is enabled by deploying the Super Node in the core network, mainly providing service control capability. Furthermore, the DSN function is gradually extended to access node, user node and external node to maximize the computing capability, storage capacity and bandwidth of access nodes and termination nodes. These nodes can provide network capability and service capability.

One point that needs to be emphasized is that DSN test-bed is also conjunctional and consistent with DSN architecture, which shares the same physical resources.

5.3.2 Software Architecture and Building Blocks of DSN

Figure 5.2 depicts our design on software architecture and building blocks of DSN. At the bottom is the infrastructure platform layer where a virtual-

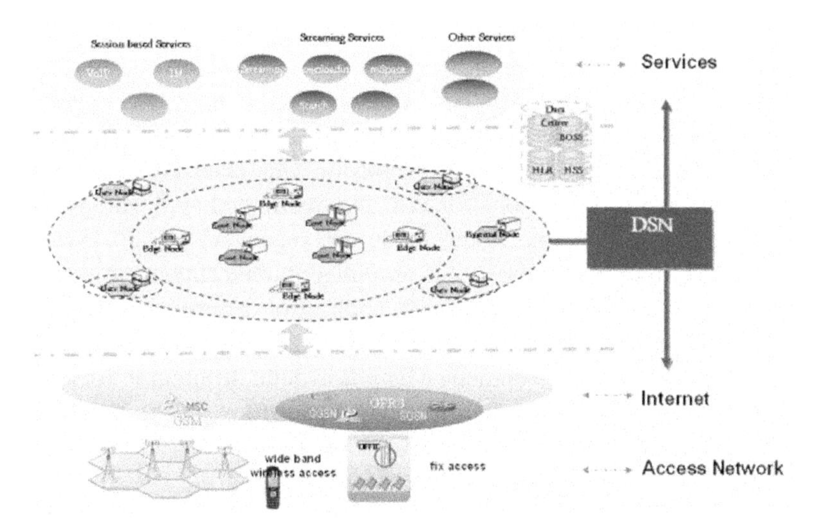

Figure 5.1 DSN system architecture.

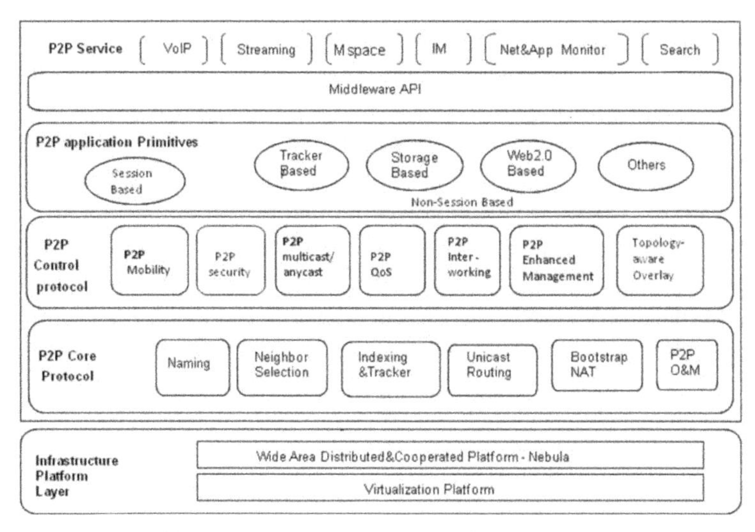

Figure 5.2 Software Architecture and building blocks of DSN.

ized platform is built to support different P2P overlays and coordinate their utilization of the machines. Based on the virtualization platform, we build our large-scale distributed and cooperative test-bed called Nebula. Nebula is used for distributed algorithm/technology experiments and DSN service deployment in real Internet. Above the infrastructure layer is the P2P core protocol layer, where the basic operation of P2P systems is abstracted. P2P core protocols include naming, neighbor selection, index and tracker building, unicast routing as well as bootstrap and P2P overlay. All these features are prerequisite functions provided by P2P systems.

The next layer is P2P control protocols. This layer provides more flexibility and enhanced functions for P2P and distributed applications, which is the key to realize operable, controllable and manageable P2P service architecture. P2P mobility component is to realize mobile support as discussed in Section 5.2. P2P QoS offers to provide application-layer QoS guarantee using overlay QoS technology. P2P security module addresses the security problem raised by distributed environment where it is quite different to identify, trace and ensure the security in peer-to-peer connections and transactions. P2P enhanced management deals with efficient meta-data collection and dissemination [30], which is a good reference for making the system work well. P2P inter-working module deals with setting up efficient and simple interconnections and relations between different P2P overlays since DSN targets for different P2P applications. Topology-aware overlay makes P2P overlay consistent with physical topology to reduce the overhead caused by the mismatch. The detailed mechanism used in the above module can be found in [2, 3, 10, 27, 28, 30]. Above this layer are P2P application primitives, or service enablers, including both session based P2P enablers and non-session based P2P enablers. In non-session based P2P enablers tracker based, storage based, Web 2.0 based service enablers are involved.

The next layer is middleware APIs for calling P2P functions and modules are open to developers. This can simplify their development of P2P applications and make more developers avoid spending too much time in P2P basic mechanisms instead of the application itself.

At the top of the DSN software architecture is the rich P2P and distributed services and applications, such as P2P VoIP, P2P streaming, file sharing, IM, gaming, etc.

5.3.3 Application Scenarios

With DSN, carriers are able to build scalable telecommunication network platforms that deliver multimedia applications and content applications:

- Multimedia Telephony (MMTel) Scenario: With DSN, carriers can build a large-scale cost-effective operable distributed MMTel service system.

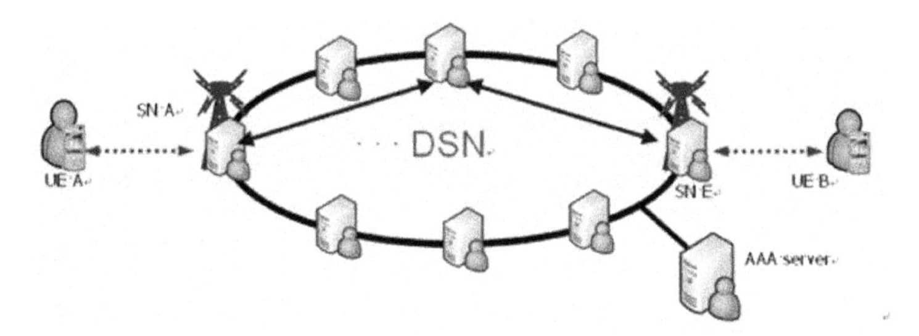

Figure 5.3 P2P VoIP system.

- Content storage and delivery: DSN enables very large data storage. In a P2P-enabled system, network resources are distributed across peer nodes in the system.
- Streaming: In DSN based streaming architecture, system capability is improved along with a large amount of peers; this also solves the scalability problem. For the system provider, the load of the server will be reduced with the increase of users. For users, it is faster to access the resources and the media is played more fluently.
- Large-Scale High Bandwidth Multi-Media Service Scenarios: Bandwidth exhausting multimedia service in the future may become the prime application of carriers. The present system requires 3–10 M network bandwidth, less than 100 ms delay, and 10 ms jitter, and also has high requirements on route setup processes and QoS. DSN provides a QoS guarantee for large capacity multimedia service.
- Other Service Scenarios: DSN can flexibly support all kinds of wireless Internet service scenarios in the future, such as IM, web2.0, and online games, etc.

5.4 Implementation of DSN Based Applications

5.4.1 VoIP

We have implemented effective P2P based VoIP applications based on DSN architecture using well-known P2P SIP protocols in the China Mobile IP Private Network, which provides mobile users with high quality and cheap voice services. In Figure 5.3, SN-A and SN-E represent two super-nodes in DSN overlay who act as the relay nodes; UE-A and UE-B are the caller and the callee [21, 29, 32] in a VoIP calling. We have developed a large-scale P2P streaming system based on DSN architecture. This is a tracker based sys-

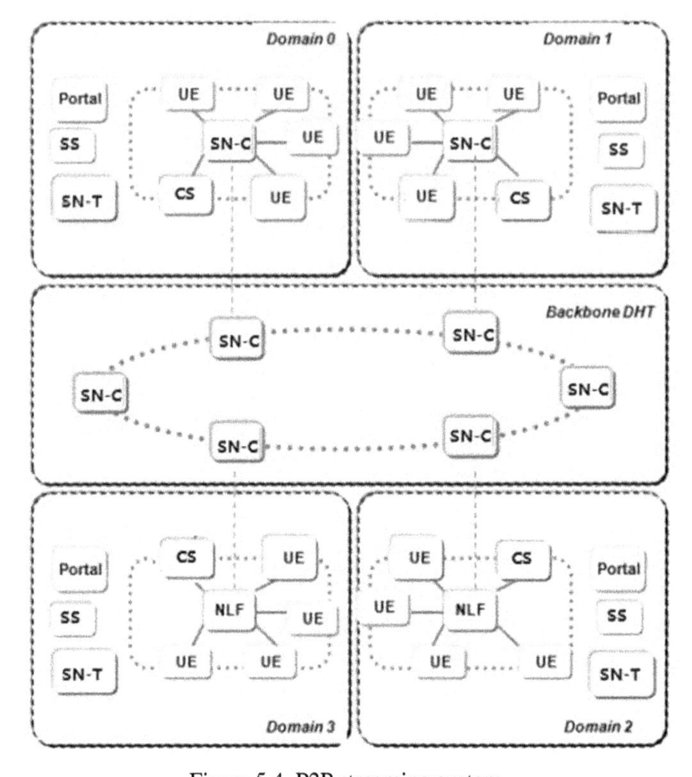

Figure 5.4 P2P streaming system.

which involves three screen interactions: PC, set top box and mobile phone. The streaming program can be shared among heterogeneous networks and terminals. We use adaptive coding H.264 SVC for different environment adaptations. This system can be easily extended to fit into a file sharing scenario since the latter has less stringent requirements for timely delivery.

In Figure 5.4 the core part is the tracker overlay by the super-nodes called SN-C. The edge parts are delivery networks consisting of operator deployed cache and CDN nodes called CS as well as user nodes called UE. SS is the source node and SN-T is the bootstrapping node.

Figures 5.3 and 5.4 show that the core architectures used in P2P VoIP and P2P streaming are just the same. Note that we can deploy the central part of the P2P VoIP (i.e., SIP server) and Streaming system (i.e., tracker) in the same super-nodes overlay.

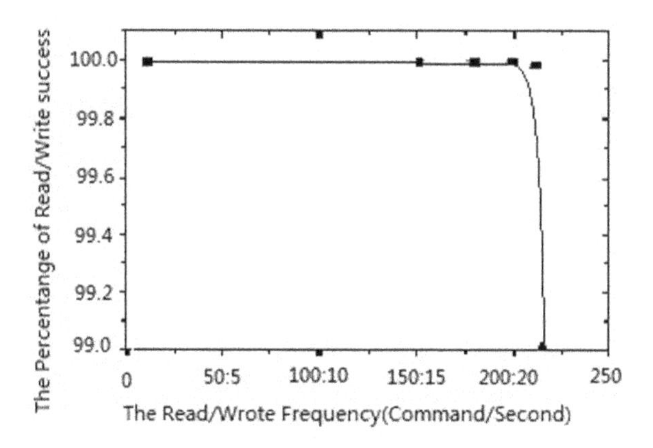

Figure 5.5 Read/write frequency and success rate.

5.5 Preliminary Local Test on the Implementations [33]

5.5.1 VoIP Test Environment and Result Analysis

5.5.1.1 Test Environment

The DSN VoIP test system has been set up consisting of 40 PC servers (DELL2950 with quad-core CPU and 8 G RAM) in three geographic areas.

5.5.1.2 Test Result

The testing results show that each NP can support 10,000 subscriber profile data storage and can reach up to 200 queries per second and 20 updates per second. Given the design capacity of each NP (e.g. 40 CPS), up to 40 queries per second need to be implemented accordingly. So the distributed P2P subscriber data storage system will not be the bottleneck of the whole VoIP switching system. According to the requirements of some subscriber profile storage systems, HSS (Home Subscriber System) needs to support 1,000 queries per second and 100 updates per second for every 1 million subscribers, i.e. 10 queries per second and 1 update per second for 10,000 subscribers. Therefore the current implementation can satisfy the requirements of carrier-grade VoIP subscriber storage systems. Regarding system capacity testing, the testing results show that each peer can reach 40 CPS without failure. According to the typical softswitch VoIP traffic model (400 CPS per 1 million subscribers), it is roughly estimated that each peer can support 100,000 subscribers. Considering it is a demo system, the system capacity can be optimized further.

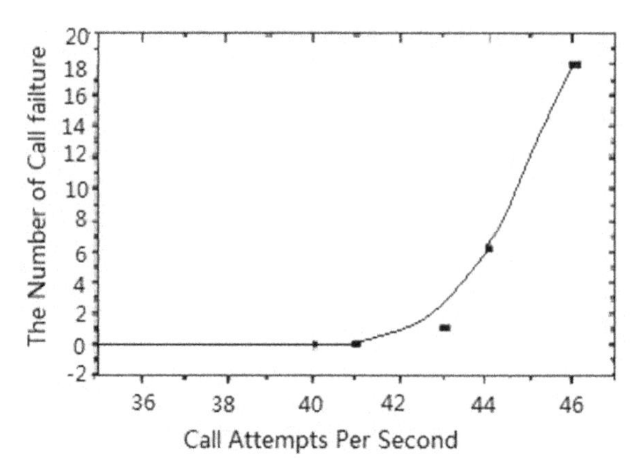

Figure 5.6 Call frequency and success rate.

5.5.2 Streaming Test Environment and Result Analysis

5.5.2.1 Test Environment

The test environment of DSN streaming systems include two domains to test the real operation in different zones (e.g., province). Every domain includes one SN-C server, SN-T server, Portal server, SS server, CS server and some PC UEs. TD-SCDMA UEs (including TD laptops and TD handsets) can access the servers and PC UEs.

5.5.2.2 Test Scenario and Result

Test scenario: Start an SN-C server, an SS server and a CS server, then use four machines to act as PC UEs. UE simulation software is run on the machine, so as to simulate 100 users which are viewing the same streaming media program simultaneously. The streaming content is distributed to CS by SS in advance. Then we start these 400 users on four machines in turn.

1. Maximum Numbers Supported by SN-C
   When the number of users which an SN-C server (Dell 360DP with Dual Core 2.5 GHz and 4 GB RAM) supports is 400, the CPU occupancy rate is around 3%, the memory occupancy is about 300 MB.
2. Maximum Numbers Supported by CS
   When the number of users which a CS server (Dell 360DP with Dual Core 2.5 GHz and 4 GB RAM) supports is 400, the CPU occupancy rate is around 20%, the memory occupancy is about 400 MB, the serving bandwidth is about 85 Mbps.

Table 5.1 Measurement results: Start-up latency.

File name	Bit rate	File format	Video format	Average start-up latency in local domain	Average start-up latency between domains
body_of_lies	968 Kbps	ASF	WMV	3 s	4 s
Star_Trek	930 Kbps	ASF	WMV	3 s	4 s
Ice_Age	769 Kbps	ASF	H.264	3 s	4 s
friends	721 Kbps	ASF	WMV	3 s	4 s

Table 5.2 Measurement results: Drag latency.

File name	Bit rate	File format	Video format	Average start-up latency in local domain	Average start-up latency between domains
body_of_lies	968 Kbps	ASF	WMV	4 s	4 s
Star_Trek	930 Kbps	ASF	WMV	3 s	3 s
Ice_Age	769 Kbps	ASF	H.264	3 s	4 s
friends	721 Kbps	ASF	WMV	4 s	4 s

3. Start-up Latency and Drag Latency
   We use four video files with different formats to act as a streaming input source. The measurement result is shown in Tables 5.1 and 5.2.

From Tables 5.1 and 5.2 we can see that the watching start-up latency and drag latency of PC UEs are 3–4 s. The latency is basically to meet the requirements of streaming media service of operators.

5.6 Simulator and Test-Bed of DSN

To evaluate the performance of the services developed based on DSN architecture, a large-scale DSN simulation environment and a large-scale distributed test-bed on the Internet are necessary.

5.6.1 Large-Scale P2P Simulator: HiFiP2P

Most of current P2P simulators have common drawbacks including the poor scalability with massive simulated peers and the lack of realistic underlay network layer support, which largely reduces the feasibility for DSN simulation. In this subsection, we present some approaches to overcome such defects, and the design considerations of HiFiP2P, our novel large-scale parallel DSN simulator with a measured realistic Internet data as its network layer support. Comparison experiments show that HiFiP2P simulator outperforms the existing simulation platform J-Sim [12] and PlanetSim [22] in both aspects of scalability and efficiency [14].

5.6.1.1 Architecture Overview

HiFiP2P is a message-driven, large-scale JAVA DSN simulator with underlay network layer support, aimed at providing a more realistic and dynamic simulated environment. To achieve this goal, HiFiP2P partitions the measured Internet data and/or the whole overlay topology into several groups, each topology group will be assigned to an independent computer.

The HiFiP2P is composed of five main components: topology partitioning, underlay network layer simulation (NLS), overlay layer simulation (OLS), distributed messaging and the simulation clock synchronization. Topology partitioning runs as a pretreatment procedure. The latter four components constitute the HiFiP2P runtime.

After topology partitioning, NLS reads partitioned network data to establish routers and links with performance metrics (delay and/or bandwidth), and communicates to different NLS only when a cross-partition link encountered. HiFiP2P has two types of NLS: lightweight and full functional. Lightweight NLS satisfies a majority of P2P simulation scenarios, it implements the interface to query the delay or bandwidth between any two routers, and it always ensures that the delay is in proportion with the link bandwidth consumption. Lightweight NLS can also import the dynamic behavior in these two metrics, and record the resource consumption of each link. Full functional NLS offers more capability than the lightweight one, including simulated TCP/ UDP and ICMP utilities like PING and Trace-route, almost all the basic IP communication methods which a real P2P program can use on a normal computer.

OLS is a discrete event simulation framework for overlay networks which can run with or without NLS. P2P protocols are implemented as OverlayNodes in OLS, such as Chord, Kademlia, Symphony, etc. HiFiP2P OLS inherits the architecture of PlanetSim and achieves the high scalability by two methods.

When running with the NLS, the OLS invokes the interface of NLS to connect the peer and routers, so as to distribute the simulations depending on the distributed NLS. Otherwise, when peers belong to different hosts want to exchange information, the NLS will directly send/receive messages by the messaging module.

HiFiP2P uses the Apache mina component to implement the messaging module. It provides a direct host-to-host communication mechanism in a messagebased driven method by the two main interfaces: RemoteMsgSender and RemoteMsgReceiver. It is the bridge to exchange the NLS messages, OLS messages and the simulation synchronization messages.

HiFiP2P has its simulated global clock advancing by events. The simulated clock is critical to calculate the dynamic link delay according to its capacity and traffic. HiFiP2P entities on different hosts run concurrently unstill the synchronous checkpoint, and then they send/receive synchronization messages to adjust their simulated clock to the same value. Sterlind et al. [34] describe the details of the single node architecture.

5.6.1.2 Overlay Layer Simulation

- Discrete Event
  HiFiP2P OLS is an extended version from the PlanetSim which is also based on the discrete event model. Every event includes a timestamp indicating when it should be expired. The difference between HiFiP2P and PlanetSim is that our events can be dynamically generated with the integration of DESMO-J [7] package.
  The effort needed for integration is to make them use the same time wheel. HiFiP2P implemented a globally consistent simulation clock for this goal. All the DSN P2P paradigm events such as "join", "leave", and “lookup” can also be recorded in a scenario file, thus this file can be used as the same test case with different network layer configurations.
- Bind Peers to Routers
  When a new peer initializes in OLS, NLS will bind it to a simulated router. In HiFiP2P, two binding policies are defined, “random” and “edge”. With the “random” policy, the overlay peer connects itself to a randomly selected router. The only constraint is that this router is hosted on the same computer. This constraint is introduced for avoiding the additional messages among computers. The “edge” policy impels the NLS to bind peers to the edge routers. The edge router is determined by the two conditions: a router with degree one or a router has a betweenness smaller than a predefined parameter. The rationale is that most P2P applications reside on computers linking to the backbone Internet with kinds of access networks. The performance parameters of the link between the peer and its first hop router are configured in this phase.
- OLS Runtime
  OLS can run with or without NLS. Running on top of the NLS, HiFiP2P takes the discrete events as the trigger, and then all the messages gener-

ated by P2P logic will be dispatched by the NLS. Thus the distribution of OLS is exactly completed by the NLS.

When the OLS runs independently, each HiFiP2P entity will have a global view mapping the peer IDs to their hosted computers. If the destination peer is on a different computer, then the two HiFiP2P simulator entities will exchange the messages through their RemoteMsgSender and RemoteMsgReceiver interfaces. Similar methods as in NLS, when the "current" packet passes through a peer, the peer increases the accumulated delay, considering the traffic and bandwidth of the peer. While the "future" packet will remain in the pending queue and not be perceived by the destination peers until such "future" packets are expired by the advancing time wheel.

5.6.1.3 Simulations and Analysis

We have made extensive simulations to evaluate the performance of HiFiP2P system and choose J-Sim and PlanetSim as comparative technique [14]. By performing the comparison experiments, we show that HiFiP2P outperforms the existing simulation platforms in both aspects of scalability and efficiency.

5.6.2 Large-Scale DSN Test-Bed for Heterogeneous Internet: Nebula

5.6.2.1 Requirements and System Overview

While simulation is a valuable tool for understanding new technologies, only testing in simulation is not enough. Just as argued by researchers, there may be over-simple assumptions for the simulation model, lack of live traffic and large-scale supporting, resulting in a credibility crisis of the simulation [20, 34]. One approach to overcome these shortfalls is to employ DSN test-beds called Nebula to support large-scale services test with hundreds of thousands of coordinated nodes to participate. The meaning of Nebula is the combination of stars and clouds in the networks, i.e., there are wide distributed machines, even mobile terminals like stars spread over the network, and there are super-nodes gathering clusters in the center of the network like clouds. Such a test-bed should meet the following requirements:

- R1: Heterogeneous network connections, including mobile and wireless connections, fixed Internet connections for different operators.
- R2: Easy coordinated nodes management. The node can be easily configured, managed and monitored with good access control and per-

mission management. Different tasks can be separated with different participating nodes and different permissions.

- R3: Good services deployment. The deployment should be automatic, fast and easily.
- R4: DSN collaboration platform can easily scale to some thousand or bigger magnitude of nodes.

PlanetLab is a well-deployed distributed test-bed for research work and MyPLC is an open source PlanetLab control center implementation, which is a good starting point to build our test-bed. However, there are some big disadvantages that cannot be used in a DSN environment:

1. Supporting fixed IP address only: Planet only supports fixed IP address access. This is quite different from the real Internet environment for which DSN is targeted. Actually, the feature of traversing NAT is necessary for our DSN test-bed since in the current Internet about 60–80% of P2P peers are found to be behind NAT [24]. Further, PlanetLab lacks mobile and wireless nodes access support. We need more types of mobile and wireless nodes to participate in to test the feasibility of the platform. WLAN nodes or 3G cellular nodes like TD-SCDMA nodes should be added in the platform. However, mobile nodes often have a dynamic and private address. So this is a similar case as the nodes behind NAT.
2. Supporting Linux only: The Windows operating system is the most popular operating system in China. According to the report [6], most of the Chinese Internet users run the Windows operating system. And for many P2P applications, e.g., PPLive, there exists only Windows-version software. However, in PlanetLab the hardware is dedicated PlanetLab nodes, as opposed to client-owned desktop machines, and it necessarily runs on the Linux operating system [31].
3. Overqualified bandwidth in PlanetLab: the network condition in PlanetLab is overqualified. because most of the participants are campus LANs with better-than-average bandwidth.

Overqualified bandwidth may lead to a totally contrary conclusion compared to real Internet. We have carried out experiments in PlanetLab to validate the effectiveness of the real nodes for P2P VoIP applications [5]. Results show few transactions benefit from the relay functionalities, which is a totally opposite conclusion compared to Skype's measurement paper [26]. We attribute the main reason for this conflicting result to the overqualified bandwidth in PlanetLab. In that environment the peer pair bandwidth is good enough so that

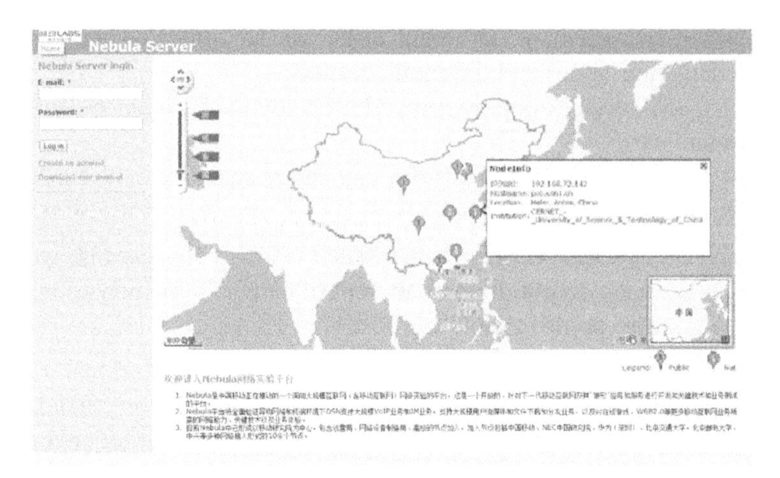

Figure 5.7 The current interface of Nebula administration.

we need not relay functionality at all. Nebula targets building a real network reflecting the true access conditions, e.g., in China over 70% of the access network is ADSL.

The technical details to realize these functionalities in Nebula are outside the scope of this paper. Interested readers can find detailed descriptions on Nebula from [8].

5.6.2.2 The Running of Nebula

Currently we have dozens of nodes on Nebula running in the Internet. We are going to enlarge the scale of Nebula and plan to have more than 1000 nodes in 3–5 years. When the scale is big enough, Nebula can connect with the leading distributed test-bed for better cooperation in distributed computing research.

Figure 5.7 shows the current interface of Nebula administration where Nebula nodes are shown with different colors. The red nodes are with public IP address while the blue ones are with private IP address. Detailed node information is provided while clicking the node icons.

5.6.2.3 Services Deployment and Key Mechanisms Measurement in Nebula

We have already deployed our P2P VoIP and P2P streaming applications in Nebula. We are doing wide range and large-scale measurements of these systems following with the increase of Nebula nodes amount to verify the

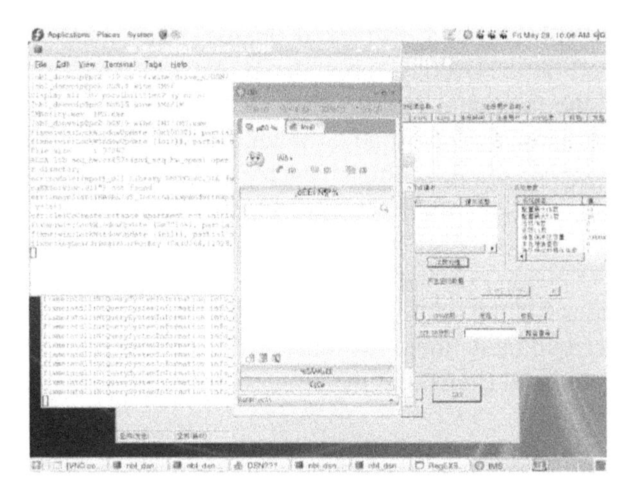

Figure 5.8 Snapshot of P2P VoIP GUI.

DSN architecture performance besides the local test. Figure 5.8 is a snapshot of P2P VoIP GUI.

With regard to key mechanisms used in DSN functional entities, we are importing P2P QoS and P2P mobility developed and simulated by HiFiP2P to Nebula test-bed for Internet-scale measurement for the performance and tuning.

5.7 Conclusions

We presented DSN, a novel carrier-level P2P and distributed services architecture for heterogeneous and converged networks. DSN is an open and promising distributed services architecture towards next generation mobile network, which effectively adopts P2P based distributed computing technology for cost effective and high quality services. Based on this unified platform, VoIP, streaming, file downloading and other popular Internet services can be supported. We gave a detailed description of DSN architecture and the related functions of core components in software architecture. We introduced the implementations of P2P VoiP and P2P streaming systems based on DSN, their local test results validate the preliminary feasibility of this architecture.

In order to further evaluate the Internet-scale performance of the services developed based on DSN architecture, a large-scale DSN simulation environment HiFiP2P and large-scale DSN test-bed Nebula on the Internet are proposed. HiFiP2P is a distributed parallel P2P simulator with a high scalability, both at the overlay and underlay network level. It is possible to make use of the measured Internet datasets as the underlay network layer model, thus the simulator can simulate a large-scale P2P application in a high fidelity method. Nebula is a complementary tool for HiFiP2P to supplement the shortcomings of simulation by real Internet experiment. It can also be used for DSN services deployment and performance measurement on the Internet.

We consider that the suite DSN methodologies and approaches discussed in this chapter will incite more excellent P2P technical proposals and useful applications, and will even boost the progress from theory to reality.

References

[1] A. Lakshminarayanan Subramanian et al. OverQoS: Offering Internet QoS using overlays. *ACM SIGCOMM Computer Communication Review*, 33(1):11–16, January 2003.

[2] BJTU, China Mobile RD. Report on real P2P system measurement under mobile IP environment, December 2008.

[3] China Mobile RD. Research report on P2P Mobility mechanisms, December 2008.

[4] China Mobile. Nebula platform design specification. Technical Report, October 2010.

[5] China Mobile. Relay effect on P2P VoIP using PlanetLab. Technical Report, October 2009.

[6] CNNIC, The 21st survey report on the Internet development in China, June 2008.

[7] DESMO-J. http://desmoj.sourceforge.net/, 2008.

[8] Yichuan Wu, Gang Li, Jing Chi, Hongluan Liao, Lifeng Le, Yunfei Zhang, Jin Peng, Xiaodong Duan, Bing Wei, and Feng Cao. DSN: A unified distributed service network system for mobile internet. Technical Report, 2009.

[9] Bhanu Krushna Rout. Extending JXTA for P2P file sharing systems. BS Thesis, ethesis.nitrkl.ac.in/306/1/10506012 and 10506034thesis.pdf.

[10] http://ietf.org/internet-drafts/draft-zhang-altotraceroute-03.txt.

[11] http://planet-lab.org/doc/myplc-3.3.

[12] http://tools.ietf.org/html/draft-ietf-alto-problem-statement-01.

[13] Introduction to JXTA. http://code.google.com/p/ppcast/wiki/WhyJXTAFailed.

[14] J-Sim. http://www.j-sim.org/, 2008.

[15] JXTA. http://en.wikipedia.org/wiki/JXTA.

[16] P2PSIP WG. www.ietf.org/dyn/wg/charter/p2psip-charter.html.

[17] PPSP WG. www.ietf.org/dyn/wg/charter/ppsp-charter.html.

[18] Telepresence. www.cisco.com/en/US/products/ps7060/index.html.

[19] China Mobile. Whitepaper of P2P technologies, December 2008.

[20] T. Anderson, L. Peterson, S. Shenker, and J. Turner. Overcoming the Internet impasse through virtualization. *Computer*, 38(4):34–41, 2005.

[21] D. Bryan, P. Matthews, E. Shim, D. Willis, and S. Dawkins. Concepts and terminology for peer to peer SIP. draft-ietf-p2psip-concepts-02. Internet Engineering Task Force, 2008.

[22] P. Garcia, C. Pairot, R. Mondéjar, J. Pujol, H. Tejedor, and R. Rallo. Planetsim: A new overlay network simulation framework. In *Software Engineering and Middleware*, pages 123–136, 2005.

[23] E. Halepovic and R. Deters. JXTA performance study. In *Proceedings of the 2003 IEEE Pacific Rim Conference on Communications, Computers and Signal Processing (PACRIM)*, Vol. 1, pages 149–154. IEEE, 2003.

[24] Y. Huang, T.Z.J. Fu, D.M. Chiu, J. Lui, and C. Huang. Challenges, design and analysis of a large-scale P2P-VOD system. In *Proceedings of the ACM SIGCOMM 2008 Conference on Data Communication*, pages 375–388. ACM, 2008.

[25] W. Kellerer, Z. Despotovic, M. Michel, Q. Hofstatter, and S. Zols. Towards a mobile peer-to-peer service platform. In *Proceedings of the International Symposium on Applications and the Internet Workshops, SAINT Workshops 2007*, page 2. IEEE, 2007.

[26] W. Kho, S.A. Baset, and H. Schulzrinne. Skype relay calls: Measurements and experiments. In *Proceedings of the IEEE INFOCOM Workshops 2008*, pages 1–6. IEEE, 2008.

[27] C. Li and C. Chen. PPStream Measurement and Data Analysis in Mobile IP Environment. In *Proceedings of the International Conference on Scalable Computing and Communications; Eighth International Conference on Embedded Computing (SCALCOM-EMBEDDEDCOM'09)*, pages 263–268. IEEE, 2009.

[28] C. Li and C. Chen. Real P2P system measurement under mobile IP environment. In *Proceedings of the International Conference on Networking and Digital Society (ICNDS'09)*, Vol. 1, pages 77–80. IEEE, 2009.

[29] Y. Li, Y.C. Wu, J.Y. Zhang, J. Peng, H.L. Liao, and Y.F. Zhang. A P2P based distributed services network for next generation mobile internet communications. In *Proceedings of the 18th International Conference on World Wide Web*, pages 1177–1178. ACM, 2009.

[30] Z. Li, G. Xie, Z. Li, Y. Zhang, and X. Duan. DHT-aid, gossip-based heterogeneous peer-to-peer membership management. In *Proceedings of the 5th IEEE Consumer Communications and Networking Conference (CCNC 2008)*, pages 284–288. IEEE, 2008.

[31] L. Peterson, T. Anderson, D. Culler, and T. Roscoe. A blueprint for introducing disruptive technology into the Internet. *ACM SIGCOMM Computer Communication Review*, 33(1):64, 2003.

[32] J. Rosenberg, H. Schulzrinne, G. Camarillo, A. Johnston, J. Peterson, R. Sparks, M. Handley, and E. Schooler. SIP: Session Initiation Protocol. RFC3261, RFC Editor United States, 2002.

[33] G. Shi, Y. Long, H. Gong, C. Wan, C. Yu, X. Yang, H. Zhang, and Y. Zhang. HiFiP2P: The simulator capable of massive nodes and measured underlay. In *Proceedings of the 17th Euromicro International Conference on Parallel, Distributed and Network-based Processing*, pages 285-292, 2009.

[34] F. Sterlind, A. Dunkels, T. Voigt, N. Tsiftes, J. Eriksson, and N. Finne. Sensornet checkpointing: Enabling repeatability in testbeds and realism in simulations. In *Wireless Sensor Networks*, pages 343–357, 2009.

6

Evolution of P2PSIP Architectures, Components and Security Analysis

Christos Tselios[1], Konstantinos Birkos[1], Christos Papageorgiou[1], Tasos Dagiuklas[2] and Stavros Kotsopoulos[1]

[1]*Department of Electrical and Computer Engineering, University of Patras, Patras, Greece; e-mail: tselioschristos@gmail.com*
[2]*Department of Telecommunication Systems and Networks, TEI of Mesolonghi, Nafpaktos, Greece*

Abstract

Peer-to-peer (P2P) networking is a distributed application that partitions resources and tasks between equally privileged, equipotent peers. P2PSIP offers peer-to-peer session management without the need for centralized servers. The distributed nature of peer-to-peer networks introduces new challenges in terms of security. Attacks can target the structure and/or the functionality of the overlay network which is the cornerstone of a P2PSIP communication system. In this chapter, the most common types of attacks that the P2PSIP protocol is prone to, along with the respective countermeasures taken in the context of the protocol, are examined. Furthermore, we present some novel mechanisms that enhance the security strength of P2PSIP, while at the same time adapting its operation in distributed networking environments. We propose a formation and maintenance scheme for the P2PSIP RELOAD overlay network, using cryptographically protected messages between the peers that can be used either complimentary to or independently of the existing security mechanisms of P2PSIP.

Anand R. Prasad et al. (Eds.), Future Internet Services and Service Architectures, 119–138.

Keywords: P2PSIP, RELOAD, overlay networks, peer-to-peer, security, VoIP.

6.1 Introduction

Peer-to-peer (P2P) is a distributed architecture, where all nodes manage and share network resources like disk space and bandwidth. The huge benefit of such an arrangement is that no expensive infrastructure such as servers is necessary. In addition, a service provider is able to handle large amount of data and users, making the whole system much more scalable and reliable than the classic client-server setup. The cost of P2P architectures is higher with a delay in lookup service, especially when the number of peers rises. The necessity of reducing that delay led to the proposal of cluster-based solutions [8]. In this context, the most reliable peers act as proxy servers for their neighbor nodes, resulting in a hybrid system architecture.

The evolution towards P2P networking caused an analogous shift in signaling protocols. The original Session Initiation Protocol (SIP) [18] that had a definite client-server orientation, was transformed into the P2PSIP protocol [2]. One of the most important aspects that the IETF P2PSIP working group [2] focused on was security. However, due to its distributed nature P2PSIP still poses several issues regarding security that remain to be resolved.

In the following sections we will first discuss on the current state of the P2PSIP protocol outlining the operation of its mechanisms and paying special attention to security provisioning. Next, we describe the most common types of attacks that the P2PSIP protocol is prone to, along with the respective countermeasures taken in the context of the protocol. Furthermore, we present some novel mechanisms that enhance the security strength of P2PSIP, while in the same time adapting its operation in distributed networking environments. We propose a formation and maintenance scheme for the P2PSIP RELOAD overlay network, using cryptographically protected messages between the peers that can be used either complimentary to or independently of the existing security mechanisms of P2PSIP. According to the proposed scheme, the nodes' keying material is periodically refreshed to mitigate the impact of attacks towards the security credentials in use. We also introduce mechanisms that adjust the P2PSIP's operation to the special properties of the wireless networking environment. A hierarchical approach in building the overlay network is presented along with a variation that operates in a more distributed way for a limited time window. Finally, we recap with the conclusions drawn from this discussion.

6.2 Current State of P2PSIP

REsource LOcation And Discovery (RELOAD) Base Protocol [10] is the leading Internet Draft in the P2PSIP Working Group [2] of IETF. It defines the structure and the operations of a peer-to-peer system that can support P2PSIP-based voice communication. Since the main difference between SIP and P2PSIP is the way each system performs storage and handles queries, RELOAD constitutes in fact a peer-to-peer overlay network suitable for users' registration information lookup and session initiation.

RELOAD consists of the following components:

1. Usage Layer: the services provided by RELOAD can be used by different applications in different ways. The Usage Layer describes the data kinds and behaviors related to each usage and it is equivalent to the application layer in the OSI model. The most common type of this layer is SIP Usage [18].
2. Message Transport: this ensures reliable end-to-end message delivery and forwards certain types of RELOAD messages.
3. Storage: RELOAD can be seen as a distributed data base in which each peer stores a portion of the necessary information for SIP communication. The Storage component is responsible for operations regarding data storage and retrieval.
4. Topology Plugin: this implements a specific peer-to-peer overlay algorithm. It defines the structure of a peer-to-peer overlay and the processes related to the overlay formation and maintenance. Although other peer-to-peer overlay algorithms are applicable, RELOAD is mainly based on Chord [24].
5. Forwarding and Link Management Layer: this handles packet forwarding between peers and implements NAT traversal.
6. Overlay Link Layer: this is the lower layer in the RELOAD architecture and it is directly related to the Transport Layer of the OSI model. It implements TLS [4] and DTLS [16] for TCP and UDP respectively.

Figure 6.1 shows the RELOAD architecture and the interactions among the different components of the protocol.

A peer who wishes to join the P2PSIP overlay network first downloads a set of overlay configuration parameters and a set of bootstrap peers from a know location. If the configuration document contains an Enrollment Server, the peer contacts this server and requests a certificate. The Enrollment Server authenticates the Joining Peer (JP) – usually by means of a username and a password – and provides JP with the requested credential as well as with a

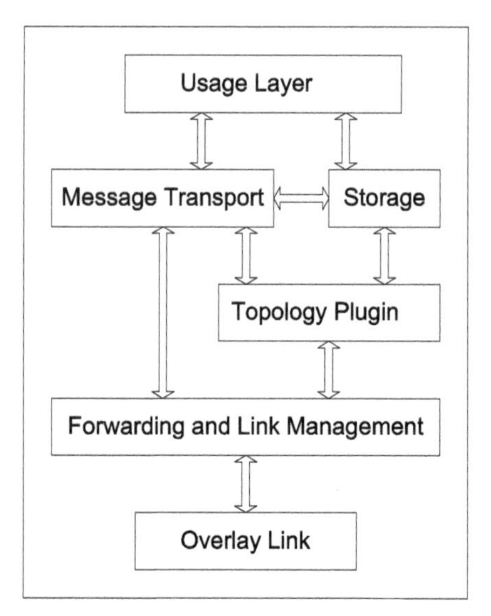

Figure 6.1 RELOAD system architecture.

unique Peer-ID which is JP's identifier in a logical identifier space. In the absence of central enrollment, JP must generate its own self-signed certificate. After a successful enrollment and authentication process, JP must contact one of the bootstrap peers, form a TLS/DTLS connection with it and send an Attach request. The bootstrap peer forwards the request to JP's Admitting Peer (AP), i.e. JP's immediate successor in the overlay. JP is then attached to AP and forms a TLS/DTLS connection with it.

After that, JP populates its routing table. The routing table consists of the finger table and the neighbor table. A peer's finger table includes all the other peers that are logically connected to this peer. It is the elements of this table that will be considered by the overlay routing mechanism to forward P2PSIP messages. The logic behind the formation of logical connections is overlay-specific and in structured overlays like Chord it is directly related to the Peer-IDs. A peer's neighbor table includes all the other peers that are logically close to that peer. By 'logically close' we mean that Peer-IDs of the neighbors are numerically close in the identifier space. The role of the neighbor table is to store a local view of the overlay that is used for maintenance and data replication purposes. The next step for JP is to explicitly request from AP to become part of the overlay. After a series of Join, Store and Update messages JP becomes a member of the P2PSIP overlay and it can initiate VoIP sessions with any other peer. According to the distributed nature of P2PSIP and contrary to conventional SIP, there is no central server to store and retrieve the necessary information for session initiation. Therefore, each peer is responsible for storing certain data items. Data storage and data retrieval is performed according to the Distributed Hash Table (DHT) functionality that is inherent to the structured peer-to-peer overlay which is the basis of RELOAD. Each data item in the DHT is identified by a data key which is a hashed value of the data. Data keys are mapped to the same identifier space with the Peer-IDs and peers are responsible for storing data items according to their Peer-ID. The purpose of the Store messages exchanged during the Join process is to transfer to JP the data items JP will be responsible for storing.

The Update messages inform other peers about JP's presence in the overlay so that they reconsider their logical connections and renew their local view of the overlay. When an Update message designates a change in the neighbor or finger table, the peer performs the necessary actions (deletes outdated entries/inserts new ones) and informs its neighbors by propagating the Update message. The propagation of Updates is the primary mechanism for overlay maintenance.

The SIP Usage [18] in the Usage Layer of RELOAD defines the functionality that substitutes the SIP Registrar. In order a peer to be accessible by other peers who wish to establish a session with it, it must register its SIP Address-Of-Record (AOR) and location in the overlay. In other words, it has to store the mapping between its SIP AOR and its Peer-ID. Consequently, when peer A needs to call peer B it actually uses the overlay routing mechanism to send a SIP INVITE to the Peer-ID of peer B that corresponds to peer B's AOR.

Security provisioning in RELOAD mainly lies in the use of public key certificates and TLS/DTLS. Each peer is identified by means of a unique Peer-ID. The Peer-ID is contained in the certificate that is usually assigned to the peer by an offline Certificate Authority (CA) before the peer joins the overlay. There is also the option of self-signed certificates under relaxed security requirements. The certificate (a) entitles the peer to store data at specific locations in the overlay, (b) entitles the user to operate a peer with the designated Peer-ID, and (c) entitles the user to use a specific user name that is also contained in the certificate. The overlay storage can be used to store peers' certificates. The advantage of this practice is that once a peer has stored its certificate in the DHT, it does not need to provide it in transactions afterwards.

Each RELOAD message exchanged between any two peers must be digitally signed. This enables the recipient to verify the origin of the sender. The digital signature also offers data integrity. Data items in the DHT are

Table 6.1 Signaling threats and countermeasures.

Threat	Attack Method	Countermeasure(s)
Caller ID Spoofing	SIP URI falsification	PKI
Put/Get Message Compromise	SIP message eavesdropping	Overall Message integrity
Squatting and Drowning	DHT Key falsification	Source Integrity

also digitally signed by the peer who stored them. Apart from public key certificates, a shared secret admission control scheme is also possible. It is based on a common symmetric key which is shared among all the peers and which is used in TLS-PSK [7] or TLS-SRP [25] connections.

6.3 Security Threats and Countermeasures in P2PSIP

For having an accurate security analysis on the P2PSIP protocol, two distinctive categories should be considered. The first one is *Signaling Security* provided that P2PSIP together with the original SIP are based on the IP infrastructure, therefore inherit most IP security vulnerabilities and the second is *Overlay Security* where more general attacks in the P2P network are described.

6.3.1 Signaling Security

As mentioned previously, P2PSIP derives from the original IP communication protocol. This fact offers many benefits but also a few disadvantages especially when it comes to security vulnerability. There is a certain number of attacks aiming to the actual signaling process, trying to compromise the overlay by causing problems to the session establishment and if possible to the DHT lookup functionality [19]. Some of these attacks are listed in Table 6.1 and are going to be presented in the following paragraphs.

6.3.1.1 Caller ID Spoofing

Caller ID spoofing is a basic attack in all networks based on IP communication protocol. The main method for identifying a specific user is the SIP Uniform Resource Identifier (SIP-URI) the user holds which is visible in all inbound and outbound messages such as *Invite* or *Request*, messages inherited to P2PSIP from earlier client-server versions of SIP. However in P2PSIP overlay consisted by untrusted nodes it is crucial that the integrity of content retrieved from the network can be verified for instance with a SIP-URI location binding method [21]. Otherwise an attacker is able to establish a call or a session to another party with a false caller ID, provided that no actual control on obtaining the SIP URI exists, manipulating the victim into believing that he is having a conversation with a friendly node. Especially in P2PSIP the adversaries have the additional ability of redirecting phone calls to malicious nodes under their control by replacing the returning IP address, totally destroying the lookup process in both nodes and the DHT. Caller ID spoofing is a rather serious attack used in all kinds of frauds.

A very good method for dealing with this category of attacks is Public Key Infrastructure (PKI) [11], systems based on public key cryptography for authentication and data integrity. PKI issues and manages digital signatures and public key certificates to each entity located in the protected network therefore an attacker is rendered unable to falsify and use this digital signature for illegitimate purposes. The process of digitally signing a message includes the use of hash functions to create a unique identifier that is placed inside the actual content. A specific algorithm called Digital Signature Algorithm (DSA) [13] and a Secure Hash Algorithm (SHA) [14] are commonly used for digital signature purposes. Especially in P2PSIP case when PKI is used, retrieving nodes becoming able of verify the caller's ID and the integrity of a data item retrieved. An alternative solution to ensure content integrity in P2PSIP is based on self-certifying SIP-URIs [1]. With this method a public key hashing generates SIP-URIs in which the owner stores additional information regarding his identity in the overlay. Such additional information might be the node location. By obtaining the aforementioned public key during the registration process, any node retrieving information stored in the DHT verifies both identity and data integrity.

6.3.1.2 Put/Get Message Compromise

In order to preserve user privacy, the network overlay should be able to provide confidentiality in all messages that are forwarded amongst nodes. A malicious node might easily take advantage possible security gaps, eavesdrop some SIP *Request* or *Response* messages and gain knowledge of a nodes' presence status, its SIP-URI address or even parts of the DHT stored in the specific node. Additionally, the attacker is possible to monitor traffic and keep track of possible overlay changes. Although eavesdropping is considered dangerous, the real threat for P2PSIP overlay comes from message tampering. The only parameters whose integrity is totally protected is of those stored in the overlay storage. All others such as key or time-to-live might be able to get modified by an attacker causing the service to function improperly. For instance, if one node's registration key is modified and stored in a fake URI,

Table 6.2 Overlay threats and countermeasures.

Threat	Attack Method	Countermeasure(s)
Routing Attack	DHT lookup request manipulation	Reputation-based routing
		ID update process
Sybil Attack	Fake node insertion	Intrusion Detection System
		Reputation-Based routing
Eclipse	Overlay Division	Central Authority
Denial of Service	Packet/Message Flooding	DHT Data Replication
		Reputation Scheme Introduction

the original URI and the corresponding user become unreachable. Therefore it is important for the system to provide integrity protection for the whole put/get message, rather than just the record value.

6.3.1.3 Squatting and Drowning Attack

A squatting attack occurs in networks that only use a unique key for each record in the overlay storage. When many user simultaneously try to put their records under this key, only the first one is going to complete the procedure successfully. An attacker might be able to put false records under well-known keys to make legitimate users unable to insert their own records. The DHT appears full and this leads to a denial-of-service regarding additional storage requests. Also a potential lookup within the overlay is high likely to return false or even no results at all since malicious nodes are not supposed to be trusted in any case. The overlay storage should also be able to protect itself against the drowning attack in which the adversary inputs large amount of mangled, destroyed or simply useless data under a target key. The large amount of junk files makes the correct value difficult to be retrieved. Possibly the optimal solution for such an attack is an implementation of source integrity protection of all records in the overlay storage.

6.3.2 Overlay Security

Some of the most common threats against Overlay structure together with the necessary countermeasures are presented in Table 6.2. A detailed analysis can be found in the following paragraphs.

6.3.2.1 Routing Attacks

The routing attack [23] is the action of the lookup request manipulation that takes place within the routing path from a malicious node. The term manipulation might mean that the node delays, alters, drops or even forwards the

Request to a wrong destination. Once a *Request* message is lost, a potential action establishment within the overlay freezes and the service supposing to be initiated becomes unavailable. Routing attacks might affect the overlay in many ways. Provided that from a technical aspect P2PSIP replaces all servers used in regular SIP signaling for locating the desired node within the overlay with a DHT algorithm, a routing attack that causes trouble in SIP messaging is highly likely to have a serious effect on the DHT as well [21]. The actual role of a DHT is to provide a basic functionality to the network nodes. It acts as a storage inventory of the whole overlay, an inventory that can provide any kind of information regarding stored data in every one of the available nodes upon request. When the overlay is under routing attack the previously described lookup service is compromised since many *Request* messages that might have been able to initiate a potential search are diverted or lost. Adversary nodes located inside the recursive routing path could easily prevent a query node from reaching its actual destination. There is no way to prevent routing attacks when lookup requests are forwarded over untrustworthy nodes.

One of the possible solutions to this thorny problem is the establishment of a central routing entity responsible for Secure node-ID assignment that could prevent a serious attack on DHT [3]. Unfortunately, the acknowledgement that even under the supervision of such an entity, malicious nodes can significantly degrade lookup service still exists [20]. This leads to the adoption of an alternative possible elucidation. If instead of recursive routing, iterative routing is used, the node from which the initial *Request* started would be able to check if the message is forwarded correctly, to a node closer to the one that should be the final destination. When a node does not issue a *Reply* message after a certain time frame is considered to be malicious. Although these tactics could render the network overlay relatively safe from routing attacks, there is no guarantee that a benevolent node will never be infected with malicious software. Therefore a reputation-based routing together with an ID *Update* procedure might be the optimal solution.

6.3.2.2 Sybil Attack

The Sybil attack [5] is commonly used against systems that do not include any authentication mechanism, allowing users to freely generate identities. Especially in pure P2P networks where no central certification authority exists, a Sybil attack might prove really dangerous. In a Sybil attack, a malicious node may present one or more fake identities to the other network nodes. This manipulates all nodes leaving them to assume that the attacker is benevolent and actually part of the network topology, and therefore treat him as such. Sybil attacks are also known as impersonation attacks and may have quite a number of variations such as:

- The attacker might present the identity of an existing note already located in the network. This way an actual impersonation of virtually any node is possible.
- The attacker might present a different identity from any other existing node. By presenting multiple non-existent identities which are indistinguishable from the existing legitimate identities an attacker avoids collision that might reveal his fraud. This trick aims on degrading multipath routing protocols since there is going to be a constantly increasing number of errors when the optimal route calculation takes place. It will be virtually impossible for the protocol to establish the shortest path between two nodes and the benefits of a multipath protocol cease to exist.

Adversarial nodes may also employ impersonation to flood the network with faulty traffic in order to force all other nodes to drop legitimate packets. This behavior leads the network's quality of service mechanism to malfunction even if the network was able to provide and ensure QoS in the first place.

P2PSIP Overlays are vulnerable to Sybil attacks since anyone is able to sign up for a SIP URI and join the network. The only reason the attack does not completely destroy the infrastructure is that not all intermediate nodes are untrustworthy. In case a DHT is used as in many structured P2P networks, the Sybil attack is able to ruin the data replication mechanism simply because all *Request* messages from a certain user towards other parts of the network is likely to be intercepted by the same malicious node. When the adversary generates multiple pseudonymous identities it is more likely that a greater amount of *Request* messages are going to end up in the wrong receiver, thus providing a higher success to the attack.

The most common solution for Sybil attacks is the implementation of a central trusted authority such an Intrusion Detection System (IDS) or a reputation scheme to certify the identity of each participating entity. Unfortunately such as solution is not able to be successful in pure P2P Overlays since it is rather contradicted to the principles of peer node architecture. On the other hand, hybrid P2P structure seems more appropriate, having the bootstrap node acting as that certified authorization unit. Another form of defense is making joining the actual network more difficult, by implementing a method of resource trade-off by the newcomer in order to establish its identity in the overlay. The alternative is, instead of an initial exchange of information, some sort of *Refresh* process to take place in order for the DHT to be updated and a kind of identity check to be possible.

6.3.2.3 Eclipse Attack

During an Eclipse attack on a P2P network overlay, the adversary will first try to gain control over a certain percentage of the whole domain. Therefore he tries to place malicious nodes or penetrate legitimate ones along strategic routing paths [22]. Once this is realized, the next step is to separate the network into different subnetworks. This strategy enables the attacker to have total control over all communication between the newly created subnetworks. If a node wants to communicate with a corresponding node in a different subnetwork its message at some point will be transferred through one of the attacker's nodes. Such a high scale attack involving strategic targeting is rather serious provided it leads to a total control of a part of the overlay. After a successful Eclipse attack the attacker becomes capable of attacking the control plane by rerouting each and every message at will, becomes capable of separating all subnetworks by dropping all receiving messages and finally attack the data plane by injecting polluted files or forwarding such files in distant legitimate nodes. In P2PSIP this leads to Denial-of-Service (DoS) and possible DHT destruction [9]. When no SIP message is forwarded correctly on the destination node, the lookup process is interrupted and DHT access is limited due to false *Request* messages. Although carrying information for every node, DHT becomes unable to answer thus returning the corresponding node identity.

There are three methods for preventing an Eclipse attack. The first one is to simply use a pure P2P network model. In this case, the bigger the network overlay, the more malicious nodes it needs to be compromised, because no actual strategic points exist for the adversary to place its nodes. Therefore no division into subnetworks can be made and P2PSIP signaling like *Request* and *Response* messaging thus DHT updates and data are forwarded normally. The second method is also based on the same principal, preventing overlay division by using an additional randomization algorithm to determine the nodes location. The third method is using a trusted centralized membership service. This service keeps track of the overlay membership and provides unbiased referrals among nodes that wish to acquire overlay neighbors. The only drawback is that it requires dedicated resources and raises concerns about availability, scalability and reliability. It is more than obvious that a form of authentication and *Refresh* process taking place in the overlay would also solve many of the existing problems like giving the bootstrap node the ability of immediate steering a newcomer towards a legitimate node rather than a suspicious one.

6.3.2.4 Denial of Service

The main goal of a Denial of Service (DoS) attack is service disruption by exhausting the network resources in order to render a node, a function or any other entity useless for its legitimate users. According to the most common strategy the adversary sends towards the victim a stream of packets for consuming the bandwidth, the processing capacity or the traffic ability, all leading to degrading services to the authorized members of the network topology. These attacks do not necessarily damage the resources they access but only occupy them and intentionally compromise their availability. The most common target of a DoS attack is no other than the bandwidth. In this case, a high volume of traffic floods the network, legitimate users are unable to get through and the result is merely frustration and degraded productivity. Something similar takes place in the case of connectivity attack. Instead of high volume traffic, the network is flooded with a high amount of connection requests towards all nodes thus rendering the system unable to distinguish and process legitimate ones [27]. In P2P overlays that use P2PSIP as signaling protocol, the most common DoS attacks are those inherited by the SIP protocol originally used in the client-client network topology. An improved version of the simple DoS is the Distributed Denial of Service (DDoS) which uses more than one terminal to launch a coordinated DoS attack versus one or more targets. By adding a multi-dimensional factor using client/server technology the adversary is able to proportionally increase the impact of his attack and make prevention towards it a much more difficult thing to do.

Refusal of Action or Service The P2P overlay consists of nodes having some data objects related to its node ID stored in the overlay inventory, the DHT. Since the DHT is a distributed table stored partially on all nodes participating in the network topology [26], there is quite a big possibility that some particular user data will be stored at a malicious node, if such exists. It is also likely that in case of some *Request* on behalf of another node, this malicious node is able to deny access to that information. In order to prevent the specific attack that makes query incomplete, the overlay storage should not only store the previously mentioned DHT part on a single node, but duplicate the record to the neighboring nodes [23]. Data replication solves denial of service issues as it eliminates the s called single point of failure. Additionally, replication also reduces lookup latency [15]. Most recent DHT implementations [17] provide some redundancy mechanism to counter this kind of attacks and also offer a much faster access to all stored data.

Flooding Attack For launching a DoS flooding attack the adversary uses a high amount of well formatted P2PSIP messages towards target nodes such as *Invite* or *Register*. The huge amount of traffic the network encounters makes impossible all alternative message exchange, leading to huge delay in communication. Depending on the exact resource the attacker is planning to consume, a different P2PSIP message might be used. For instance a fast, constant stream of *Register* messages leads to an increase of the bootstrapping nodes' processing load, while a high count of *Invite* messages with different session identifiers might cause the victim node to run out of memory. The simple act of generating SIP-compliant, unencrypted and unauthenticated messages remains a dangerous threat that P2PSIP overlays have to be protected from.

RFC 4987 [6] defines some basic defensive methods against the flooding attack. One of the most immediate solutions is a timer implementation that simply inputs a time interval between all incoming messages and the ACK the entity is supposed to send back. If the actual load rises, no ACK is sent back and the entity responsible for resource usage takes action. Alas, this solution is not very effective when the attacking packet rate is high. Introducing a reputation scheme is a far more advanced and trust-worthy solution since makes the network overlay capable of excluding from the DHT all suspicious nodes that only cause dubious traffic.

6.4 P2PSIP Extensions

In the following we present two extensions of RELOAD which aim at improving the provided security level. Towards extending the security mechanisms in RELOAD, a Public/Private Key Infrastructure (PKI) scheme is used along with a symmetric cryptographic mechanism. For this purpose each node is preconfigured with a key generation function for its Public/Private Key (PPK) pair and a network-wide shared key. Furthermore, each peer periodically refreshes its PPK pair in order to further improve the security level of the system. Public key cryptography is used to encrypt the communication between the peers, while symmetric cryptography is used primarily for authentication purposes. The above mechanisms can be used either complimentary or independently of the TLS mechanism existing in the current RELOAD draft, since they handle security threats at different levels and they offer enhanced protection against insiders.

In any case however, the VoIP sessions established by the P2PSIP protocol are guarded against a series of attacks that are possible using RELOAD. More specifically, the authentication provided by the network-wide key prevents a malicious peer from disrupting the protocol's operation by trying to act as admitting or even Bootstrap peer. The encryption provided by the public key cryptography ensures that the peers' public keys and local view of the overlay network transmitted with the *StoreReq* and *UpdateReq* messages respectively, are protected against eavesdroppers. Finally, the refresh mechanism puts a temporal limit in the operation of a malicious peer that managed to intercept a valid set of security credentials and peer ID, in contrast to RELOAD where neither a certificate revocation scheme or any mechanism whatsoever exists as countermeasure against a malicious peer operating within the network.

In the proposed mechanisms, we maintain the notion of higher hierarchy level peers (Bootstrap peers) that are also present in RELOAD. The Bootstrap peers are responsible for building and maintaining the overlay network, and implementing the key distribution and refresh processes.

Two mechanisms are proposed in order to extend the RELOAD protocol towards the above directions. A hierarchical approach, to be referred to as Hierarchical RELOAD (HR), is first presented, according to which the Bootstrap peers are central to all the operations implemented within the protocol. We also propose a variation of the HR protocol, Semi-Hierarchical RELOAD (SHR) that follows the same logic but also permits for a limited time window a more distributed operation for the protocol. In the following the HR and SHR protocols are described in detail.

6.4.1 Hierarchical RELOAD

In the Hierarchical RELOAD (HR) protocol, the joining peer (JP) first initiates a neighbor discovery mechanism in order to detect if there is someone within its communication range. In doing so, the joining peer JP broadcasts a *Hello* message that is acknowledged by every peer already connected to the overlay network that receives it with an *AckHello* message. The joining peer JP then transmits a *JoinReq* to the sender P of the first *AckHello* message that

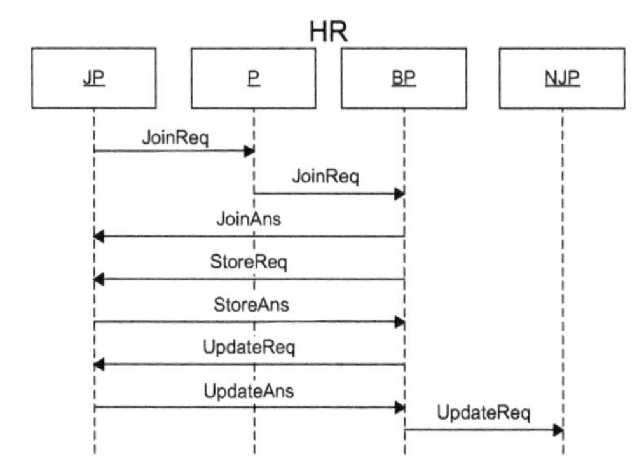

Figure 6.2 The join process in the HR protocol. We denote the joining peer as JP, the first peer discovered by JP as P, the Bootstrap peer as BP and the logical neighbors of JP as NJP.

arrived, ignoring all the subsequent ones. Upon receiving *JoinReq*, peer P forwards it towards the Bootstrap peers of the overlay. The *JoinReq* message contains the joining peer's public key encrypted with the network-wide secret key in order to authenticate the joining peer and prevent eavesdropping of the transmitted keying material.

The Bootstrap peers then send a *JoinAns* message to the joining peer JP acknowledging the *JoinReq* message and a *StoreReq* message containing the public keys of the joining peers' logical neighbors as defined by the overlay construction algorithm. Both *JoinReq* and *StoreReq* messages are encrypted using the joining peer's public key, as above, for authentication and protection purposes. The joining peer confirms the successful reception of *StoreReq* with a *StoreAns* message.

To complete the join process *UpdateReq* messages are sent from the Bootstrap peers to the joining peer and the peers selected as its logical neighbors informing them about the fact that they are logically connected with each other. These messages contain the local view of the sender about the state of the peer-to-peer network. Again as above, the Bootstrap peers encrypt the *UpdateReq* messages using the public keys of their recipients. These *UpdateReq* messages are acknowledged by respective *UpdateAns* messages following the opposite direction. This update process is also initiated by any peer when there is a change in its logical connections, i.e. the peers affected of such a change are informed by *UpdateReq* messages. Figure 6.2 illustrates the aforementioned message exchange sequence.

The leave process is also implemented in an analogously hierarchical manner with the Bootstrap peers receiving the *Leave* message sent by the leaving peer informing the affected peers by this exit with *UpdateReq* and *StoreReq* messages as in the join process described above.

An important security extension to the RELOAD protocol is a key refresh mechanism, according to which the public/private key pairs of the participating peers are periodically refreshed in order to strengthen the system against attacks based on the retrieval of users' public or private keys. When a peer joins the overlay, it can use its assigned keying material for a limited time period. The Bootstrap peers maintain the time when the public/private key pair of every peer must be regenerated and periodically check if the time interval until this time falls below a certain predefined critical margin. In this case, the Bootstrap peers send a *RefreshReq* message to the peer whose credentials are to be refreshed, to be referred to as refreshing peer (RP). The *RefreshReq* message contains a new seed for the key generation function of the refreshing peer so that a new public/private key pair is created. Upon receiving *RefreshReq*, the refreshing peer first acknowledges this reception to the Bootstrap peers by sending them a *RefreshAns* message and then restarts a new join process with its newly generated public key (Figure 6.2). If the Bootstrap peers do not receive a *RefreshAns* within a certain timeout period, they retransmit the *RefreshReq* message. The maximum number of retransmissions is a parameter in the protocol's operation.

6.4.2 Semi-Hierarchical RELOAD

The Semi-Hierarchical RELOAD (SHR) protocol aims at relaxing the strictly hierarchical nature of Hierarchical RELOAD in order to enable a fast integration of the joining peer to the overlay, at least for a limited time window until the full update process as described in Hierarchical RELOAD is completed.

The joining peer is first discovered through the same *Hello* beacon protocol described in Hierarchical RELOAD peers that are located within its transmission range. Let P be the first peer to reply with a *AckHello* message. As in Hierarchical RELOAD, the joining peer sends P a *JoinReq* message containing its public key and P apart from forwarding it towards the Bootstrap peers, also handles it as if it were a Bootstrap peer. This means that P, judging from its local view about the state of the peer-to-peer network, temporarily places the joining peer in the overlay and thus sends the corresponding *StoreReq* and *UpdateReq* messages containing the public keys of the peers selected as the joining peer's logical neighbors.

The rest of the join process is implemented concurrently as described in Hierarchical RELOAD and when the *UpdateReq*/*StoreReq* messages arrive

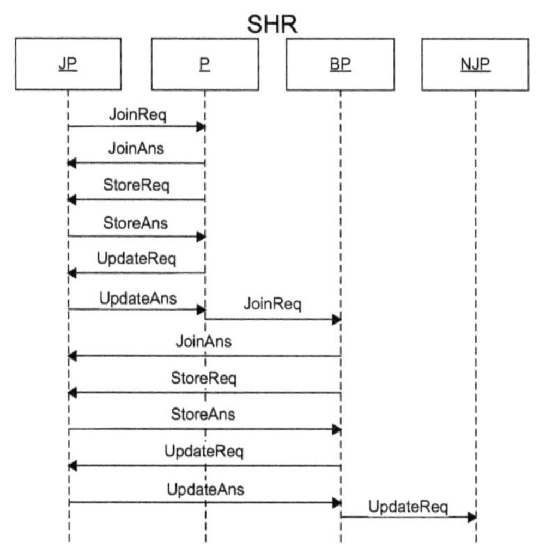

Figure 6.3 The join process in the SHR protocol. We denote the joining peer as JP, the first peer discovered by JP as P, the Bootstrap peer as BP and the logical neighbors of JP as NJP.

at the joining peer from the Bootstrap peers the joining peer's view about the peer-to-peer network and its position in it is correspondingly updated. If this does not happen within the time window given for this temporary placement of joining peer in the overlay, its entrance is canceled. Figure 6.3 depicts the operations performed during the join process in Semi-Hierarchical RELOAD.

The above relaxation in the join process enables the faster integration of a peer in the network without compromising the provided system security level, since the use of both the public/private and the symmetric cryptography in the message exchange among the peers is not sacrificed. Especially in the case of sparsely connected wireless ad hoc networks, this variation in the protocol's operation could prove beneficiary when the round trip time of a message to the Bootstrap peers grows too large.

The Semi-Hierarchical RELOAD (SHR) protocol is characterized by the majority of the security features included in the Hierarchical RELOAD (HR). with some relaxations in favour of fast deployment and flexibility. The peer P at which JP addresses its *JoinReq* cannot check whether the joining peer's ID maps into the specific public key. Peer P can only confirm that the joining peer (JP) holds the network-wide key. However, the joining peer (JP) is permitted to operate normally within the overlay for a limited amount of time until it gets final permission from the Bootstrap peers. Thus the join process as described in Hierarchical RELOAD continues concurrently.

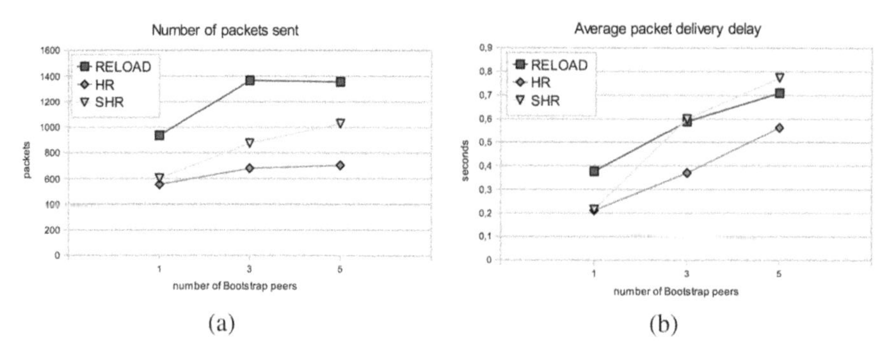

Figure 6.4 (a) The total number N of messages sent. (b) The average message delivery delay D in a simulation in order to build and maintain the peer-to-peer network over the number of Bootstrap peers in the case of 25 nodes participating in the network.

6.4.3 Security Extensions Evaluation

We implemented and evaluated the performance of HR, SHR and RELOAD in Network Simulator (ns-2) [12]. We compared the proposed protocols to RELOAD focusing on how they scale when the network size grows and on the effect the number of Bootstrap peers has on their performance. The results are depicted in Figure 6.4.

As the number of Bootstrap peers grows, all protocols seem to be affected uniformly, except the Hierarchical RELOAD that produces the smallest total number N of messages sent and the lowest average message delivery delay D, for all cases of Bootstrap peers and remains stable regardless of the number of Bootstrap peers. The Semi-Hierarchical RELOAD is almost identical to HR when there is only one Bootstrap peer but the distance between them widens as more Bootstrap peers are added. This is due to the fact that the cost of the parallel temporary admission is higher when there are more Bootstrap peers in the network, since a different series of messages is exchanged between the joining peer and each one of them. Finally, RELOAD in all scenarios produces significantly higher N than the rest due to the more complex signaling it defines.

6.5 Conclusions

In this chapter we examined P2PSIP as a distributed approach for session initiation in IP networks. PSPSIP relies on a structured overlay network and uses a distributed storage functionality to store and retrieve the users' registration information. Since, like in most peer-to-peer systems, the deployment of P2PSIP generates several security concerns, emphasis was given to security aspects and how the different components of the P2PSIP system architecture can be used to address certain threats.

Security threats can affect the signaling as well as the overlay network. Signaling threats include Caller ID Spoofing, Put/Get message compromise, Squatting and Drowning. Countermeasures for this category of threats are based on PKI and also on the adoption of integrity protection mechanisms. Overlay threats include overlay routing attacks, the Sybil attack, the Eclipse attack and DoS attacks. DoS attacks form a greater category of threats with refusal of action or service, flooding attack and DDoS attacks being more common to P2PSIP systems. Iterative routing, reputation-based routing, central trusted authorities, randomized node location, data replication and IDSs are among the most prominent solutions against attacks on the P2PSIP overlay.

Moreover, we proposed two extensions of P2PSIP, namely Hierarchical RELOAD and Semi-Hierarchical RELOAD. These extensions exploit a two-level hierarchy among the participating peers, symmetric/asymmetric cryptography and a key refresh mechanism in order to protect the structure of the P2PSIP overlay and alleviate the effects of malicious activity. The proposed extensions differ in that SHR relaxes the strictly hierarchical nature of HR in order to enable a fast integration of joining peers to the overlay.

References

[1] P2PP prototype implementation: http://www1.cs.columbia.edu/~salman/peer/. PhD Thesis.

[2] P2PSIP, http://www.p2psip.org/ietf.php.

[3] OpenChord. PhD Thesis, Otto-Friedrich-Universität, Bamberg, http://www.unibamberg.de/en/pi/bereich/research/software-projects/openchord, 2005.

[4] T. Dierks and E. Rescorla. The Transport Layer Security (TLS) protocol, Version 1.2. IETF RFC 5246, 2008.

[5] John R. Douceur. The sybil attack. In *Proceedings of 1st International Workshop on Peer-to-Peer Systems (IPTPS'02)*, 2002.

[6] W. Eddy. TCP SYN flooding attacks and common mitigations. IETF RFC 4987, 2007.

[7] P. Eronen and H. Tschofenig. Pre-shared key ciphersuites for Transport Layer Security (TLS). IETF RFC 4279, 2005.

[8] L. Garcés Erice. Hierarchical peer-to-peer systems. PhD Thesis, Polytechnic University, 2005.

[9] Cheevarat Jampathon. P2PSIP security. PhD Thesis, Helsinky University of Technology, http://nordsecmob.tkk.fi/Thesisworks/Cheevarat-final.pdf, 2008.

[10] C. Jennings, B.E. Lowekamp, E. Rescorla, S. Baset, and H. Schulzrinne. REsource LOcation And Discovery (RELOAD) base protocol. IETF, internet draft, http://tools.ietf.org/id/draft-ietf-p2psip-base-11.txt, 2010.

[11] National N. Nazareno. Federal public key infrastructure (PKI) technical specifications (version 1) Part B: Technical security policy. NIST PKI Technical Working Group, 1996.

[12] The Network Simulator NS-2. http://www.isi.edu/nsnam/ns/.

[13] National Institute of Standards and Technology (NIST). Digital signature standard. FIPS PUB 186, 1994.

[14] National Institute of Standards and Technology (NIST). Secure hash standard. FIPS PUB 180-1, 1995.

[15] V. Ramasubramanian and E.G. Sirer. Beehive: O(1) lookup performance for power-law query distributions in peer-to-peer overlays. In *Proceedings of USENIX Symposium on Design and Implementation (NSDI)*, 2004.

[16] E. Rescorla and N. Modadugu. Datagram transport layer security. IETF RFC 4347, 2006.

[17] Sean Christopher Rhea. OpenDHT: A public DHT service. PhD Thesis, University of California, Berkeley, 2005.

[18] J. Rosenberg, H. Schulzrinne, G. Camarillo, A. Johnston, J. Peterson, R. Sparks, M. Handley, and E. Schooler. SIP: Session Initiation Protocol. IETF RFC 3261, 2002.

[19] H. Schulzrinne, E. Marocco, and E. Ivov. Security issues and solutions in peer-to-peer systems for realtime communications. IETF RFC 5765, 2010.

[20] J. Seedorf and C. Muus. Availability for DHT-based overlay networks with unidirectional routing. In *Proceedings of WISTP 2008*, 2008.

[21] Jan Seedorf, Frank Ruwolt, Martin Stiemerling, and Saverio Niccolini. Evaluating P2PSIP under attack: An emulative study. In *Proceedings IEEE GLOBECOM*, 2008.

[22] Atul Singh, Tsuen-Wan Ngan, Peter Druschel, and Dan S. Wallach. Eclipse attacks on overlay networks: Threats and defences.

[23] E. Sit and R. Morris. Security considerations for peer-to-peer distributed hash tables. In *Proceedings of 1st International Workshop on Peer-to-Peer Systems (IPTPS'02)*, 2002.

[24] I. Stoica, R. Morris, D. Liben-Nowell, D. R. Karger, M. F. Kaashoek, and F. Dabek. Chord: A scalable peer-to-peer lookup protocol for internet applications. *IEEE/ACM Trans. on Networking*, 11(1), 2003.

[25] D. Taylor, T. Wu, N. Mavrogiannopoulos, and T. Perrin. Using the Secure Remote Password (SRP) protocol for TLS authentication. IETF RFC 5054, 2007.

[26] Guido Urdaneta, Guillaume Pierre, and Maarten van Steen. A survey of DHT security techniques. *ACM Computing Surveys*, 2009.

[27] D. Wallach. A survey of peer-to-peer security issues.
PhD Thesis, http://www.cs.rice.edu/~dwallach/pub/tokyo-p2p2002.pdf, 2002.

7

Conferencing Services in P2P Networks: Trends and Challenges

Andre Rios[1], Alberto J. Gonzalez[1,2], Jesus Alcober[1,2], Javier Ozon[1] and Kayhan Z. Ghafoor[3]

[1]*Department of Telematics Engineering, Universitat Politècnica de Catalunya, Esteve Terradas 7, 08860 Castelldefels, Spain; e-mail: andre.rios@upc.edu*
[2]*i2cat Foundation, Gran Capità 2–4, Edifici Nexus I, 08034 Barcelona, Spain*
[3]*Faculty of Computer Science and Information Systems, Universiti Teknologi Malaysia, 81310 Skudai, Johor D.T., Malaysia*

Abstract

Conferencing systems allow interactive communications and facilitate the joint work among users regardless of their geographical location. The vast majority of those are usually implemented by means of either centralized client-server architectures or high performance devices called Multipoint Conference Unit (MCU). To date, few researchers have carried out the implementation of these services in a fully decentralized and multipoint scheme. However, the main trend, in order to overcome the scalability issues, is to use the Peer-to-Peer (P2P) paradigm, despite the fact that in this type of systems, its implementation is at an early stage of development. Currently proposed P2P conferencing solutions generally employ the same elements and mechanisms as P2P media streaming systems, however, from the operational point of view, they are subject to additional or greater constraints such as to maintain minimum delay (bounded) at distribution level and media mixing. This chapter analyzes all these issues including some architectures for developing P2P conferencing and the use of enhanced video coding such as source and network coding to offer robustness and scalability, both at transmission and reception in heterogeneous environments, floor control characteristics and

Anand R. Prasad et al. (Eds.), Future Internet Services and Service Architectures, 139–160.

management. Moreover, based on real experiences and projects, a prototype giving some recipes for implementation considering later aspects such as possible business models and operational limitations will be described.

Keywords: peer-to-peer, conferencing, video, audio, coding, mixing, ALM.

7.1 Introduction

Currently, streaming applications arouse a great interest both at commercial level and academic research. Live streaming introduces new challenging problems different to ordinary file-sharing. In general, P2P media streaming solutions have different features that determine the operation of the applications, i.e., a large volume of media data along with stringent timing constraints, the dynamic and heterogeneous nature of P2P networks and the unpredictable behavior of peers. Additionally, if the media streaming service is a conference, then there is a "big challenge" to maintain the interactive experience, especially when there are many participants (large-audience).

Conference services are generally deployed for open group applications in order to distribute public events, such as lectures, conferences, concerts or Internet TV. Multiparty conferencing systems with a reduced and controlled membership group (close groups) are better suited to set up meetings, discussions, seminars or workshops. To design a system that supports both schemes in order to have a more generic and flexible platform is a challenging task. In general, collaborative tools on the Internet include discussion boards, bulletin boards, chat, P2P file-sharing networks, electronic newsletters, blogs, newsgroups, instant messaging, and video conferencing through Internet webcams and software such as Skype. Mainly, a conference is a real-time application that in its point-to-point basic scheme involves bidirectional communication with stringent requirements for end-to-end delay. Moreover, the development of a multipoint conference in a P2P scheme has other critical challenges such as bandwidth and processing limitations of peers and the delay associated with multicast among peers.

There are several existing solutions handling P2P conferencing, such as using IP multicast, full-meshed conferencing and Application-Layer Multicast (ALM) [1]. IP multicast is a first traditional approach based on network layer. Currently, its deployment is limited due to its complexity and security problems. The second approach is to use a full-meshed scheme, but clearly its main problem is the scalability and applicability, especially when the number of users is high. Finally, the third approach is the most widely used currently:

Multicast at Application Layer, which reduces the network usage and balances the links in the underlying network. In [2–6], some P2P systems for conferencing services are described.

In relation to the media, the existing approaches assume that all participating devices support audio mixing but not video mixing. This assumption is often based on the proliferation of devices, especially IP phones or softphones running on PCs, with abundant CPU power and large bandwidth. However, mixing at all the participating nodes is usually not scalable. Besides, in terms of signaling, there are several works that try to mix P2P techniques with the Session Initiation Protocol (SIP), designed for the establishment and maintenance of multimedia sessions. Mainly, there are two approaches: SIP using P2P (replace SIP location service by a P2P protocol) and P2P over SIP (implementation of P2P using SIP signaling). Currently there is an IETF working group, which efforts to standardize P2PSIP [7].

Taking into account these above mentioned antecedents, in this work we explore ALM techniques that allow to improve the existing solutions in order to offer a system with real-time support and adding features like media mixing (centralized or distributed) depending on the scheme selected: pure conference or multiparty. In a similar way, it is understood that a good conferencing service should besides meet certain basic requirements. It should be easy to set up and maintain and it should dynamically add and drop participants to the conference. In this sense, SIP-based conference control framework (SIPPING) offers a suitable environment to work while allowing interoperability with other SIP compliant systems.

Moreover, in media distribution, there are two important aspects that may be relevant to the implementation of video conferencing. At transmission level it is key to minimize packet loss and to maintain a bounded delay. Therefore, it is necessary to have a media coding scheme that allows a robust communication and efficiently uses the network resources. MDC, SVC and NC are schemes that can be useful in this sense and that will be analyzed further in this chapter. In [8] a system using SVC is described. Finally, this work proposes an overview about P2P multipoint conference systems based on ALM and SIP, which is a first approach to overcome some problems associated to P2P conferencing. P2P conference systems are well suited to setup spontaneous conferences, because they do not depend on a certain infrastructure.

The remainder of this article is organized as follows. Section 7.2 describes the typical P2P architectures used in media streaming and the requirements for offering conferencing services. In Section 7.3, some media coding tech-

Table 7.1 Main constraints.

Constraint	Description
Scalability	The system must scale according to the number of users who are connected to the service (VoD, live streaming and multiconference service).
Bandwidth constraint	The video streaming rate should not exceed the channel capacity.
Real-time constraint	The delay in video packet delivery should not exceed the play-out deadline of a video frame at reception time.
Quality of Service (QoS)	QoS must guarantee a minimum decoded video quality and a maximum transmission error rate over the duration of the streaming session despite the variation in channel conditions.

Table 7.2 A taxonomy of typical P2P applications.

Category	Bandwidth-sensitive	Delay-sensitive	Scale
File download	No	No	Large
On-demand streaming	Yes	Yes	Large
Audio/Video Conferencing	Yes/No	Yes	Small
Live Broadcast	Yes	Yes	Large

niques that can be used in a conference system are presented. Section 7.4 describes a network model for P2P streaming for both traditional conference and multiparty services using SIP protocol and ALM mechanism. Section 7.5 analyzes the use of ALM trees for the model described in the previous section, in order to determine for example the number of participants and recommended level in the trees. Finally, a testbed and conclusions are presented.

7.2 P2P Media Streaming Architecture

Media streaming applications can be classified according to the delay tolerance just as shown in [1]. Real-time applications need to have low delay tolerance because of the interaction among end-to-end users. Therefore, delays longer than 400 milliseconds are not accepted (ITU-T recommendation G.114), in order to guarantee fluid interactivity among them. A one-way transmission delay of 150 ms is considered to offer the same experience as using PSTN. However, live broadcast applications typically have no interactivity requirements and, consequently, longer delays are tolerated, commonly up to 30 seconds. This delay cannot be detected without interactivity or without a reference point. In the end, on-demand media applications present greater delay tolerance because the existent interactivity is limited to change the channel or due to VCR-like control. Table 7.1 presents the following main constraints in video streaming.

Combined, the above characteristics yield a unique application scenario that differs from other typical peer-to-peer applications, including on-demand streaming, audio/video conferencing, and file download (see Table 7.2).

The key problem in a P2P video broadcast system consists of organizing the peers into an overlay for disseminating a video stream. The main criteria to be considered in overlay construction and maintenance operations can be found in the following list:

1. Overlay Efficiency: The construction of the overlay network must be efficient, because the streaming video requires high bandwidth and low latencies. However, if these applications are not interactive, then a start-up delay can be tolerated.
2. Scalability and Load Balancing: The overlay must be scalable in order to support a large amount of receptors, and the associated overhead must be reasonable at these large scales.
3. Self-organization: The overlay must be built in a distributed manner and it also must be robust enough to support dynamic changes of the peers, which take part in the overlay. Moreover, the overlay should continuously adapt to changes in the network, such as bandwidth and latency variances. The system should be self-improving, i.e., the overlay should evolve towards a better structure as more information becomes available.
4. Bandwidth constraints: The system depends on the bandwidth contribution of the present peers, so it is important to ensure that the total contribution of the bandwidth of a user does not exceed its bandwidth capacity.
5. Delay: The overlay should minimize the delay between the source and each receiver in the system. This is a critical parameter, specially considered in interactive applications such as videoconference.
6. Failure Recovery: The overlay must have robust mechanisms to react to failures. Typically these failures are caused by peer disconnections without any kind of control. Concretely, in peer-to-peer applications, peers appear and disappear arbitrarily. This is known as churn. The consequence of churn is very important as the overlay needs to be re-organized efficiently and quickly in order to keep the continuity of the established communications.

7. Other System Considerations: The selection of a suitable transport protocol which allows to overcome connectivity restrictions, such as NAT and firewall traversal must be considered.

7.2.1 Types of Overlay

The distributed interactive applications have stringent latency requirements and dynamic user groups. These applications may benefit from a group communication system to improve the system support for such applications. Using tree or mesh structure becomes an important decision because it is crucial to develop low-latency overlay networks for ALM. Next, tree and mesh overlay are briefly described.

7.2.1.1 Tree-Based Overlay

In tree-based systems the overlay is hierarchically organized. Data is sent from the source node to the rest of nodes of the tree. This approach is known as source-driven. The technique used to send data all along the tree is push-based, that is, when a node receives a data packet, it forwards the packet to each of its children.

There are several requirements for tree-based systems. The system has to build an efficient tree that matches the underlying network, and also it has to provide a shallow tree, since when a node leaves or crashes all the branches of this node will stop receiving packets. This failure may interrupt data delivery to an important part of the tree, momentarily dropping the overall performance. A shallow tree minimizes this problem, but also decreases the upload rate, since most of nodes are leaves and the upload bandwidth of leaves is not used.

7.2.1.2 Mesh-Based Overlay

The construction of a mesh-based overlay is an unstructured approach, where no explicit structure data delivery is constructed or maintained. Normally, this mesh-based overlay uses a data-driven approach for exchanging data. Data-driven approach is guided by data availability, which is used to route the data in the overlay. With pull-based techniques, e.g. CoolStreaming, each node keeps a set of neighbor peers (partners) and periodically exchanges its data availability. Later, one node may retrieve unavailable data from one or more partners and, at the same time, will supply its available data to its partners. Another technique that can be used is the push-pull based approach, where each node is autonomous and pushes data to its partners without the pull request. The push is performed on time-based prediction, but a wrong prediction leads under flow or duplication issues.

This approach is robust to failures, because available data is redundant (kept in several partners), and the departure of a node simply implies that its partners will use other nodes to receive segments of data which only have local impact. The potential bandwidth of partners can be totally used for exchanging data between partners. In mesh-based applications, the scheduling algorithm is a key component which must schedule the segments that are going to be downloaded from various partners to meet playback deadlines. It is the brain or core component of the system. So, is there a future for mesh-based live video streaming? The answer to this question is found in [9] where the authors conclude that mesh-based systems can achieve near-optimal rates in practice with negligible chunk misses (1 percent) at 90 percent maximum stream rate, and comparable or better rate than other tree-based systems, for example, AQCS, GridMedia. Also, mesh-based overlay is still an attractive choice since it follows a simple unstructured overlay (no tree-like structure is difficult to maintain), it is scalable (the delay grows slowly with overlay size), and it has high rates and chunk-tolerance. The main disadvantage, in comparison with tree-based, is that the delay remains higher than in some tree-based systems, e.g. GridMedia.

7.3 Advanced Media Coding in P2P Conferencing

There are basically two techniques that currently are being studied to be applied to a P2P Conference system: source coding and network coding. In general, the P2P systems have well known advantages in terms of scalability, robustness and fault-tolerance. However, if all users receive and serve data, the probability that one stream breaks is higher because of the replication rate of the video streams. Furthermore, the connectivity to the network and the different paths used is strongly variable. MDC, SVC and NC are coding techniques that can be applied in media streaming for situations where the quality and availability of connections vary over time. Next, all these will be described.

7.3.1 Multiple Description Coding and Scalable Video Coding

MDC [10] is a source coding technique, which encodes a signal (audio/video) into a number of N different sub-bitstreams (where $N \geq 2$). Each bitstream is called descriptor (or description), as shown in Figure 7.1(a). The descriptors,

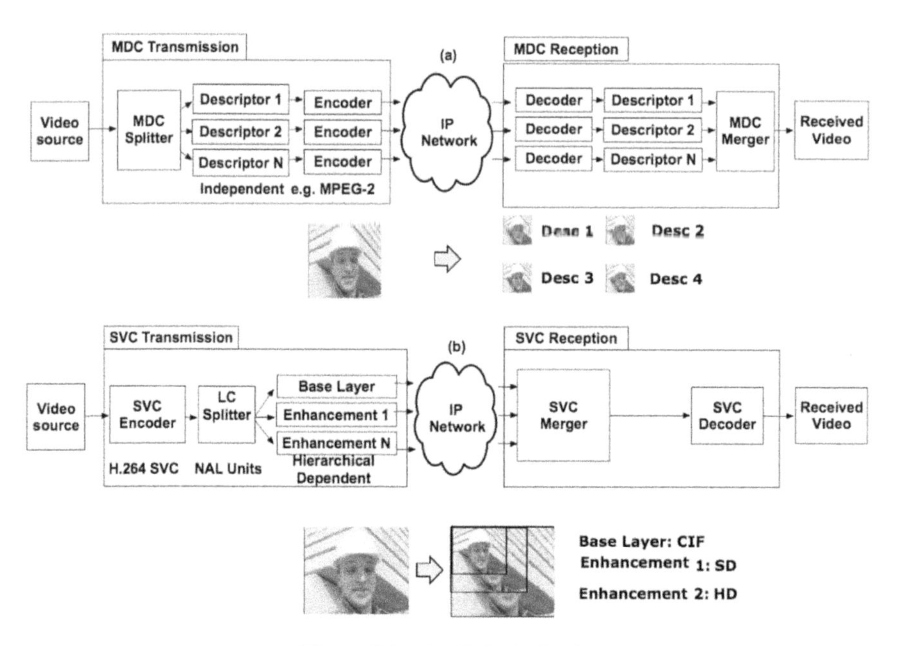

Figure 7.1 (a) MDC. (b) SVC.

which are all independently decodable, are meant to be sent through different network paths in order to reach a destination. The receiver can play the media when any of the descriptors is received. The quality of the reproduced media is proportional to the number of descriptors received; that is, the more descriptors are received, the better the reconstruction quality is. Since an arbitrary subset of descriptors can be used to decode the original stream, network congestion or packet loss, which are common in best-effort networks such as the Internet, will not interrupt the playback of the stream (continuity) but will only cause a (temporary) loss of quality. SVC [11], also called Layered Coding (LC), adapts the video information to the network constrains splitting the images into different layers (similar to MDC). These layers represent the quality of the image, so from the base layer each successive layer improves the image quality, getting the full picture quality with the total amount of layers used (see Figure 7.1(b)). Specifically, SVC is the name given to an extension of the H.264/MPEG-4 AVC video compression standard. It must be noticed that the main difference between MDC and SVC is that MDC creates independent descriptors (can be balanced or unbalanced) while SVC creates dependent descriptors (unbalanced). According to SVC,

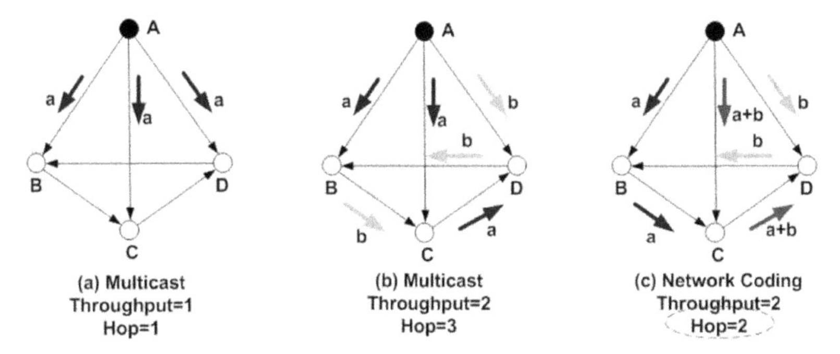

Figure 7.2 Minimizing delay with network coding.

we can apply different techniques to make the video data scalable, in the same way as MDC. Currently, real-time coding using SVC technique is still a challenge due to the high computational requirements (to date, there is no SVC real-time software encoder, there are just a few real-time hardware encoders below HD resolution), which supposes an important limitation at implementation stage.

7.3.2 Network Coding

Network coding [12] is a mechanism proposed to improve the throughput utilization of a given network topology. The principle behind network coding is to allow intermediate nodes to encode packets. Compared to other traditional approaches (e.g. building multicast trees), network coding makes optimal use of the available network resources. Each time a client needs to send a packet to another client, the source client generates and sends a linear combination of all the information available to it. After that, a client receives enough linearly independent combinations of packets and it can reconstruct the original information. In [13] the authors proposed the use of network coding to minimize the maximal transmission delay during a multicast session, while maintaining high throughput at the same time. Figure 7.2 shows an example where using NC is possible to reduce the number of hops and, therefore, the delay.

7.4 Network Model for P2P Video Conferencing

Next, a basic scheme for having a P2P Conferencing service based on SIP signaling and ALM protocol, is presented.

7.4.1 SIP Conferencing

SIP [14] is a protocol that provides mechanisms based on dialogs (SIP messages) for the initiation, modification and termination of media sessions between user agents. While the original SIP scheme was developed for a two party communication, SIP communication sessions with multiple participants introduce other technical complications which were not solved in the initial specification. In order to overcome these issues, different conferencing frameworks have been proposed: SIPPING and XCON. This work is based on the former because it does not define new conference control protocol to be used and it provides mechanisms for inviting, expelling and configuring media streams during the conference. SIPPING workgroup has defined some new functional entities: FOCUS, Conference Notification Service, Conference Policy Server and Mixer. FOCUS is a central controller entity, which receives all *INVITE* requests to a conference. Moreover, it maintains SIP signaling relations with each and every participant and takes the responsibility of sending media to all the participants in the conference. The Conference Notification Service maintains a Subscription/Notification service with the participants for the modification of the conference primitives. On the other hand, every conference is governed by a certain set of rules, which are enforced by Conference Policy Server. Finally, media transmission and distribution is done by the Mixer. It receives a set of streams of the same type and combines their media in a type specific manner and redistributes the resultant media to every participant in the conference. Besides, it includes transport protocols such as RTP for communications. SIPPING framework is open to define the physical realization. For instance, it is possible to mention some models: centralized server, endpoint server, media server component and distributed mixing. In this work we try to exploit all of these functionalities into a P2P network in order to allow large-audience conferencing. Figures 7.3(a, b, c) depict different models of multiparty communications supported by SIP and Figure 7.3(d) shows a basic SIPPING scheme.

7.4.2 P2P Conferencing System

This section explains a prototype system that allows the creation of a large-audience conference by taking advantage of P2P techniques for distributing the conference stream among all the participants and uses SIP as standard signaling and messaging protocol. The system discerns two specific roles of the peers that join the system: the leader and the participants. The leader is the

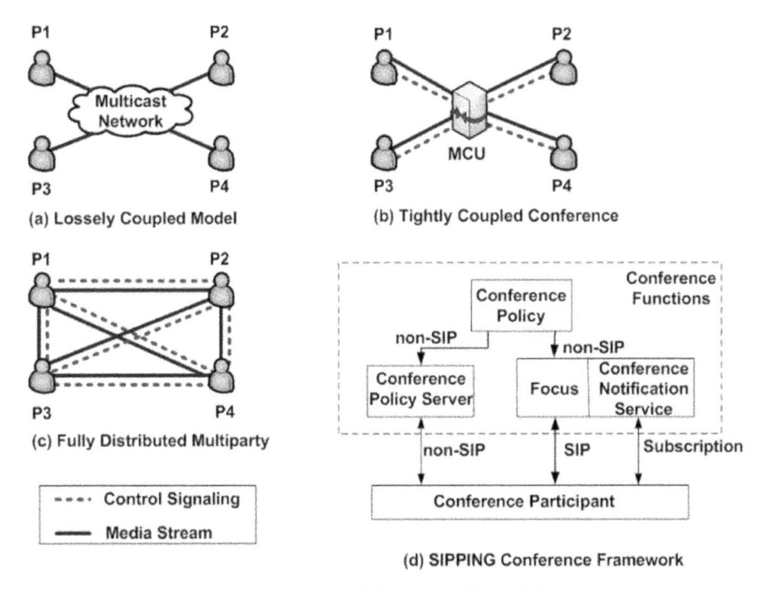

Figure 7.3 Multiparty models and SIPPING framework.

peer that creates the conference, the one in charge of creating the virtual conference room and announcing to other users the event. The peers that receive the announcement of the conference will become participants when joining the virtual room. Once the conference has started, a participant can adopt two behaviors: actively participate on the conference or passively participate on it. This yields two types of participants, the first one is called speaker, which has video capturing capabilities, and the second one is known as listener, which only wants to be present at the conference and does not take part in it.

The proposed conference system allows a limited number of speakers. In this case we define a limit of four simultaneous speakers on the conference. Note that this number cannot be very high, otherwise the conference would be not useful. In order to allow the participation of many peers in an ordered way, the leader manages a list of participation events. The list consists of an ordered queue that allocates the queries of the different participants that want to become speakers. This list applies a First-In-First-Out (FIFO) policy. The leader has also the ability to reject some participants of the conference. Moreover, the proposed conferencing system borrows ideas from the framework IETF SIPPING Working Group. In Figure 7.4(a) a general SIP conferencing architecture implemented by each peer present in the system

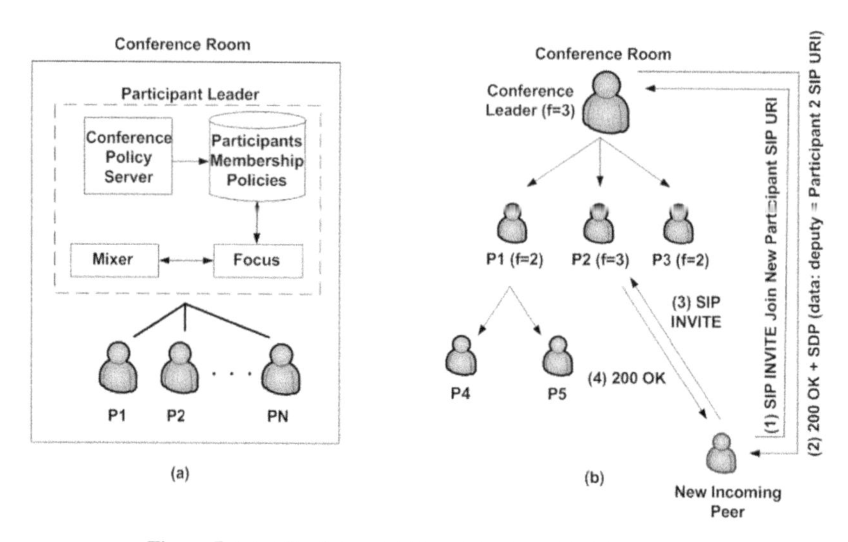

Figure 7.4 (a) Participant SIP architecture. (b) Join process.

is shown. This architecture allows any peer in the system to become leader, speaker or listener.

The conference media stream will be delivered to the audience by constructing an ALM Tree rooted at the leader peer. The leader peer will create a map of the conference connected peers by creating a logic tree structure. The tree is constructed using the data gathered from new joining peers. The new incoming peers must specify the number of supported children, by calculating its own fanout f. The fanout can be obtained by dividing the peer current available connection speed by the stream bitrate. A new peer, which tries to join a conference room, sends a *INVITE* Join SIP message indicating its fanout in the attached SDP. The leader looks in its structure for the first peer in the system with an available connection (deputy). Once this has been obtained, the deputy peer sends its SIP URI to the new incoming peer through the SDP attached to the *200 OK* SIP message. Then, the new peer just has to send an *INVITE* SIP message to the indicated deputy in order to start joining the conference. The described process can be seen in Figure 7.4(b).

We define failure as the situation where a peer cannot receive the media stream because the communication with its parent peer is broken (i.e. when a peer accidentally leaves or there is a problem in the link between parent and children). A peer detecting a failure tries to establish a new connection with another peer. If a peer fails in receiving data from its parent, then it will ask

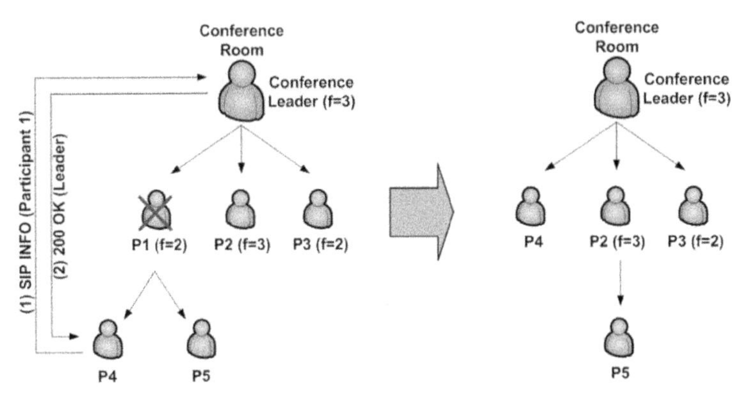

Figure 7.5 Peer reconnection.

the leader for another peer with enough available fanout in order to become its new parent. The peer also communicates to the leader which peer has failed in order to update its peer-map. Once the peer establishes a new connection with its new parent, the communication can continue. When the peer detects the failure, it uses an *INFO* message in order to inform the leader source that its parent is unavailable. Then, the leader peer deletes the leaving peer and looks for a new parent for the peer and notifies it to the peer on the *200 OK* response SIP message, as shown in Figure 7.5. The leader peer is able to mix the different streams generated by the speakers (see Figure 7.6(a)). The participants send *REFER* SIP messages in order to add its stream to the conference. If the maximum number of simultaneous speakers is reached, then the candidate speaker is added to a queue and can wait till its turn arrives. Once the speaker can participate in the conference, it sends its video stream to the leader who mixes the streams generated by the speakers (S1, S2, S3 and S4) and creates a single stream (S) that will be distributed (see Figure 7.6(b)) all along the network from the leader to the participants. In [15] it is described how it is possible to improve the mixer scalability.

7.5 Analysis of the ALM Tree

In this section a brief analysis of the performance of the ALM tree depicted in former section is presented. We calculate the maximum number of levels of the multicast tree and the number of participating peers, as a function of the stream delay and the fanout of the peers. We show that there is an exponential relation between them. We also calculate the robustness of the multicast tree.

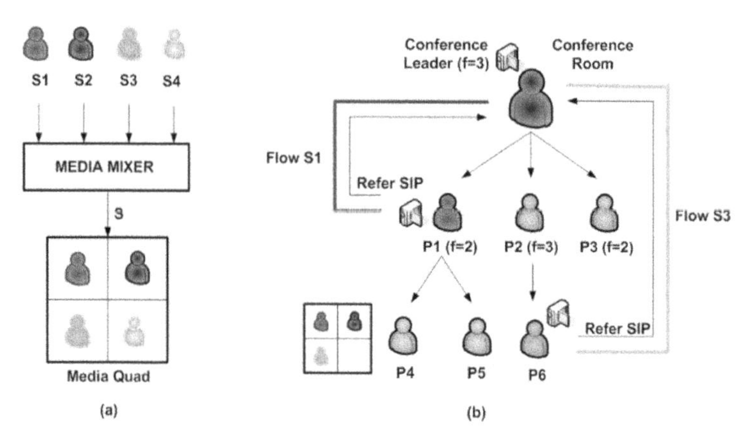

Figure 7.6 Mixing 4 speakers, S1, S2, S3 and S4 in to single stream S.

7.5.1 Number of Participants

We denote by f_p the fanout of peer p; that is, the number of peers, here called children, which can simultaneously receive a message from p. As we have seen before, the fanout may be calculated as the connection speed of a peer divided by the stream bitrate. Moreover, we define the latency parameter λ_{pq} between peer p and peer q, as the time elapsed since the message starts to be sent by parent p until it is completely received by child q. Note that λ_{pq} is equal to transmission time μ_p plus the propagation delay between p and q, as shown in Figure 7.7. For simplicity's sake, we assume at the moment that f_p and λ_{pq} are the same f and λ for all the peers. Let l be the maximum number of levels of the multicast tree, t_p the time taken by the leader to process and mix the streams of all the speakers, and D the maximum delay allowed at the multicast tree. We assume that the processing time of each different peer from the leader is negligible. Then the maximum delay of the stream must be by definition equal or lower than D:

$$\lambda + t_p + \lambda \cdot l \leq D \Rightarrow l = \left\lfloor \frac{D - \lambda - t_p}{\lambda} \right\rfloor = \left\lfloor \frac{D - t_p}{\lambda} \right\rfloor - 1 \tag{7.1}$$

We have calculated the maximum delay (for the peers of the last level) as follows. First each speaker sends a stream to the leader with a delay of λ. Then the leader mixes in t_p time units the streams of each speaker into one puzzle stream (mosaic) and sends it to the peers of the first level in λ time units (note that by convention the leader is said to be at level 0). Now for each level we have an additional delay of λ which gives us the maximum delay for

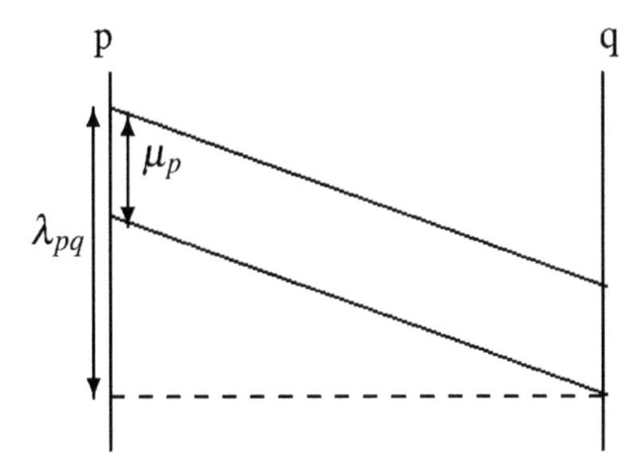

Figure 7.7 Latency parameter λ_{pq}.

the last level l of $\lambda + t_p + \lambda \cdot l$ as reflected in equation (7.1). Note that t_p depends on the number of speakers and that D must be lower than the well-known bound of 150 milliseconds for reasons of comprehension. We also assume that $(D - t_p)/\lambda$ is equal or higher than two. Otherwise the leader could not send the stream to any peer and the multicast transmission would not be possible. Now we calculate the maximum number of participating peers, including the leader, as

$$n = 1 + f + f2 + \cdots + f^l = \frac{f^{l+1} - 1}{f - 1} = \frac{f^{\lfloor \frac{D - t_p}{\lambda} \rfloor} - 1}{f - 1} \tag{7.2}$$

We remark the exponential relation between the allowed number n of participants and both the processing time t_p of the leader and the latency λ between two peers. If we consider a particular case in which each peer has a connection of 2048 Kbps and the bitrate of the leader is 512 Kbps, we will have a fanout of $f = 2048/512 = 4$. Then, if we consider a maximum delay D of 150 ms, a processing time t_p of 10 ms and a latency λ of 26 ms, the maximum number of levels will equal to 4. In this case, the maximum number of peers will be of 341. If now we divide the latency λ by 2 down to 13 ms, the number of levels will increase to 9 and the maximum number of peers will equal to 349525, more than one thousand times larger than for $\lambda = 26$ ms.

In a general case we may find bounds for the number of participants, if we take the maximum and the minimum values of f and λ, respectively:

$$\frac{f_{\min}^{\lfloor D - t_p/\lambda_{\max} \rfloor} - 1}{f_{\min} - 1} \leq n \leq \frac{f_{\max}^{\lfloor D - t_p/\lambda_{\min} \rfloor} - 1}{f_{\max} - 1} \tag{7.3}$$

Table 7.3 Assignment of each peer to a level according to its decreasing fanout.

peer	leader	1	2	3	4	5	6	7	8	9	10	11	12	13	14	15	16
F	2	3	2	2	2	1	1	1	1	1	1	1	1	1	1	1	1
level	0		1				2						3				4
$\sum F$	2		5				6						6				

In the case that the fanout is not the same for all the peers – and if we still consider λ as a uniform parameter – we could minimize the mean delay by assigning to the first levels of the multicast tree the peers with higher fanouts. This can be done by ordering the peers according to its decreasing fanouts and then choosing them in groups until we fill out each level, as shown in Table 7.3. In that case, we begin (as always) with the leader, which in the table has a fanout of two. Thus, the leader sends the stream simultaneously to two peers at the first level. These two peers hold the maximum fanouts, in this case of 3 and 2 respectively, so at the second level we will have five peers $(3 + 2)$. We repeat this schedule until we reach all the peers of the network. In a more general case where both f and λ are not the same for all peers, the list from Table 7.3 should be reordered considering not only the fanout of each peer but also its latencies to the other peers of the network. That is, we should find a balanced rule between the fanout of each peer and the sum of its latencies, depending each case on the topology of the network.

7.5.2 Robustness of the Multicast Tree

Frequently, real-time applications use unreliable transport-layer protocols such as UDP. That means that it is not always possible to ensure the complete arrival of the stream at the destination peers. We analyze at this point, under the assumption that there are no retransmissions, the robustness of the application layer multicast tree as described in previous sections.

As we have seen before, for a fanout of $f = 1$ our protocol depicts a linear topology. In this case, when a stream arrives at a given peer, it is only forwarded once to another peer. On the other hand, for a general fanout of f the protocol depicts f divergent paths from each peer, as shown in Figure 7.8. Hence, for $f = 1$ the probability that a stream arrives at l peers is always lower than the probability of arriving at $l - 1$ peers. This is because a stream will arrive at the lth peer only if it arrives first at the preceding $(l - 1)$th peer in the multicast tree, and then it is successfully transmitted through the edge which joins them. This undesirable characteristic for $f = 1$ does not appear for $f > 1$ due to the thicker multicast tree that depicts the algorithm, with

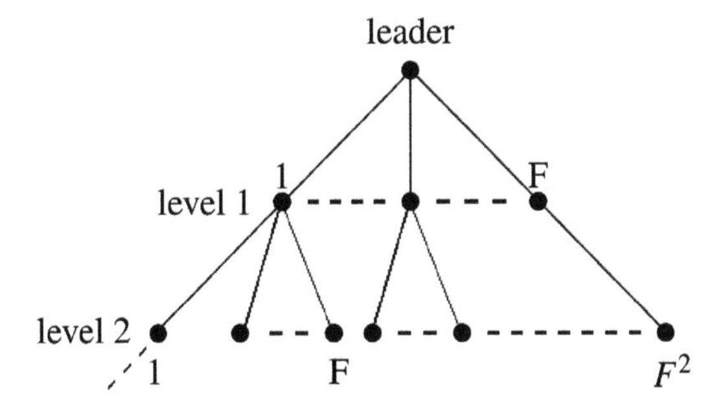

Figure 7.8 Levels at multicast tree.

divergent paths. For $f = 3$, for example, the probability of arriving at three peers is in general much higher than the probability of arriving at only one, an issue which does not state for $f = 1$.

Let $P_c(p, q)$ be the probability that peer q receives correctly a stream sent directly by its parent p. For the sake of simplicity we consider that $P_c(p, q) = P_c$ for all pairs of peers. Actually, if we consider $P_c = \max\{P_c(p, q)\}\ \forall p, q$, we will obtain a lower bound of the robustness of the multicast tree. We call a peer which has correctly received a stream a "visited peer". We denote by s the number of speakers and by n_r the average number of peers that receive the stream. To calculate n_r we divide the peers into levels, according to Figure 7.8. We call E_i the average number of peers that receive the stream at level i. For the first level we have f peers and then

$$E_1 = E(r_{11}+r_{12}+\cdots+r_{1f}) = f\cdot E(r_{11}) = f\cdot(0\cdot p_1(0)+1\cdot p_1(1)) = f\cdot P_c^s\cdot P_c \tag{7.4}$$

By definition, r_{ij} is 1 if the peer j at level i has received the stream and 0 otherwise. Thus $r_{11}+r_{12}+\cdots+r_{1f}$ is equal to the number of peers that have received the stream at the first level. Moreover, $p_i(1)$ is the probability that the stream arrives at a peer which belongs to level i. Thus, for the first level we calculate $p_1(1)$ as the probability that the leader correctly receives the stream from each one of the s speakers, that is P_c^s, multiplied by the probability P_c that the stream is correctly send from the leader to the peer at the first level. For the f^2 peers of the second level we have

$$\begin{aligned} E_2 &= E(r_{21}+r_{22}+\cdots+r_{2f2}) = f^2 E(r_{21}) \\ &= f^2(0\cdot p_2(0)+1\cdot p_2(1)) = f^2 P_c^s P_c^2 \end{aligned} \tag{7.5}$$

And in general for the level i

$$E_i = E(r_{i1}+r_{i2}+\cdots+r_{if^i}) = f^i E(r_{i1}) = f(0\cdot p_i(0)+1\cdot p_i(1)) = f^i P_c^s P_c^i \tag{7.6}$$

Finally, we calculate n_r as the sum of the averages of each level. We denote by l the number of levels

$$\begin{aligned} n_r &= E_0 + E_1 + \cdots + E_l = P_c^s(1 + fP_c + \cdots + f^l P_c^l) \\ &= P_c^s \frac{(fP_c)^{l+1} - 1}{fP_c - 1} \end{aligned} \tag{7.7}$$

As in previous sections, we assume that the number of peers is $1 + f + f2 + \cdots + f^l = (f^{l+1} - 1)/(f - 1)$. For $f = 1$ the assumptions of equation (7.7) are valid. In this case, since we have $l = n - 1$, the following results:

$$n_r = P_c^s \frac{1 - P_c^n}{1 - P_c} < \frac{P_c^s}{1 - P_c} \tag{7.8}$$

Thus if $P_c^s/(1 - P_c)$ is much smaller than the number n of peers, the average number n_r of visited peers with $f = 1$ will also be much smaller than n. But if, on the contrary, $P_c^s/(1 - P_c)$ is higher than n then the average number of visited peers may be close to n. In Table 7.4 we see some values of the average n_r for $f = 1$ and $s = 4$, depending on n and P_c. For $P_c = 0.9$ we have $P_c^s/(1 - P_c) = 6.56$ and then the average may not be greater than 6.56, no matter how high n is. However, for the usual values of $P_c = 0.999$ and $n = 10$ or $n = 100$ we have good averages, close to n. For other values the average is much smaller than n, which means that the stream is not received by a large percentage of peers. In this case, the minimum accepted fanout should automatically change to $f = 2$. For instance, for $n \approx 1000$ and $P_c = 0.999$ we would arrive at only the 62.97% of the peers with $f = 1$ whereas for $f = 2$ the percentage would be of 98.80%. In this case, the percentage for $f = 3$ would be 99.05%, only a little higher than for $f = 2$. In a general case the robustness of $f = 2$ will be acceptable.

Therefore, when we define the multicast tree in a real network, the algorithm should estimate the average number n_r of peers that will receive the stream for $f = 1$ and if it is not large enough then it should estimate n_r for $f = 2$ and so forth until it finds a value of $f = f_r$ such that n_r is large enough. At this moment, the algorithm should reject any peer with a fanout lower than f_r. We should remark, nevertheless, that the assumption that P_c is equal for all the links has greater implications for $f = 1$ than for $f \geq 2$.

Table 7.4 Number n_r of peers that receive the stream for $f = 1$ and $s = 4$.

n	$P_c = 0.9$	$P_c = 0.99$	$P_c = 0.999$	$P_c = 0.9999$
10	4.27	9.19	9.92	9.99
100	6.56	60.90	94.83	99.47
1000	6.56	96.06	629.78	951.29
$P_c^s/(1 - P_c) = \max n_r$	6.56	96.06	996.01	9996.00

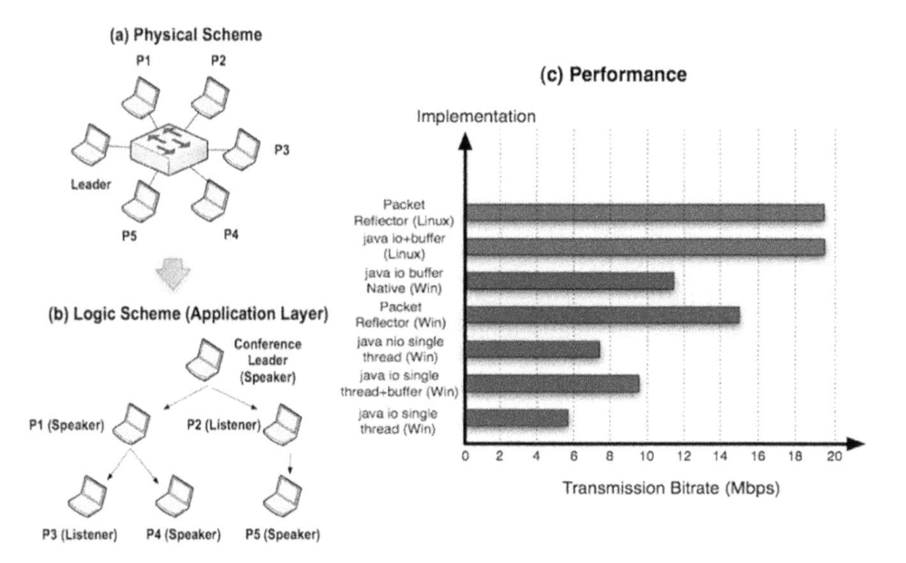

Figure 7.9 Testing scenario.

For topology reasons, there are usually a larger percentage of short end-to-end transmissions for $f = 1$ than for $f \geq 2$ (we assume that if a peer has a fanout equal to one, it will choose a close peer as a child). This means that for $f = 1$ there will be in general more transmissions than for $f \geq 2$ with a probability of success larger than P_c. Thus, the results of equation (7.7) can be more pessimistic for $f = 1$ than for $f \geq 2$.

7.6 Testbed

As proof of concept, a prototype application has been developed to test the distribution of four mixed media streams and to demonstrate the feasibility for building a P2P conference system. The mixing function has been realized using VideoLAN Client (VLC) 0.8.5, which allows to create a mosaic of different video/audio streams. The main goal was to mix four video streams obtained from video cameras connected to four participants and then distribute it on the network using a single MPEG-TS stream. Note that the four sound tracks from the four streams will also be streaming in the same MPEG-TS stream. The deployed scenario can be seen in Figure 7.9(a). The peer that acts as leader configures the transmission network interface to receive the desired stream from the speakers and also to send the stream to its children. The connected participants receive the media stream from their parent peer and forward it to their respective children (see Figure 7.9(b)). The transmission layer of the developed prototype was implemented using different programming languages (Java and C), different socket implementations (java.io blocking socket, java.nio non-blocking socket, Packet Reflector C based) and different operating systems (Windows XP and Ubuntu Linux). When observing the gathered results, the best performance was achieved when using the Packet Reflector implementation running under Linux. It was achieved a transmission bitrate of 20 Mbps when sending four mixed streams. The test results are graphically represented in Figure 7.9(c).

7.7 Conclusions

Usually, studies about P2P media distribution are mainly focused on media file download or video streaming for P2PTV or P2PVoD. On the contrary, P2P Conferencing is a service technically more challenging than the previous ones and only in recent years it has begun to be studied in more detail. The main goal of this chapter was to present an overview about conferencing systems over P2P networks describing the basic elements and related technologies (e.g. MDC, SVC, NC), indicating some technical considerations to take into account and delivering an analysis to determine its operational feasibility. Moreover, in a proof of concept, we have presented a prototype system, which allows to create large-audience media conferences by applying ALM techniques for distributing the media stream and which makes an efficient use of the network resources. The system uses SIP as signaling protocol providing flexibility to the solution and allowing the development of new value-added services. Moreover, this prototype is based on the mixing of the different speaker streams in order to deliver the resulting conference stream to all the participants in the conference.

Acknowledgements

This work was supported by the MCyT (Spanish Ministry of Science and Technology) under Project No. TSI2007-66637-C02-01, which is partially funded by FEDER, and the i2CAT Foundation.

References

[1] W.P. Ken, X. Jin, and H. Gary Chan. Challenges and approaches in large-scale P2P media streaming. *IEEE Multimedia*, 14(2):50–59, April 2007.

[2] X. Wu, K. Kishore, and V. Krishnaswamy. Enhancing application-layer multicast for P2P conferencing. In *Proceedings of IEEE Consumer Communications and Networking Conference (CCNC)*, Las Vegas, NV, pages 986–990, January 2007.

[3] M. Dowlatshahi and F. Safei. Multipoint interactive communication for peer to peer environments. In *Proceedings of IEEE International Conference on Communications (ICC)*, pages 317–322, June 2006.

[4] Ch. Luo, J. Li, and S. Li. DigiMetro: An application-layer multicast system for multi-party video conferencing. In *Proceedings of IEEE Global Telecommunications Conference, GLOBECOM*, Vol. 2, pages 982–987, December 2004.

[5] H. Horiuchu, N. Wakamiya, and M. Murata. A network construction method for a scalable P2P video conferencing system. In *Proceedings of IASTED European Conference on Internet and Multimedia Systems and Applications*, Chamonix, France, pages 196–201, March 2007.

[6] M. Brogle, D. Milic, and T. Braun. Supporting IP multicast streaming using overlay networks. In *Proceedings the Fourth International Conference on Heterogeneous Networking for Quality, Reliability, Security and Robustness*, pages 1–7, August 2007.

[7] P2PSIP. http://www.p2psip.org/.

[8] M. Ponec, S. Sengupta, M. Chen, J. Li, and P.A. Chou. Multi-rate peer-to-peer video conferencing: A distributed approach using scalable coding. In *Proceedings of the IEEE International Conference on Multimedia and Expo*, pages 1406–1413, July 2009.

[9] F. Picconi and L. Massoulie. Is there a future for mesh-based LiveVideo streaming? In *Proceedings of the Eighth International Conference on Peer-to-Peer Computing*, pages 289–298, 2008.

[10] V.K. Goyal. Multiple description coding: Compression meets the network. *IEEE Signal Processing Magazine*, 18(5):74–93, September 2001.

[11] H. Schwarz, D. Marpe, and T. Wiegand. Overview of the scalable video coding extension of H.264/AVC. *IEEE Transactions on Circuits Systems for Video Technology*, 17(9):1103–1120, September 2007.

[12] C. Fragouli and E. Soljanin. *Network Coding Fundamentals, Foundations and Trends in Networking*. Now Publishers, 2007.

[13] C. Fragouli and E. Soljanin. *Network Coding Applications*. Now Publishers, 2008.

[14] J. Rosenberg, H. Schulzrinne, G. Camarillo, A. Johnston, J. Peterson, R. Sparks, M. Handley, and E. Schooler. SIP: Session Initiation Protocol. IETF RFC 3261, June 2002.

[15] E. Cheung and G.M. Karam. A novel implementation of very large teleconference. In *Proceedings of IPTComm 2010 Principles – Systems and Applications of IP Telecommunications*, pages 89–95, 2010.

8

Overlay Multimedia Streaming Service for Consumer Electronic Products

Ettikan Kandasamy Karuppiah[1], Eiichi Muramoto[2], Boon Ping Lim[1], Thilmee Baduge[2] and Jonathan Tan Ming Yu[1]

[1]*Panasonic R&D Centre Malaysia, 63000 Cyberjaya, Malaysia;*
e-mail: ettikan.karuppiah@mimos.my
[2]*Video Communication Development Office, Panasonic Corporation, Osaka, Japan*

Abstract

Application Layer Multicast (ALM) enables packet replication and forwarding at the application layer (as opposed to other techniques, for example, Internet Protocol (IP) layer multicast) through an overlay tree. This packet forwarding capability is suitable for applications such as real-time audio-video (AV) content distribution since it provides independence from the lower layer (e.g. IP layer) while maintaining better end-to-end packet delivery times. ALM requires all the participating nodes in a given active session to be collectively responsible for forwarding the group's streaming data among the members via a set of chosen paths which make up the overlay tree.

This chapter discusses the unique characteristics of ALM when used in consumer electronics, the design considerations that need to be thought about when creating such a system, and the considerations required when testing the system.

Keywords: application layer multicast, tree construction, bandwidth fairness, high definition audio/video streaming, testbed.

Anand R. Prasad et al. (Eds.), Future Internet Services and Service Architectures, 161–181.

8.1 Application Layer Multicast: A Consumer Electronics Perspective

ALM is one of the overlay multicast methods [1] which handle group membership, overlay tree and packet forwarding independently without any additional multicast support from Internet routing protocols. This provides independence for applications to control data forwarding and group membership based on their specific needs, without being constrained by the features available or unavailable in the lower layers. Small and medium-sized groups such as real-time AV content sharing can benefit from such independence as contents can be streamed in a simple, trusted and fast manner.

8.1.1 Unique Characteristics of Consumer Electronics ALM

Even though many ALM applications have been developed and deployed in the Internet to date, hardly any of them were intended as actual consumer electronics (CE) products. However, there are projects still in the research stage to use overlay methods for broadcasting information to CE devices over the Internet [2].

As for actual products, network-enabled CE devices that include real-time AV content sharing have been announced [3, 4], and it is expected that such functionality will be more widespread. ALM technology can be used in such devices to introduce or enhance these functions.

ALM as used in CE devices is different from typical content distribution methods in the following ways:

- *Usage of ALM is ad hoc.* Typically CE devices are kept switched off or in standby mode until their use is necessary, and users may not schedule usage of the CE device in advance. For example, a user may impulsively decide to transmit live video material of a party without much planning in advance. System startups taking a few minutes (or more) may be acceptable in other devices, but not in CE devices as user demands fast action.
- *CE devices are less powerful than comparable content distributing devices.* Users utilising other distribution methods may have access to routers or special-purpose nodes with great processing power, but CE device users who use ALM only have the processing power of their device at hand. Also, due to cost-efficiency reasons, CE devices are typically less powerful than a standard personal computer.

- *CE devices are used on networks with less capability.* Users utilising other distribution methods may have access to higher-bandwidth, lower-latency and more reliable networks, for example, by obtaining a lease line. CE devices may be used in homes or small offices with no access to such infrastructure, and must still be useable in lower-bandwidth networks compared to corporate networks. It is also possible that CE devices are used in bandwidth-heterogeneous situations, where some nodes have access to a high-bandwidth connection to the network and some do not. The system should be able to make use of the nodes with high-bandwidth connection access to improve the quality of service, while still including the other nodes in data transmission.
- *CE devices may be used by people with less training.* The users of CE devices may not be specialists in the computer network field, therefore the operation of the system must be robust and easy to use. For example, the system itself should recover from network interruptions without user intervention.

These differences must be kept in mind when designing an ALM system for consumer electronics. In this chapter, we relate the special needs of consumer electronics to ALM in the following aspects:

- overlay tree construction (Section 8.2),
- packet forwarding (Section 8.4.1),
- group membership (Section 8.4.2).

8.1.2 Sample Use Case: Real-time Video Streaming

We believe real-time AV content streaming among a group of recipients will be the prime application which capitalizes on ALM's capability. Non-real-time data streaming such as Video on Demand is already popular and best done using P2P architecture since timeliness of data delivery is not an issue.

An example of real-time AV content streaming is in ad hoc video conferencing during personal events or business meetings. HD (High Definition) cameras and HD televisions (HDTV) with network capability are becoming common in today's households. Commoditization of bandwidth in many parts of the world, on the other hand, provides a great opportunity for exchanging HD content in real-time among a group of friends and family members during personal events or even among colleagues for formal meetings. Connecting these devices with an application that uses ALM will instantly enable rich AV and personal experience exchange. Capability of the AV application to include more trusted members instantly into the session in a scalable manner without overloading the HDTV's processing power is vital. It is also notable that bandwidth heterogeneity will exist amongst the CE devices. Thus, such an application must handle this bandwidth heterogeneity in making real-time AV sharing common in daily life. In the near future, inclusion of mobile devices into HD Visual communication will enable cross-device HD content streaming.

8.2 Design Considerations for Overlay Tree Construction

The special nature of consumer electronics ALM gives certain considerations to take into account when designing algorithms for tree construction. This section discusses three of them: timeliness, bandwidth limitation and heterogeneity, and the effect of group membership changes even during data streaming.

8.2.1 Timeliness

Due to the ad hoc nature of the use of ALM in this scenario, and the fact that the CE devices may operate on poor networks, the algorithms for tree construction must not be too complicated while ensuring that an acceptable end-to-end delay overlay tree is produced. For example, the user will not want to wait a long time for a tree construction algorithm that requires traversal of many hops over the network, which will take a long time if the nodes in the network have high latency links to each other.

8.2.2 Bandwidth Limitation and Heterogeneity

The overlay tree construction method must support low bandwidth capacity nodes while effectively utilising any high bandwidth capacity nodes available. Figure 8.1 illustrates this advantage among four nodes located in four different locations with different bandwidth capability. The AV content flow from the source node (e.g. originating from Vietnam) is duplicated and forwarded at application layer of an end host (e.g. Japan), with separate paths for content distribution to other two receiving nodes (e.g. Singapore and Malaysia). This allows a bandwidth-limited node (e.g. Vietnam) to stream high bandwidth data to all receiving nodes capitalizing on the resources of bandwidth abundant nodes.

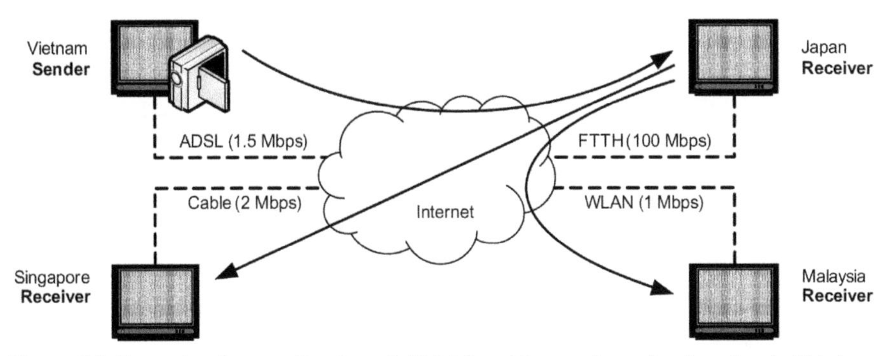

Figure 8.1 Example of an application of ALM for video conferencing in a bandwidth heterogeneous situation. The arrows represent the transmission of the video stream through ALM.

8.2.3 Group Membership Dynamism

In tree-based ALM, nodes not only receive data streams but duplicate and forward the received stream to other nodes. Even though terminal-based content duplication and forwarding facilitates load distribution, this load distribution feature complicates the adaptation of the overlay tree due to occurrences of nodes joining or leaving since specific node dynamics' impact must be minimized to other nodes.

Support for group membership dynamism is even more important in CE ALM, because the system cannot assume that all the devices involved have high availability. It is possible that nodes may join the session late (after the streaming has started) or leave the session early due to user unavailability (or network disconnection).

Therefore, the dynamic joining and leaving of participating ALM nodes need to be considered during overlay tree construction or reconstruction. These mechanisms should consider route changes that have low impact to existing paths while maintaining overall stream timeliness. The additional resources needed to handle joining nodes or the "orphans" of leaving nodes must be obtained from existing nodes without increasing their bandwidth and AV processing load beyond allowed capacity. For example, a joining participant's requirements for the stream must be met without impacting other nodes' needs. A leaving participant may have responsibilities in forwarding the stream to other nodes; these responsibilities must be shifted to other nodes without overburdening them. In addition, all nodes involved in any tree reconstruction steps must be informed and synchronized during the tree reconstruction process.

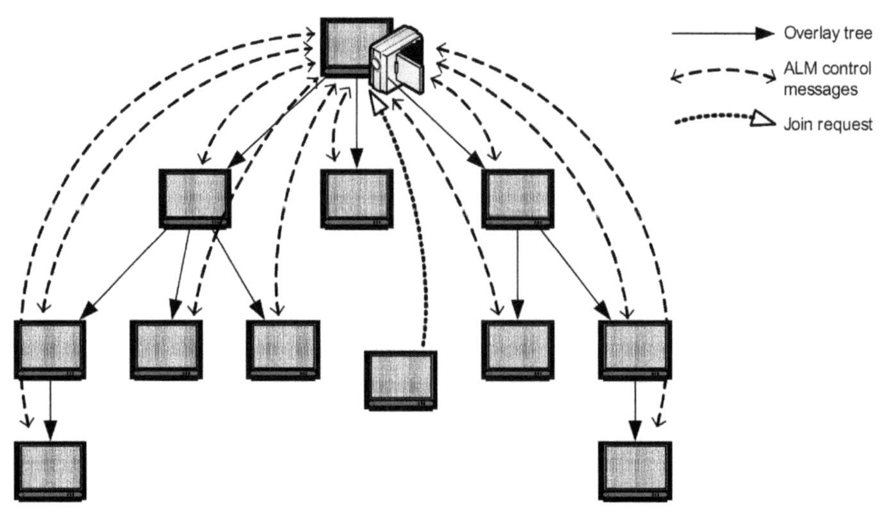

Figure 8.2 ALM-based video streaming with central session control. A node not in the group is requesting to join in the above illustration.

The tree reconstruction mechanism must operate within given resource and time constraints. Various design options must be considered for overlay tree construction and reconstruction mechanism. For example, when a node joins, a parent node must be found for the new node that has enough residual upload bandwidth to be responsible for relaying the stream. Minor rerouting may be triggered if a node has greater bandwidth and shorter latency than the existing nodes. The high capable node should be placed closer to the source in order to reduce the latency from source for nodes that join in the future. Most importantly, all these activities must happen in a timely manner for responsiveness in the overall user experience and with a close-to-zero interruption in the data streaming process.

8.3 Overlay Tree Construction Algorithms

There has been considerable work [5–7] undertaken which can produce overlay trees using various different policies and protocols for different purposes. This section discusses examples of algorithms specifically tailored to the consumer electronics requirements that follow the considerations in Section 8.2.

In general, this section suggests a simple approach, controlled centrally (as opposed to distributed control), that works in heterogeneous-bandwidth environments and with dynamic group membership. Central control is chosen to reduce the time taken during tree construction or reconstruction, and because bandwidth may be limited. With central control, a joining node (or orphans of a leaving node) need not probe many other nodes to find a new parent to receive the stream from; it needs only to connect to the server which performs the central control. (In the case of orphans of a leaving node, the server will perform the overlay tree reconstruction automatically.) The probing will be especially time-consuming in poor high-latency and low-bandwidth networks (each probe will take a long time and has a high chance of being unsuccessful, as many nodes already receiving the stream will not have great upload capacity and will refuse more children), and timeliness is important in CE applications.

The proposed method for overlay tree construction and reconstruction builds on the N-tree algorithm [8]. The N-tree algorithm considers bandwidth as its primary (most important) input, to enable effective utilization of the available bandwidth capacities in heterogeneous-bandwidth networks. Latency is a secondary input. Bandwidth is prioritized here to ensure the highest bit rate possible is used, because the algorithm cannot assume that all nodes have more than enough bandwidth available to support an arbitrarily constructed overlay tree.

The algorithm assumes that bandwidth and latency measurements between every pair of nodes are provided or estimated. The administrator may optionally configure clusters of nodes based on geographical location (or other factors) so that the overlay tree construction algorithms will avoid paths which redundantly use physical links.

Examples of the overlay trees generated initially and upon group membership change can be found in Table 8.3.

8.3.1 Initial Overlay Tree Construction

An initial overlay tree consisting of active participating nodes is constructed during the session startup. The tree construction algorithm will first ensure that all clusters have a node represented close to the source in the initial tree by adding one node with the greatest bandwidth from each cluster to the tree. After that, all nodes are added one by one to nodes of their own clusters, from the highest bandwidth nodes to the lowest. If two nodes have the same bandwidth, the lowest latency (from the source) node is added first. After all nodes are added, a tree optimization step is triggered by moving high-latency nodes to lower latency positions.

8.3.2 Node Join during Streaming

A node that was not present when the tree was constructed initially may join the group after streaming has started. Bandwidth and latency metrics between the new node and all current participants in the tree are assumed to be present (or estimated). If the new node does not have greater upload bandwidth compared to other nodes of the same cluster, it must join as a leaf node to a parent with these characteristics:

- The parent must have remaining upload bandwidth.
- The edge from the parent to the new node must have remaining bandwidth.
- If possible, the new node and the parent must belong to the same cluster.

If the new node has greater upload bandwidth, it can join as a leaf node with the constraints stated above, or it can reroute an edge between two existing and adjacent nodes through itself (that is, split the existing edge into two and insert itself at the free endpoints of the resulting two edges) with the following constraints:

- The two edges added as part of the rerouting must have remaining bandwidth.
- The child of the rerouted edge must have less upload bandwidth than the new node.
- If possible, either the new parent or the new child (or both) must belong to the same cluster as the new node.

Given the above constraints, the algorithm will compute the position that gives the new node the lowest latency from source node. If there is more than one such position, joining as a leaf node is preferred over rerouting.

Rerouting is considered for high capacity nodes so that the nodes which are capable of transmitting to greater numbers of nodes are closer to the source. If nodes closer to the source transmit to more nodes, the overlay tree will be shallower and thus more nodes will receive the stream at lesser latency.

8.3.3 Node Leave During Streaming

A node in the group may leave after streaming has started. During node leave, a cleanup step is first done where all "dead" edges (i.e. those that are incident on the node that left) are removed, which results in (among other things) all of the leaving node's descendants no longer being in the content distribution tree. Then, one by one, the leaving node's immediate children are joined (in the same way as a new node would join) to the existing tree. When an immediate child joins, all its descendants automatically become part of the tree as well. When all immediate children have rejoined the tree, the tree reconstruction is complete.

8.4 Overlay Middleware

The tree construction module is supplemented by, among others, a packet forwarding module and group membership module.

8.4.1 Packet Forwarding

Unlike the tree construction module, ALM packet forwarding can be performed in the same way in CE devices as in non-CE applications, but an example protocol is included for completeness.

8.4.1.1 Example

Packet forwarding is performed using a proprietary protocol named ALMcast. ALMcast follows the application layer packet relay concept, where it is the end nodes (as opposed to intermediate routers) that perform packet forwarding by referring to their respective locally stored forwarding tables. The local forwarding table is constructed external to this module (for example, the server could construct the overlay tree centrally as in Section 8.3 then inform the participant nodes of their respective forwarding tables via control messages). ALMcast relies on ordinary unicast IP packets without needing any packet header modification or extension (at the IP layer). Therefore no additional modification is required to handle the network sockets at the end nodes or at any intermediate routers.

In order to ensure maximum flexibility and interoperability, ALMcast operates using a "port forwarding" strategy. Any packets received by ALMcast on a specific User Datagram Protocol (UDP) port are multiplexed to other ports or nodes according to the locally stored forwarding table. The number of the local ports that ALMcast listens to are not fixed but can be configured at run time, allowing each node to support multiple content streams at any one time. The packets multiplexed by ALMcast can be retrieved at receiver nodes by reading from a preconfigured (UDP) port.

In order to perform the application layer multicast, the applications at sending nodes only need to write packets to the local port using methods

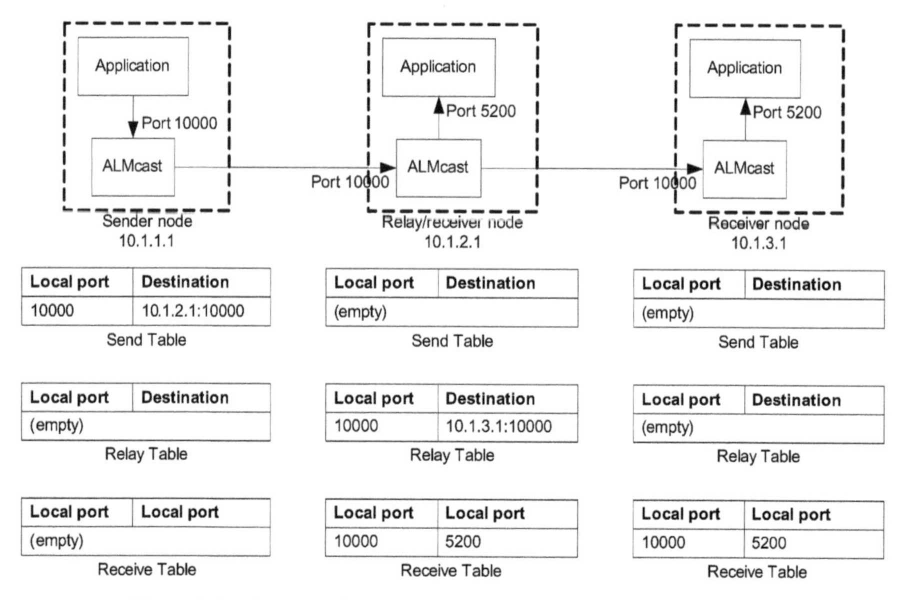

Figure 8.3 ALMcast virtual overlay and interaction with application.

provided by the operating system. The applications at receiving nodes only need to read packets from a local port using similar methods.

As shown in Figure 8.3, each node contains three forwarding tables to send, relay and divert packets. The sender side uses only the Send Table, leaving the other two tables empty. Any application that requires the transmission of packets will write the required packets to the preconfigured local port. Upon receiving a packet at the preconfigured port, ALMcast will replicate the packet and transmit it to the preconfigured destinations.

At relay nodes, the ALMcast Relay Table must be configured with a local ALMcast port and the correct IP addresses and ports that any packet received should be forwarded to. Any packet received on the local ALMcast port will be forwarded to all the nodes in the Relay Table.

The reason for separating the Send and Relay tables is to allow optimizations in implementations that can take advantage of the fact that Send Tables handle locally generated packets and Relay Tables handle packets received from the network, but this separation is not mandatory.

At receiving nodes, the ALMcast Receive Table must be configured with the local (application) port that ALMcast should forward its packets to. The application should not read from the ALMcast port directly which is handled by ALMcast. A node may both be a relay and receiving node, in which case any packet arriving at the configured ALMcast port is both forwarded to other nodes in the Relay Table and to the local (application) port to be read by the application.

8.4.2 Group Membership Management

As described earlier in Section 8.2.3, group membership dynamism is important to the use of ALM in CE devices, hence the method used for group membership must support nodes joining or leaving at any time. Because a node may leave without prior notification (for example, from a forced shutdown or network interruption), and because nodes are responsible for not only themselves but also for the nodes that they forward streams to, a method for detecting a so-called uninformed leave must be implemented. Typically, a heartbeat mechanism could be used where the consecutive absence of a few heartbeats would signify that the node has left and a tree reconstruction is possibly needed (if the node was forwarding to other nodes at the time of leave).

8.4.2.1 Example

Various studies have been done on methods of group management [5–7]. For CE purposes, a centralized system is used where group modifications (participants joining or leaving the group) can be done in both active (client-initiated) and passive (server-initiated via polling) methods. Central control is used for the same reasons as in Section 8.3.

Typically, the client application will establish one connection first before negotiating for ALM group membership. For example, in a video conferencing application, a conference session establishment via the Session Initiation Protocol (SIP) may be performed first. Only after that, the client will initiate a connection request to the server in order to establish the ALM session between them.

In addition, the underlying ALM communication framework also maintains the connectivity between ALM nodes using a heartbeat mechanism. A keep-alive heartbeat message is exchanged periodically between all clients with the server. Two scenarios of dynamic leave during the session are possible: informed leave and uninformed leave. Informed leave occurs when a node leaves after receiving permission from the server. Uninformed leave occurs when nodes are dropped upon network disruption, loss of power, and so on.

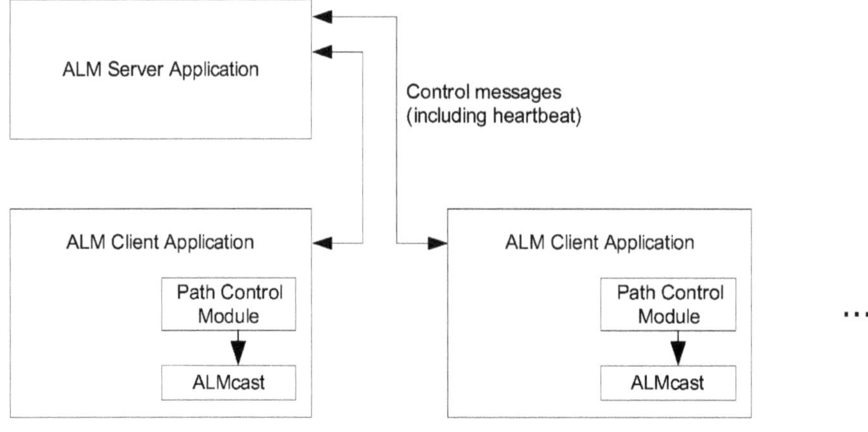

Figure 8.4 The ALM server and client.

In the case of informed leave, the system controls the timing of a node's leaving via delaying the transmission of the approval message until overlay tree reconstruction and forwarding table updates of all other nodes have been completed. This approach ensures reduced streaming disruption to the downstream nodes of the leaving node. While uninformed leave of nodes affects streaming quality tremendously, this problem is mitigated by the keep-alive messaging between the server and all the clients, which provides timely detection of uninformed leave.

8.4.3 Module Interaction

The modules discussed above are packaged into two applications, namely, the ALM server and the ALM client. Figure 8.4 shows an abridged view of their composition and relations.

The ALM server controls the module responsible for overlay tree construction and manages the overall group membership. In any ALM session, only one node performs the tasks of the ALM server.

All receiving nodes run the ALM client. The node running the ALM server may also run an ALM client if necessary, or the ALM server role may be taken up by a specialized node. All ALM clients are capable of network metrics collection, updating their respective ALMcast tables, and communicating with the ALM server for group management purposes.

Whenever a node joins or leaves the session, the ALM server is informed (either with an explicit control message or by the sudden absence of the heartbeat messages). Then the ALM server recomputes forwarding tables, and sends them to the affected ALM clients. When a client receives a new forwarding table, it updates its own ALMcast forwarding table as explained in Section 8.4.1.

8.5 Testing ALM Applications

ALM applications can be tested both on a simulated network using a (local) testbed and on a real network (the Internet). The advantages of testing on real networks are obvious, but testbed tests provide greater flexibility as different scenarios can be easily simulated even with the same physical layout of nodes.

Most of the testbed considerations for CE ALM application testing are the same as that of tests of ordinary network-enabled applications, although CE applications add their own requirements as well. For example, the test mechanisms and testbed need to support situations like bandwidth heterogeneity and the simulation of network interruption; these situations are common in normal CE use.

The purpose of this section is both to serve as an example of a testbed mechanism that can be used to test CE devices, and to validate the policies that we have selected in Sections 8.3 and 8.4 as suitable for use. The remainder of this section describes the methods used for testbed simulation and the results of the simulation, with a brief description of a live test in Section 8.5.5.

8.5.1 Description of Testbed

The testbed is modeled after the StarBED project [9, 10]. It comprises 16 nodes (15 ALM nodes and one router node) which are Intel® Pentium® 4 computers with two network cards installed. Similar to StarBED, each node participates in two networks: experimental and management. The experimental network is the network where all tests are performed on, and where network events like bandwidth limitation and network disconnection are triggered as part of tests. All IP addresses configured in the ALM software stack correspond to experimental network interfaces. On the other hand, the management network is used to coordinate the test. (Two networks are needed because control is needed over all nodes even though a test may mandate a node being disconnected.) A set of tools called SpringOS [10], operating on

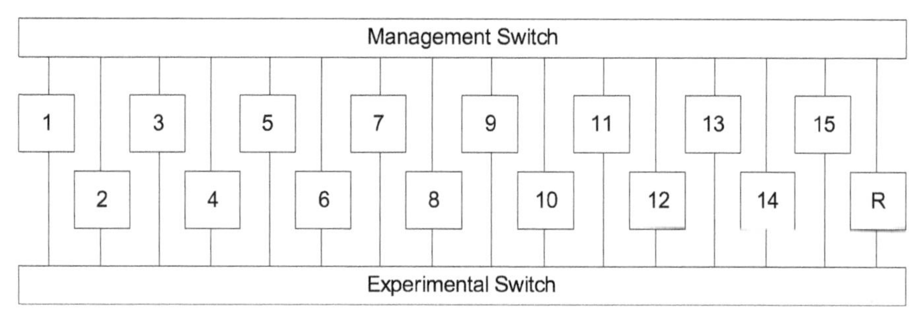

Figure 8.5 A diagram of each node's physical connections with the management and experimental switch.

the management network, is used to allow automation of tests. Figure 8.5 shows the physical layout of the testbed.

To test the software stack's capability in heterogeneous-bandwidth networks, Hierarchical Token Bucket [11] is used as a network limiter on a designated router node. The connections between all ALM nodes and the experimental network switch have been assigned individual Virtual Local Area Networks (VLANs) and the connection between the router node and the experimental network switch has been VLAN-trunked with all assigned VLANs. Also, routing tables have been configured on all nodes such that traffic is routed through the correct gateways. This forces all packets to be routed through the router node, incurring the configured bandwidth limitation, instead of bypassing the router node through Layer 2 routing on the switch. Figure 8.6 shows the logical connections made.

As in [9], the testbed has been verified to ensure that the router node is capable of routing 15 streams as once, while bandwidth limitation is performed on the router. For each limitation, the router node is configured to only permit each node to upload the given bandwidth (measured in kilobits per second). Then, 14 nodes are used to simultaneously generate a packet flow, transmitting them through the router node to the respective next nodes using a tool called Distributed Internet Traffic Generator (DITG) [12], which also returns statistics on the number of packets actually received, thus giving the measured actual bandwidth. The results are as in Table 8.1. The test shows that the router node is capable of restricting bandwidth to within 4% of accuracy.

In the scenarios (Section 8.5.2), however, only 4000 kbps streams are used.

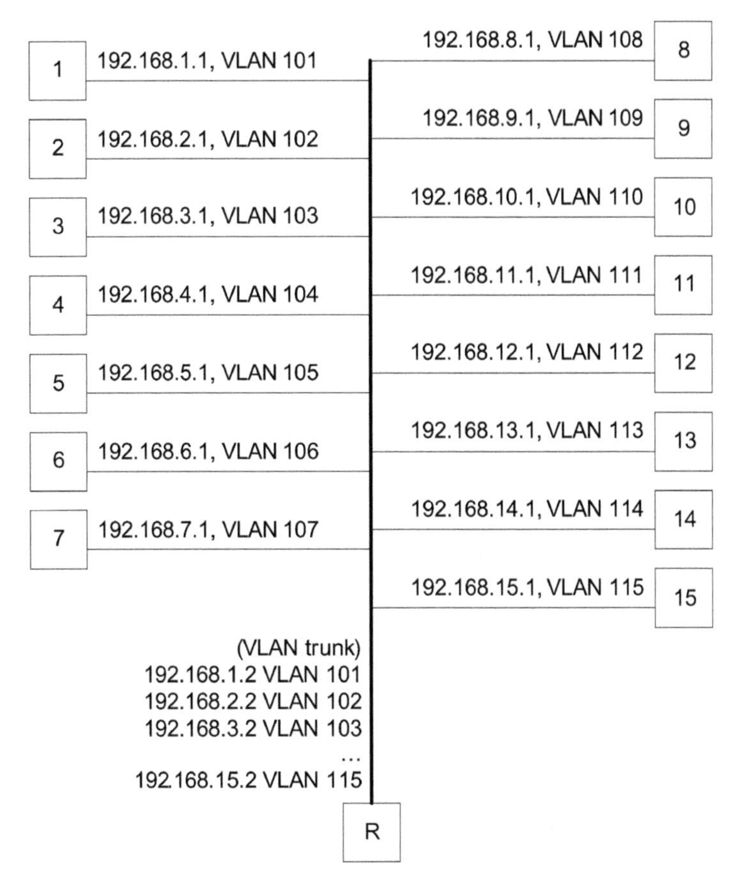

Figure 8.6 A logical diagram of the experimental network. Nodes 1 through 15 are in their own individual VLANs, and the router node (R) resides in all VLANs through VLAN trunking. For each of the numbered nodes, the router node (R) is set as the gateway for any network connections in the experimental network.

Table 8.1 Results of Hierarchical Token Bucket (HTB) verification. The theoretical bandwidth is different from the HTB bandwidth limitation because DITG reports bitrates excluding IP and UDP header overhead, whereas HTB includes them.

HTB bandwidth limitation	Theoretical bandwidth (not including IP and UDP header overhead)	Average measured bandwidth (14 nodes)
4000 kbps	3888 kbps	3820 kbps
8000 kbps	7776 kbps	7586 kbps
12000 kbps	11664 kbps	11235 kbps

8.5.2 Scenarios Tested

The algorithms and mechanisms described in Sections 8.3 and 8.4, including the ALM server and client, have been implemented in the C programming language.

The ALM server software stack is run on one node, and the ALM client software stack is run on 15 nodes. Group membership change is handled as follows: a client which wishes to join the group will send a connection request to the server (the active method), whereas the server detects a client leaving the group by the absence of heartbeat messages.

A packet generator is also run on the "sender" node that generates 4 megabits of packets per second, writing them to the ALMcast port to be transmitted to the other nodes. These other nodes run packet receivers, which detect packet loss or interruption.

The 15 nodes are divided into two administrator-defined clusters. The first cluster consists of nodes 0 through 7 (8 nodes in total), of which node 0 is the sender. All nodes in the first cluster can transmit 8 megabits of stream data per second, except for node 7 which is capable of transmitting 12 megabits per second. The second cluster consists of nodes 8 through 14 (7 nodes in total). All nodes in the second cluster are capable of transmitting 12 megabits per second. The network links between all nodes are capable of transmitting the 4 megabit/second stream and will not be a limiting factor in this scenario. The router node has been reconfigured for this heterogeneous bandwidth situation.

Initially, nodes 0 through 4 and nodes 8 through 14 participate in the ALM session. Then:

1. Node 5 (one node) joins the session,
2. Node 8 (one node) leaves the session,
3. Node 7 (one node) joins the session,
4. Node 6 (one node) joins the session.

A summary of the nodes, their bandwidth limitations and the group management scenarios is available in Table 8.2.

8.5.3 Evaluation and Analysis

During the test scenarios, the overlay trees are produced and the time taken to produce the new overlay trees are recorded. For node join scenarios, the time between the client sending the connection request and receiving the first packet through ALM is also recorded. In general, overlay tree reconstruction takes very little time compared to typical Internet latencies.

Table 8.2 The nodes and their respective characteristics and behaviours.

Node	0 (source)	1-4	5	6	7	8	9–14
Bandwidth capacity (kbps)	8000	8000	8000	8000	12000	12000	12000
Cluster node resides in	1	1	1	1	1	2	2
Group membership							
Initial scenario	In	In	Out	Out	Out	In	In
Scenario 1	In	In	In	Out	Out	In	In
Scenario 2	In	In	In	Out	Out	Out	In
Scenario 3	In	In	In	Out	In	Out	In
Scenario 4	In	In	In	In	In	Out	In

Table 8.3 shows the produced overlay trees with descriptions on how these trees are produced. The table also illustrates the trade-offs made when selecting an algorithm for use (as in Section 8.3). For example, it is theoretically possible to produce a lower-latency tree (one where the maximum and/or average source-to-node delay for all nodes is the lowest) if we allow some shuffling of the nodes in the newly reconstructed overlay tree, but we do not do this to reduce stream disruption. Instead, we allow some nodes to operate at less-than-full capacity (e.g. in Table 8.3, node 7 when it has just joined) and allow nodes that join after the reconstruction to use those locations in the tree (e.g. in Table 8.3, node 6) instead of assigning them to existing nodes.

8.5.4 Extension to Larger Groups

This software stack has been tested on 99 nodes in the StarBED facility provided by the National Institute of Information and Communications Technology [13] with similar results. Due to space constraints and similarity of results, details are omitted.

8.5.5 Live Internet Video Streaming Test Results

The software stack has also been cross-compiled to a consumer electronics board running the UniPhier® Panasonic® Processor [14], and integrated with other modules such as High-Definition Multimedia Interface (HDMI) video display, a video codec, and AV Quality of Service (including Forward Error Correction [15] handling).

This CE board with the associated software stack has been tested in a live video transmission scenario over the Internet. Five CE board nodes were used in this test, as shown in Figure 8.7: one node in Shinagawa with the video camera acting as the sender, one node in Osaka acting as a relay node

Table 8.3 Analysis of five scenarios tested.

Scenario	Distribution tree	Analysis	Join time
Initial	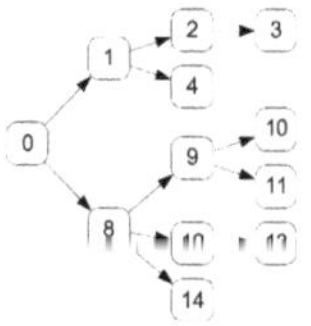	This is the initial tree produced; notice that the clusters do not intermix. Nodes in the first cluster (0-4) are capable of forwarding to 2 nodes, whereas nodes in the second cluster (8-14) are capable of forwarding to 3.	N/A
Node 5 joins	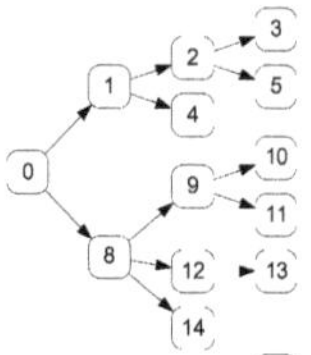	The ALM server finds the closest possible location (to the source) in the first cluster to place node 5.	< 2 ms
Node 8 leaves	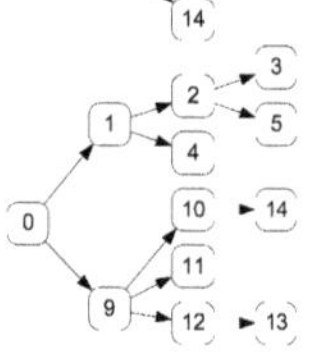	The server rejoins node 9 by connecting it to node 0, bringing along its children (nodes 10 and 11). Then node 12 is rejoined by becoming node 9's third and final child. Node 14 is rejoined by joining to node 10.	N/A
Node 7 joins	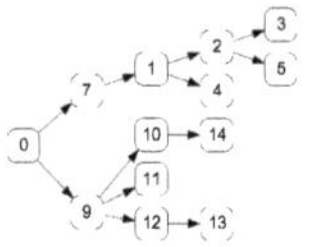	Because node 7 is a high-capacity node (see Table 8.2), the server connects it to the source node 0 directly. Node 1 is now node 7's child. The server does not do any other path shuffling to reduce stream interruptions.	< 2 ms
Node 6 joins	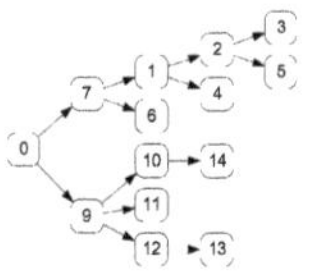	The server joins node 6 to the recently joined high capacity node 7.	< 2 ms

(receiving the stream from the Shinagawa sender and relaying it to the three Shinagawa nodes), and three more nodes in Shinagawa acting as a receiver. All relay nodes and receivers present the stream received on an attached television, connected through a HDMI cable.

In typical situations where one would use ALM, this setup is not recommended due to the redundant use of the Internet physical link. However, for the purposes of this test, we traverse the physical link over the Internet multiple times.

In this test, we are evaluating the ability of our middleware modules to handle a real-world application (video streaming) in a real-world situation.

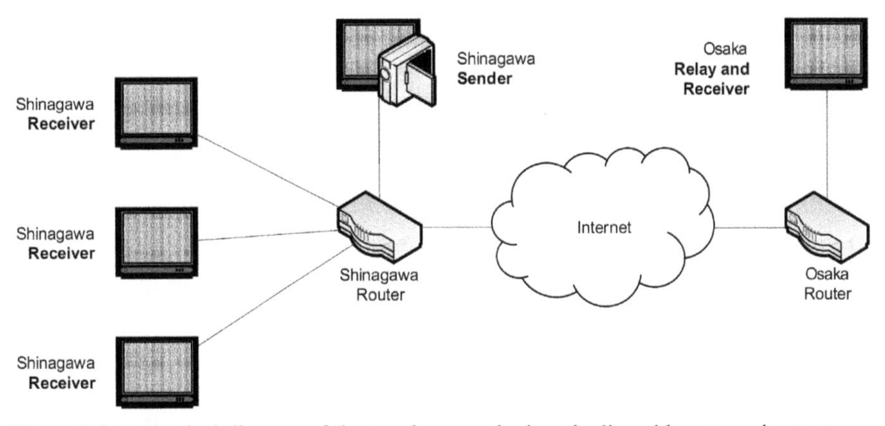

Figure 8.7 A physical diagram of the topology used when the live video streaming test over the Internet was conducted.

Table 8.4 The minimum, average, and maximum central processing unit (CPU) utilization of the sender, relay, and receiver nodes.

Node	Minimum	Average	Maximum
Sender	7%	10%	14%
Relay	18%	21%	25%
Receiver	9%	12%	14%

A 3 megabit/sec video stream was used, and 25% Forward Error Correction was activated.

Running on the CE board, there was no perceptible drop in AV quality. Central processing unit (CPU) utilisation of the board is also at low levels (see Table 8.4 for the full details).

8.6 Conclusion and Future Work

In a recent CE show event [3], a few vendors have demonstrated Skype® on television which represents one of the popular consumer applications. Even though numerous personal computer-based applications use ALM-like technology, similar adaptation in CE world has just begun. Although using ALM in CE devices brings other challenges into view, these challenges can certainly be overcome. We have described several design considerations, shown examples of the algorithms and mechanisms that can be used, and discussed options for tests on CE devices and in CE situations. We anticipate many more CE vendors will design their products embedding ALM-like solutions for better consumer experience. This is a foundation that can be built upon in the future to offer a complete real-time AV content sharing system using CE devices that can scale to large numbers of participants.

Acknowledgements

The authors would like to indicate that this research work was undertaken by numerous people within and outside the organization that made various publications possible. They are Jason Soong, Lin En Shu, and Ngo Chuan Hai from Panasonic R&D Centre, Malaysia, Truong Khoa Phan and Nam Thoai from Ho Chi Minh University, Vietnam, and Kunio Akashi and Kenichi Chinen from Japan Advanced Institute of Science and Technology.

References

[1] S.R. Narayanan et al. Peer-to-peer streaming for networked consumer electronics. *IEEE Communications Magazine*, 45(6):124–131, June 2007.

[2] http://www.nhk.or.jp/strl/open2009/english/exhibit/t25.html.

[3] Brad Stone. A venture integrating Skype into the family room. *The New York Times*, 5 January 2010. Retrieved from http://www.nytimes.com/2010/01/05/technology/internet/05hdtv.html.

[4] http://www.ivci.com/videoconferencing-sony-pcs-hg90.html.

[5] D. Pendarakis, D. Verma, S. Shi, and M. Waldvogel, ALMI: An application level multicast infrastructure. In *Proceedings of USENIX Symposium on Internet Technologies and Systems*, page 5, 2001.

[6] P. Francis. Yoid: Extending the multicast internet architecture. http://www.aciri.org/yoid/,1999,

[7] Y. Chu, S.G. Rao, and H. Zhang. A case of end system multicast. In *Proceedings of SIGMetrics*, pages 1–12. ACM, 2000.

[8] Lim Boon Ping, K.K. Ettikan, Eiichi Muramoto, Lin En Shu, Truong Khoa Phan, and Nam Thoai. Bandwidth fair application layer multicast for multi-party video conference application. In *Proceedings of CCNC2009*, pages 113–117, 2009.

[9] Thilmee Baduge, Lim Boon Ping, Kunio Akashi, Jason Soong, Kenichi Chinen, K.K. Ettikan, and Eiichi Muramoto. Functional and performance verification of overlay multicast applications – A product level approach. In *Proceedings of CCNC2010*, pages 740–744, 2010.

[10] Toshiyuki Miyachi, Ken-ichi Chinen and Yoichi Shinoda, StarBED and SpringOS: Large-scale general purpose network testbed and supporting software. In *Proceedings of Valuetools 2006*, Pisa, Italy, October 2006.

[11] Martin Devera, Hierarchical token bucket, http://luxik.cdi.cz/ devik/qos/htb/.

[12] Alessio Botta, Alberto Dainotti, and Antonio Pescape', Multi-protocol and multi-platform traffic generation and measurement. In *Proceedings of INFOCOM 2007*, DEMO Session, May 2007.

[13] http://www.starbed.org.
[14] http://panasonic.co.jp/corp/news/official.data/data.dir/en070619-4/—en070619-4.html.
[15] C. Perkins and O. Hodson. Options for repair of streaming media. RFC 2354, June 1998.

9

Improving Wide Area P2P Service Discovery Mechanisms Using Complex Queries

Jamie Furness and Mario Kolberg

Computing Science and Mathematics, School of Natural Sciences, University of Stirling, Stirling FK9 4LA, Scotland; e-mail: {jrf, mko}@cs.stir.ac.uk

Abstract

With smartphones and other network enabled consumer devices becoming increasingly popular, the number of available services and their complexity is growing considerably. With an increasingly large and dynamic environment it is important that users have a comprehensive yet efficient mechanism to discover these services. Many existing wide-area service discovery mechanisms are centralised and do not scale to large amounts of users. Peer-to-peer networks however have been prove to scale well, and can be used to provide not just a platform on which peers can offer and use services without relying on a centralised resource, but also as a means of service discovery. There are various wide-area peer-to-peer service discovery mechanisms that allow discovery of services via their attributes, however the majority are limited to keyword matching and do not support other types of complex queries.

This chapter starts with a review of complex queries and existing approaches which provide support for such queries. We illustrate the use of blind search in Distributed Hash Tables (DHTs) to provide support for all types of complex queries, such as wild-card search, range queries, and even regular expressions. Using blind search allows for processing the search query at every node within the network, supporting queries as complex as required. However due to the nature of broadcast trees search performance suffers under high churn levels; to combat this we note that data is already

Anand R. Prasad et al. (Eds.), Future Internet Services and Service Architectures, 183–203.

replicated within the network for redundancy. This can be further used to improve the success rate of blind search when under high churn. Finally, we present novel results considering churn level vs replication of data.

Keywords: P2P, DHT, service discovery, complex queries, blind search, broadcasting.

9.1 Introduction

The number of smartphones and other network enabled consumer devices is rapidly growing, resulting in an increasingly large and dynamic environment in which users wish to find services. At the same time the number of services available to users is growing at a considerable rate. For example, both the Android and iPhone application stores now have over 60,000 and 220,000 applications respectively. This shows that with these two examples alone there is a large number of users actively discovering a large number of available applications and services every day. While these centralised and closed repositories may work for giant distributors such as Google and Apple, whom already have the infrastructure to handle scalability and failures in place, a more open approach with a lower barrier of entry is important for smaller companies or individuals wishing to advertise their services.

It is important to note that while there are various local area Service Discovery Mechanisms (SDMs), such as Simple Service Discovery Protocol (SSDP) in UPnP [8], and Bluetooth Service Discovery Protocol (SDP) [1], the focus of this chapter is on wide area SDMs.

Traditionally wide area SDMs have had a centralised architecture, for example UDDI [24], consisting of multiple repositories that collect service advertisements which are then synchronised periodically. However as the number and complexity of services grows it becomes obvious that such a centralised approach is not sustainable. To solve this problem a number of decentralised approaches have been proposed based on structured peer-to-peer networks. The advantages of such approaches include:

Scalability. The performance of a SDM should not degrade as the size of the system increases. This is a common problem with centralised approaches, where the overall capacity stays fixed but the number of users, and hence load on the system, grows. Decentralised systems however tend to cope well with increasing network size because the load on each node is distributed equally amongst all nodes. This means that as the network size, and hence load, grows, so does the overall capacity.

Fault tolerance. Centralised systems by definition have low fault tolerance due to a single point of failure; if a centralised server fails the whole system is usually rendered inoperable. On the other hand, due to their highly dynamic nature, peer-to-peer systems are specifically designed to be fault tolerant, with the assumption that nodes will fail at some point. To cope with failure, data in a peer-to-peer network is automatically replicated at a set number of nodes – for data loss to occur all nodes replicating the data item would be required to fail simultaneously.

Low barrier of entry. In the traditional client/server model, to support a large number of services and users requires a server farm, that is, a data centre with sufficient network capacity and a set of server-class hosts. To achieve fault tolerance this data centre is often replicated at multiple physical locations, and requires around the clock maintenance to respond to failures. This presents a huge start-up cost to small businesses. Conversely peer-to-peer networks can handle both scalability and fault tolerance without the need for any expensive set-up costs.

However it is important to note that using a Distributed Hash Table (DHT) based approach does introduce other issues, the most limiting of which is only a basic support for search – a fundamental requirement for service discovery.

In Section 9.2 we discuss different types of search queries which are covered under the term complex queries, and examples of existing service discovery mechanisms which provide support for such queries. In Section 9.3 we discuss the advantages of using blind search when performing complex queries, and outline some specific implementations based on efficient broadcast. Due to efficient broadcasts use of broadcast trees, the blind search approaches tend to be very sensitive to node failure; in Section 9.4 we look at the effect of churn, the act of nodes leaving or joining the network, on the success rate of blind search. Finally, in Section 9.5, we discuss how load is distributed during blind search, and what effect the distribution of data within the network has on the success rate of blind search when under churn.

9.2 Complex Queries: Examples and Use in Service Discovery

Complex query is a vague term which is often used to describe any form of search query more complex than an exact match. In this section we discuss different types of queries which are often referred to as complex queries. For completeness we start with the basic exact match.

9.2.1 Exact Match

Due to the hash-table nature of a DHT, the only form of search query supported without some form of extension is the exact match. All DHTs support, either directly or indirectly, a hash-table interface providing the methods $put(key, value)$ and $get(key)$. Given the key corresponding to a service, a DHT can guarantee the return of results, usually within a specific number of hops, if any such results exist. In many cases such limited support for search is not sufficient and clearly does not support complex queries, which has resulted in many proposed systems which build support for more complex queries on top of the exact match facility.

9.2.2 Keyword Search

If exact match look up is not sufficiently powerful, DHTs can be extended to support keyword-based queries. Keyword-based search provides the ability to associate multiple keywords with a single service.

A simple approach to supporting keyword-based search, as used by INS/Twine [2], is to simply index pointers to the service at multiple locations within the network, defined by the associated keywords.

An alternative approach, called Squid [18, 19, 20], represents services in a d-dimensional coordinate space, with their position defined by a set of keywords. For example using base-26 coordinates, a service providing the weather in Scotland may be stored at the coordinates $(scotland, weather)$, as shown in Figure 9.1. This d-dimensional coordinate space is then mapped to a 1-dimensional index space using a Hilbert Space Filling Curve (Hilbert SFC), allowing the query to be routed by the underlying DHT.

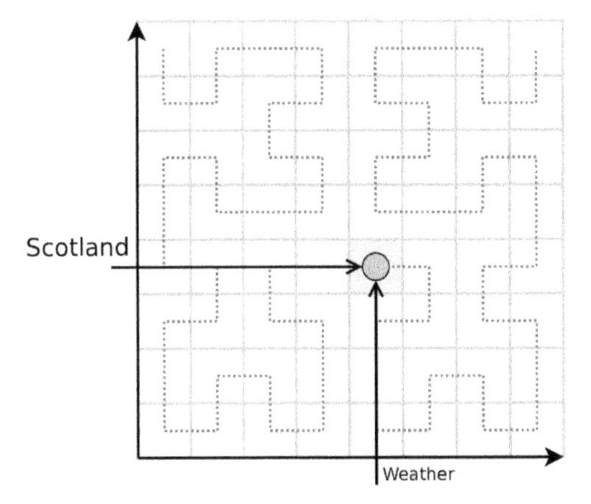

Figure 9.1 A 2-dimensional keyword space in Squid, showing the mapping of a service at $(scotland, weather)$ to a single point in the index space.

9.2.3 Range Queries

Range queries are a type of query which look for any document, possibly indexed by keywords, which lies within a given range. For example, a user may wish to find available services within a certain price range.

Range queries are made challenging by the use of consistent hashing in most DHT-based schemes, which uniformly distributes data throughout the network without preserving locality. It follows that if a hashing algorithm can preserve locality of objects, range queries could be supported, however the load balancing effect of consistent hashing is then lost.

An example SDM which implements range queries, by removing the consistent hashing algorithm to preserve locality, is PRoBe [16]. PRoBe organises peers into a multi-dimensional logical space, similar to that used by CAN, described in Section 9.3.1.4. Since the data items and queries are directly mapped onto the logical space, locality is preserved and the range queries can be processed by only visiting the relevant peers. To balance load virtual peers are used, allowing peers to manage multiple zones rather than a single zone. When a peer *P* finds that the ratio between the load of its most loaded neighbour and itself crosses a threshold, it will hand over its virtual peers to its least loaded neighbour. *P* then splits the load with the most loaded neighbour such that their loads are nearly equal.

Other approaches to supporting range queries include using multi-set DHTs [5] and distributed segment trees [26].

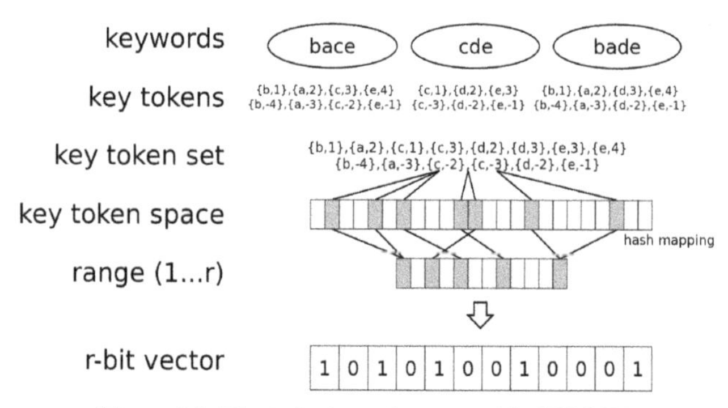

Figure 9.2 The indexing scheme used in KISS-W.

9.2.4 Wild-Card Search

The term wild-card search is used to describe a search in which part of the search term is unknown. Note that wild-card search is related to range queries, but not exactly the same. For example, assuming keywords were restricted to $a \dots z$, the wild-card search for "acm*" can be converted to a range query for everything between "acm" and "acn". However other forms, such as "*acm", "a*m", or "*acm*" are difficult to map to a range query [11].

KISS-W [11] is a system in which wild-card search is supported over a hypercube based network by extracting key-tokens from keywords. A key-token is a pair $\{c, i\}$, where c is a character in the keyword and i is the characters position. Key-tokens are also extracted counting backwards to allow queries where the wild-card is of unknown length. The keyword "acm" would result in a key-token set of {a,1},{c,2},{m,3},{a,-3},{c,-2},{m,-1}, and the search term "a*m" would result in the key-token set {a,1},{m,-1}. After extraction the key-tokens are mapped to the key-token space then converted to an r-bit vector by applying a hash function, as shown in Figure 9.2. The document, or a pointer to the document, is then stored at the node with the identifier corresponding to the calculated r-bit vector.

To perform a search the the set of nodes which may index a service matching the description need to be identified. It is easy to see that these nodes must all contain the key-token set extracted from the search term, and hence their identifiers must all have 1 at the same positions as the r-bit vector calculated from the search term key-token set. Using a spanning binomial tree [10] of the sub-hypercube induced by the node defined by the search terms r-bit vector, the appropriate nodes can be queried and matching services found.

9.2.5 Full-Text Search

Full text search refers to a technique for examining all words in all documents, to try match the search term supplied by the user. While this type of search may initially sound similar to the keyword search, keyword search solutions assume a limited number of keywords and do not scale well above a certain limit. A straight forward approach to full-text search is to use the DHT to store and retrieve inverted indices, however this approach is inefficient due to the high cost of transporting these inverted indices between peers to be intersected, since inverted indices are likely to be large. Reynolds and Vahdat [15] suggest several optimisations, such as bloom-filters and caching, to reduce the cost of transporting inverted indices.

A bloom-filter is a hash-based data structure that summarises membership in a set. Since bloom-filters are hash-based, it is possible for a bloom-filter to contain false positives, however the probability of such false positives is tunable by adjusting the number of bits used in the filter; false negatives will never occur. Using bloom-filters the inverted indices can be summarised and hence the required bandwidth is reduced during the intersection phase. Due to the false positives that can occur, the final intersection result has to be passed back along the peers holding the original inverted indices to eliminate any false positives, however in most cases, by this time the set will be much smaller and hence require less bandwidth.

9.2.6 Semantic Search

Semantic search is a content-based search, in which queries are expressed in natural language instead of keywords. The goal in semantic search is to use semantics, the meaning of words, to return results which are relevant to the search terms, without simply matching keywords.

One approach for semantic search is known as pSearch [23], built on top of a hierarchical version of CAN known as eCAN [25]. Given a document or service description, term vectors are computed using a Vector Space Model (VSM). From these term vectors, a semantic vector S is computed using Latent Semantic Indexing (LSI), and the document or service description is then stored in the DHT at the coordinates denoted by S, resulting in similar services being close together. During search the semantic vector Q of the search query is computed and the query is routed to the coordinates denoted by Q. Upon receiving the search query, the node responsible for Q floods the query to nodes within a defined radius based on the similarity threshold, as

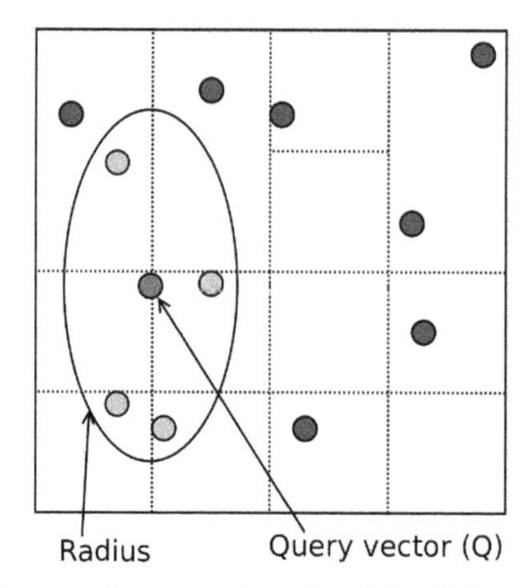

Figure 9.3 An example query using pSearch in a 2-dimensional CAN.

shown in Figure 9.3. Receiving nodes perform a local search using LSI, and return their results to the querying node.

9.2.7 Regular Expressions

Although not often mentioned, in some circumstances it may be desirable to have the ability to search for data using regular expressions. Evaluation of queries as complex as regular expressions is currently only supported using a method that processes queries locally, such as blind search.

9.3 Blind Search and Use in Service Discovery

The term blind search is used to describe a search operation in which no information about the search space is known, other than to distinguish the goal state from all others. In other words, as a query traverses through the network it has either reached the goal or not, there is no concept of distance to the goal as with regular operations in a DHT. Blind search can be thought of as the structured equivalent to flooding, providing all the flexibility of flooding, without the downside of redundant messages.

The ability for P2P service overlays to efficiently and effectively perform blind search is important as it provides the facility to discover services using any type of complex queries, such as described in Section 9.2. Without such a facility a P2P service overlay will be restricted to, at best, a specific subset of complex queries. While keyword-based search can support linking multiple keywords to a service and some implementations also support range queries and wild-card, there are various limitations, outlined below:

- Depending on the indexing scheme used, the number of keywords which can be associated with any one service may be limited. For example, in Squid [18, 20, 19], the maximum number of keywords which can be associated with any one service is defined by the dimensionality of the keyword space. While this is a parameter which can be tweaked, once the network has been set-up there is a hard limit which cannot be exceeded. Schmidt and Parashar also note that as the dimensionality of the keyword space grows, the efficiency of their algorithm falls – in simulations a dimensionality of three was the maximum evaluated, resulting in a maximum of three keywords per service. In some cases a limit on the maximum number of keywords may not be a problem, however in many cases attempting to describe a service in a limited number of keywords would result in a much less comprehensive description. The blind search approach however supports service descriptions of any length.
- To support multiple keywords some indexing schemes, such as INS/Twine [2], split service descriptions into keywords and then store the service description at every node associated with a keyword, resulting in the service description being duplicated a large number of times around the network. As the space required to store a service description is fairly minimal, the storage in itself is not a problem, however it makes updating the service description a much more expensive operation, especially when a large number of keywords are present. In a static network this may not cause a problem, however if services can only serve a limited number of uses at a time they may wish to remove their advertisement when they have no available slots, or even include the number of available slots in the service description. In such an environment with regularly changing service descriptions the overhead of maintaining accurate service descriptions at a large number of nodes could be high.
- Keyword-based search is limited, as the name suggests, to keywords. In other words there is no support for matching phrases, regular expressions, or other such queries, which are all supported by blind search.

Based on these issues it should be clear that while keyword-based search can be useful in some cases, it is not suitable for every situation. In these cases blind search can be a powerful alternative adding support for full-text search and regular expressions, albeit with increased bandwidth requirements.

Due to the lack of knowledge about the search space when performing blind search, the query must be processed locally at each node within the network, and as such the process of blind search is almost identical to that of broadcasting. Although there are a number of algorithms which can be used to perform blind search on various different overlay structures, they all stem from the idea of efficient broadcast, first published by El-Ansary et al. [6].

Efficient broadcast is performed by dynamically constructing a broadcast tree, using only information in the underlying structure of the overlay.

9.3.1 Methods for Blind Search

9.3.1.1 Efficient Broadcasting

Efficient Broadcasting [6] is an algorithm for broadcasting complex queries, or indeed any data, over DHTs, proposed by El-Ansary et al. using Chord [21] as an example. In Chord nodes form a ring structure; in a network of size N each node maintains routing state information for at most $\log_2(N)$ other nodes, namely the node succeeding it, known as the successor, and a set of finger nodes. The finger nodes are chosen at logarithmically increasing distance around the ring, the ith entry in the table at node n contains the identity of the first node that succeeds n by at least 2^{i-1} $(i \geq 1)$.

In the proposed algorithm, to initiate a query, a node n will send the query along with a limit to every (non-redundant) node in its finger table. The limit parameter given is the identifier of the next finger in the finger table, and is used to restrict the forwarding space of the receiving node to $(n, limit)$. The last node in the finger table is given a limit of the originating node. When the message is received by a node it forwards the broadcast to any fingers it has within the forwarding space, giving each one a new limit.

9.3.1.2 HyperCup

Schlosser et al. [17] propose a new network structure known as HyperCup which allows for efficient broadcast and search. The structure proposed organises peers into a hypercube, or more generally a Cayley graph, topology. Edges in the graph are tagged based on their dimension. Figure 9.5 shows a hypercube topology with base $b = 2$, which in three dimensions turns out to be a hypercube with b nodes in each dimension. A complete hypercube graph consists of $N = b^{L+1}$ nodes, where all nodes have $(b - 1) \times (L + 1)$

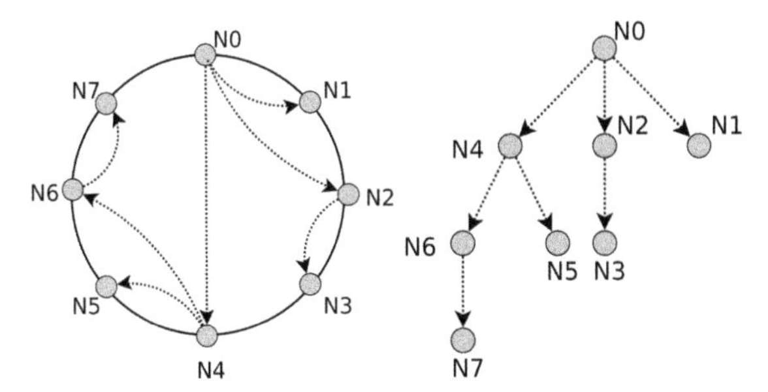

Figure 9.4 Dissemination of an example broadcast message using Efficient broadcast in a Chord network, where N is a node and L is the limit.

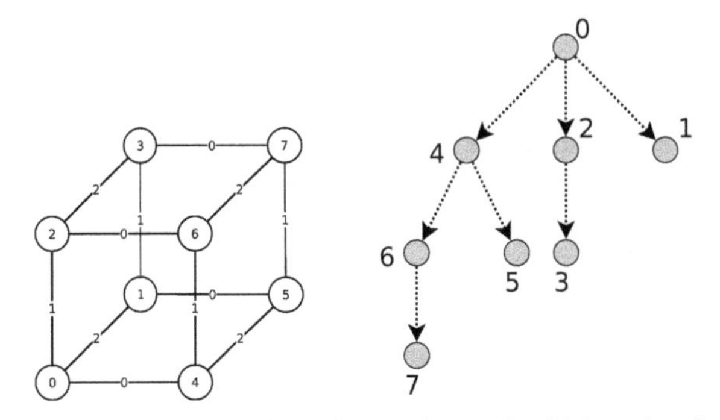

Figure 9.5 A 3-dimensional hypercube graph with base $b = 2$.

neighbours, $(b - 1)$ in each dimension – where $L + 1$ is the number of dimensions.

A broadcast algorithm is described which is guaranteed to reach all nodes with exactly $N-1$ messages. A node invoking a broadcast sends the broadcast to all its neighbours, tagging it with the edge label on which the message was sent. Nodes receiving the message restrict the forwarding of the message to those links tagged with higher edge labels.

9.3.1.3 Pastry Broadcast

Castro et al. [4] describe a method for broadcast over Pastry. While not specifically designed with search in mind, it can be used to broadcast complex queries.

Each pastry node maintains a routing table, a neighbourhood set, and a leaf set. In a network of size N using identifiers with base 2^B, each nodes routing table is designed with $\log_B(N)$ rows, where each row holds $B - 1$

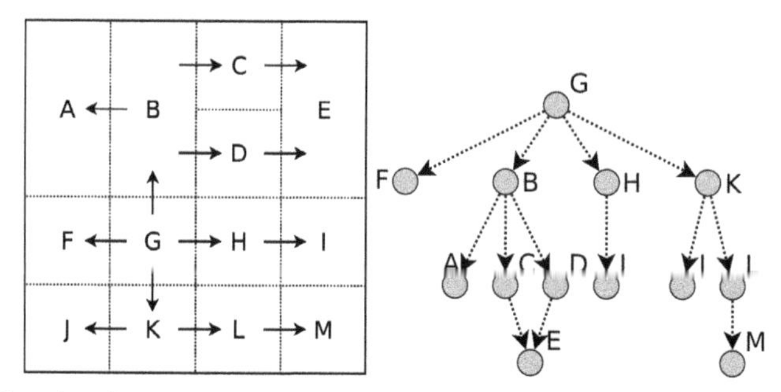

Figure 9.6 Broadcasting a message from node G in a 2-dimensional CAN overlay with 13 nodes.

entries. All the entries at row r of the routing table each refer to a node whose identifier shares the current node's identifier in the first r digits, but whose $(r + 1)$th digit does not match that of the current node.

A node broadcasts a message by sending the message to all nodes in its routing table; each message is tagged with the routing table row r. When a node receives a message it forwards the message to all nodes in its routing table with rows greater than r. This continues until a node receives a message tagged with r and has no entries in rows greater than r.

9.3.1.4 CAN Broadcast

In a solution for application-level multicast in CAN presented by Ratnasamy et al. [14], multicast is performed by creating a mini-CAN of participating users, then broadcasting over that CAN. While using this technique specifically for complex queries is not suggested, broadcasting is the basis of most complex querying techniques.

CAN is a rather unique type of overlay, designed around a virtual d-dimensional Cartesian coordinate space. The entire coordinate space is dynamically partitioned among all the nodes in the system such that every node owns an individual zone. Keys are mapped onto a point in the coordinate space using a consistent hash function, and the responsible node is the node whose zone contains that coordinate.

In a CAN with dimension d, each node has at least $2d$ neighbours; one to move forward in dimension d and one to move backwards. To initiate a broadcast the source node forwards the message to all its neighbours. When a node receives a broadcast message from a node with which it neighbours along dimension i, it will forward the message to those neighbours along dimension $1 \dots (i - 1)$ and the neighbours in dimension i on its other side.

To prevent the message from looping back around a node does not forward a message along a particular dimension if that message has already traversed at least half-way across the space from the source coordinates along that dimension. An example broadcast can be seen in Figure 9.6.

It is worth noting that for a perfectly partitioned coordinate space the algorithm ensures each node receives the message exactly once. For an imperfectly partitioned space, a node may receive the same message from multiple neighbours. For example, in Figure 9.6, node *E* would receive a message from both neighbours *C* and *D*.

9.4 Effects of Churn

Due to the tree structure used during blind search all approaches share one major weakness – susceptibility to churn. It is widely accepted that in any peer-to-peer network deployed in the real world, routing tables will always have varying degrees of accuracy; as nodes join or leave a network it takes time for changes to be reflected in other nodes routing tables. How long it takes for changes to be reflected depends on the underlying overlay chosen, for example Pastry uses reactive maintenance which means nodes attempt to update their routing tables as soon as a change is detected; in Chord periodic maintenance is used, causing the routing tables to only be updated at specific intervals, defined by a network parameter.

There are two types of failures due to churn which must be considered:

1. When routing tables are out-of-date a node may unintentionally forward a search message to a node that no longer exists in the network, expecting them to cover a partition of the network. A solution to this category of failures would be to implement an acknowledgement and timeout, allowing for message retransmission upon failure. Since the network should self-repair the message would be successfully delivered eventually. However while this would increase the success rate, it would also double the number of messages required, and increase search latency.
2. Even if the addition above was to be implemented, a second type of failure can occur, in which a node may receive a search message successfully, but leave the network before covering the partition it was assigned. In this case the acknowledgement would need to be delayed until after the nodes children had been covered; however this makes choosing a suitable timeout problematic as the length of time taken will depend on the depth of the tree, which cannot accurately be calculated.

Ghodsi suggests this could be solved by implementing a failure detection mechanism independently from the broadcast mechanism [7], in which nodes periodically send messages to their children and await an acknowledgement. Again while this would increase the success rate, it would further increase the number of messages required, and further increase the search latency.

Using the OverSim P2P network simulation framework [3] for OMNeT++, two sets of simulations were performed, with network sizes of 1,000 nodes and 10,000 nodes, comparing the performance of blind search using the Efficient Broadcast technique for Chord, and Pastry's Broadcast technique. To simulate churn the network was filled, thereafter nodes would join and leave based on times drawn from a Weibull distribution, as recommended by Stutzbach and Rejaie [22]. Lifetime mean ranged from 100 seconds to 6,000 seconds.

Data was distributed randomly among the nodes, such that each node would have on average two data items, plus any replica. The replication rate was varied from 1 (no redundancy) to 8.

Once the network reached its target size we initiated 20 searches sequentially, each from a random node and for a random data item known to be currently within the network. If the data item queried was found at least once the search was considered a success.

In the Chord network we ran two sets of simulations using parameters: $N = 8$, $S = 1$, and $D = \{120, 10\}$, where N is the number of successors, S is the stabilisation delay, and D is the finger table maintenance delay in seconds.

In the Pastry network we used parameters: $L = 16$, $M = 0$, and $B = 4$, where L is the leaf set size, M is the neighbour set size, and B is the number of bits per digit.

Results, shown in Figure 9.7, showed that under high churn the success rate for blind search is poor, dropping to under 40% success rate when the lifetime mean is reduced to 100 seconds in a 1,000 node Pastry network. With Chord it was possible to increase the success rate by reducing the maintenance delay, however even with a low maintenance delay of 10 seconds the success rate dropped as low as 51% with 100 seconds lifetime mean.

Whether or not such high churn is likely to be realistic in reality depends on the type of SDM deployed and the type of users participating. In a service overlay where all users participate in the network the churn level is likely to be higher. However some approaches, such as INS/Twine [2], suggest a

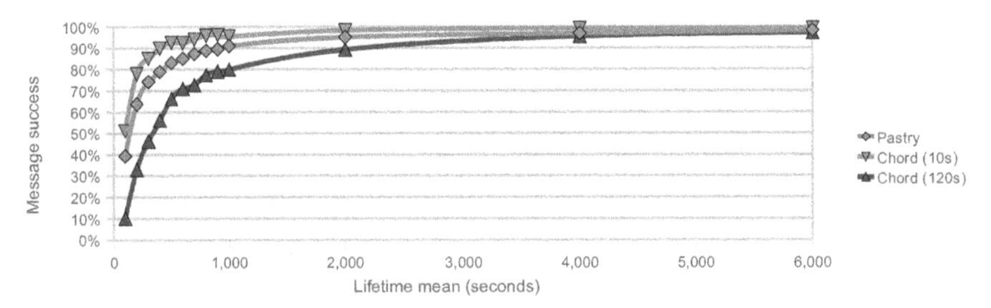

Figure 9.7 Average message success rate of Efficient Broadcast in 1,000 node Chord and Pastry networks while under churn.

hybrid approach where the network consists solely of resolvers – clients only communicate with their local resolver, which is responsible for forwarding their request to the network – in such an approach the resolvers are likely to be much more stable, and hence the churn level will be much lower. If users of the SDM are likely to be connecting from mobile devices then the churn rate is likely to be higher due to poor signal strength, movement in and out of reception areas, and interruption from other activities. Alternatively users connecting from desktop PCs are likely to stay in the network for longer periods of time without interruption.

9.5 Load Distribution

When considering load distribution in a service overlay there are two main aspects that must be considered – the distribution of messages (bandwidth), and the distribution of data (storage).

9.5.1 Bandwidth

If all nodes within a network are equally likely to initiate searches, then the overall load on the network will be balanced. However when considering a single search query, the load on the network is not balanced, and hence if searches are not distributed evenly, the load will not necessarily be balanced.

In each of the approaches for blind search, described in Section 9.3.1, the initial branching factor of the broadcast tree is different. However all approaches attempt to construct the widest broadcast tree possible by using the highest branching factor possible; In each approach the root node always sends the message to every node in their routing table, thereafter nodes always forward the message to every node in their routing table within the defined

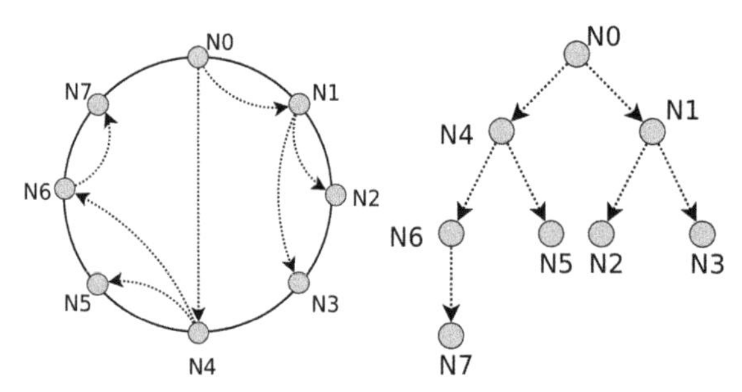

Figure 9.8 Dissemination of an example broadcast message using a binary partition tree in a Chord network, where N is a node and L is the limit.

limits. This results in a shallower broadcast tree, and hence lower average hop count and search latency, however nodes higher up in the broadcast tree are placed under higher load than those lower down.

To combat this issue, Huang and Zhang [9] proposed an adaption of the Efficient Broadcasting algorithm which works by hierarchically partitioning the identifier space into two subspaces to construct a binary partition tree. Upon receiving a broadcast message with a limitation value l, each node n partitions its limited identifier space (n, l) into two subspaces (n, f) and $[f, l)$, where f is the closest finger node to the limitation value l. Next, the node n, selects the first finger found in (n, f) as its left child, and f as its right child. Finally the node forwards the broadcast message with the limitation value f to its left child and with the limitation value l to its right child, as shown in Figure 9.8. This approach has the property that each node forwards the broadcast to at most two children, and hence the broadcast tree is balanced and the branching factor fixed at 2. However, due to the decreased branching factor this approach results in an increased average hop count, and hence increased search latency.

9.5.2 Storage

This section describes the different replication techniques used to distribute data within structured peer-to-peer networks, and the effect these techniques have on the success rate of blind search. In all the techniques discussed, assuming consistent hashing, load should be spread evenly across the network.

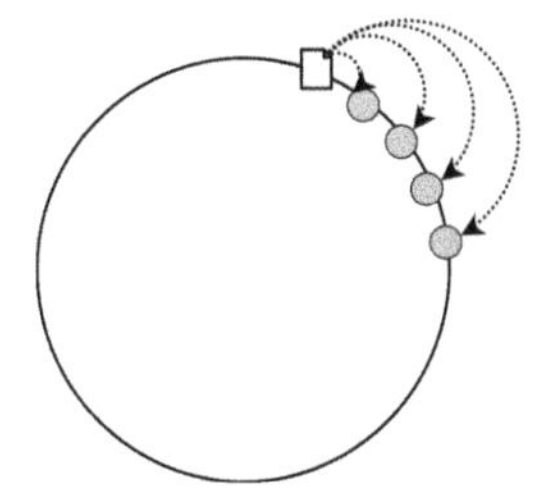

Figure 9.9 Placement of replica in the neighbour replication strategy.

9.5.2.1 Neighbour Replication

Neighbour replication is the term used to describe different replication strategies, such as successor replication and leaf set replication, in which data is replicated at neighbours of the original node, as can be seen in Figure 9.9.

Neighbour replication is the default used by the majority of existing overlays. Neighbour replication is cheap to maintain, since in most DHT nodes already store and maintain a list of their neighbours; for example as each node in a Chord network stores and maintains a list of successors, neighbour replication would mean successor replication. The main advantage to neighbour replication is that it guarantees data will always be in the right location. For example in Chord when a node leaves the network the responsibility falls to its successor – since data is replicated at the nodes successors it will already be in the right place.

However, it should be clear that replicating data at neighbouring nodes is not the best approach to data replication when considering blind search. Due to the nature of the broadcast tree built during blind search, a single node or message failure can cause a whole partition of the tree to not receive the search query, as discussed in Section 9.4. When partitions, in other words groups of neighbouring nodes, can fail, it should be clear that storing all replica in a single partition will not provide any redundancy. The effects of this can be seen in Figure 9.10.

9.5.2.2 Multi-Publication Replication

Multi-Publication replication is different from neighbour replication in that the data is published at multiple keys within the network, each of which is of equal importance. Unlike neighbour replication, because each replica has equal importance, the node inserting the data is responsible for generating each key and sending the appropriate data. There are three proposed strategies for implementing multi-publication replication [13]:

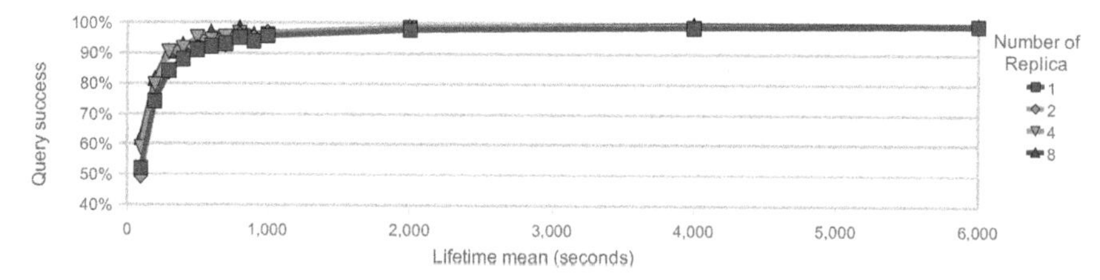

Figure 9.10 Effects of Neighbour Replication rate on the success rate of Efficient Broadcast in 1,000 node Chord network under churn.

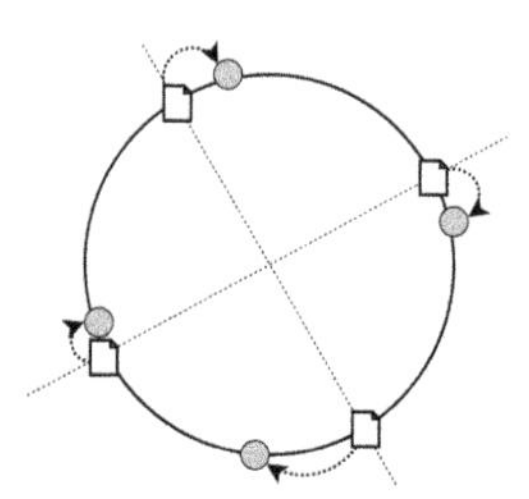

Figure 9.11 Placement of replica in the symmetric replication strategy.

Multiple hash functions. To generate R unique keys R different hash functions are chosen. How replica are spread depends on the functions.

Correlated hashing. To generate R unique keys the numbers $r = 0 \ldots R$ are prepended to the original key before hashing. Assuming consistent hashing, replica should be spread evenly throughout the network.

Symmetric replication. To generate R unique keys, first the hash h is calculated, then the keys are defined as $(h + (r * \frac{N}{R}))\%N$ where $r = 0 \ldots R$ and N is the size of the key space. Using this formula, replica will be spread perfectly evenly throughout the network. An example of replica placement using symmetric replication can be seen in Figure 9.11.

All of the multi-publication approaches have the advantage that replicas are generally spread out throughout the network, and hence partition failures are less of a problem than when using neighbour replication. In fact using symmetric replication with a replication rate of two or above it is impossible for one single partition failure to cause data loss. This property makes multi-publication replication a much better candidate when a network has support for blind search, as can be seen in Figure 9.12.

This method also has advantages that are not directly related to blind search, for example due to the method used to generate the keys, any node

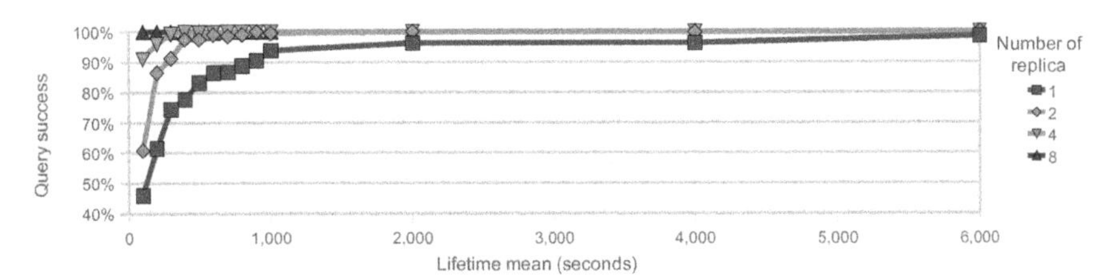

Figure 9.12 Effects of Symmetric Replication rate on the success rate of Efficient Broadcast in 1,000 node Chord network under churn.

within the network is able to generate keys for all replica of a specific data item. This allows the querying node to send multiple parallel requests for the data, or even choose a replicating node in its local proximity by choosing the numerically closest replica key.

Multi-publication replication has a higher success rate when the network is under churn [13]. However, it was also found to have by far the highest overhead, nearly 6 times that of neighbour replication with a replication rate of 8. This is partially due to the implementation of symmetric replication using multiple parallel requests. Parallel requests are required as when a node fails there will be a period of time until the failure is detected and replica is transferred to the correct replacement node. If parallel requests were not issued it would be possible for a failure to occur, requiring the requesting node to calculate the key of the next replica and sent a second request, greatly increasing the search latency. The increased overhead can also be partially attributed to replica maintenance, which for multi-publication replication is more demanding. For each stored replica, the local node must find the location of the next replica and verify if the corresponding node stores a copy of the data. If not, the node must transfer a copy of the data item. Since replica are no longer stored at neighbours, both the verification and transferring of replica become multi-hop operations, increasing their cost. By employing multicast the overhead of parallel requests can be reduced [12], however receiving multiple responses can still be costly.

9.6 Conclusions

As shown in Section 9.2 there are many different types of search queries which can be classed as complex queries, with many different approaches existing for performing specific types of queries. The main advantage to blind search is its ability to support all types of complex queries in a single service overlay, allowing the flexibility of searching for services using the most suitable query type. However it is important to note that in some cases the use of blind search also provides features which are not available using type specific approaches, such as the ability to associate an unlimited number of keywords with a service, or the ability to process regular expressions.

We have reviewed multiple techniques for performing blind search over various different DHTs, however simulation has shown that all approaches share a susceptibility to churn. By choosing an appropriate data replication technique the success rate of blind search can be increased sufficiently to make it a viable alternative even in networks under high churn.

References

[1] Specification of the Bluetooth System (Volume 3), 2007.

[2] M. Balazinska, H. Balakrishnan, and D. Karger. INS/Twine: A scalable peer-to-peer architecture for intentional resource discovery. In *Proceedings of the First International Conference on Pervasive Computing*, pages 195–210, 2002.

[3] I. Baumgart, B. Heep, and S. Krause. OverSim: A flexible overlay network simulation framework. In *IEEE Global Internet Symposium*, pages 79–84, 2007.

[4] M. Castro, M.B. Jones, A.-M. Kermarrec, A. Rowstron, M. Theimer, H. Wang, and A. Wolman. An evaluation of scalable application-level multicast built using peer-to-peer overlays. In *INFOCOM 2003: Proceedings of the 22nd Annual Joint Conference of the IEEE Computer and Communications Societies*, Vol. 2, pages 1510–1520, 2003.

[5] G.D. Costa, S. Orlando, and M.D. Dikaiakos. Multi-set DHT for range queries on dynamic data for grid information service. In *Data Management in Grid and Peer-to-Peer Systems*, pages 93–104, 2008.

[6] S. El-Ansary, L. O. Alima, P. Brand, and S. Haridi. Efficient broadcast in structured P2P networks. In *Proceedings of IPTPS'3: The 2nd International Workshop on Peer-to-Peer Systems*, pages 304–314, 2003.

[7] A. Ghodsi. Distributed k-ary system: Algorithms for distributed hash tables. PhD Thesis, The Royal Institute of Technology (KTH), 2006.

[8] Y. Goland, T. Cai, P. Leach, Y. Gu, and S. Albright. Simple Service Discovery Protocol/1.0, 1999.

[9] K. Huang and D. Zhang. A partition-based broadcast algorithm over DHT for large-scale computing infrastructures. In *GPC'09: Proceedings of the 4th International Conference on Advances in Grid and Pervasive Computing*, pages 422–433, 2009.

[10] S.L. Johnsson and C.-T. Ho. Optimum broadcasting and personalized communication in hypercubes. *IEEE Transactions on Computers*, 38(9):1249–1268, 1989.

[11] Y.-J. Joung and L.-W. Yang. Wildcard search in structured peer-to-peer networks. *IEEE Transactions on Knowledge and Data Engineering*, 19(11):1524–1540, 2007.

[12] M. Kolberg. Employing multicast in P2P overlay networks. In *Handbook of Peer-to-Peer Networking*, X. Shen et al. (Eds.), pages 861–874, Springer, 2010.

[13] S. Ktari, M. Zoubert, A. Hecker, and H. Labiod. Performance evaluation of replication strategies in DHTs under churn. In *MUM'07: Proceedings of the 6th International Conference on Mobile and Ubiquitous Multimedia*, pages 90–97, 2007.

[14] S. Ratnasamy, M. Handley, R. Karp, and S. Shenker. Application-level multicast using content-addressable networks. In *NGC'01: Proceedings of the Third International COST264 Workshop on Networked Group Communication*, pages 14–29, 2001.

[15] P. Reynolds and A. Vahdat. Efficient peer-to-peer keyword searching. In *Middleware'03: Proceedings of the ACM/IFIP/USENIX 2003 International Conference on Middleware*, pages 21–40, 2003.

[16] O.D. Sahin, S. Antony, D. Agrawal, and A. El. Abbadi. PRoBe: Multi-dimensional range queries in P2P networks. *Proceedings of WISE 2005: Web Information Systems Engineering*, pages 332–346, 2005.

[17] M. Schlosser, M. Sintek, S. Decker, and W. Nejdl. A scalable and ontology-based P2P infrastructure for semantic web services. In *P2P'02: Proceedings of the Second International Conference on Peer-to-Peer Computing*, page 104, 2002.

[18] C. Schmidt and M. Parashar. Flexible information discovery in decentralized distributed systems. In *Proceedings of the 12th IEEE International Symposium on High Performance Distributed Computing*, page 226, 2003.

[19] C. Schmidt and M. Parashar. A peer-to-peer approach to web service discovery. *World Wide Web*, 7(2):211–229, 2004.

[20] C. Schmidt and M. Parashar. Enabling flexible queries with guarantees in P2P systems. *IEEE Internet Computing*, 8(3):19–26, 2004.

[21] I. Stoica, R. Morris, D. Karger, M.F. Kaashoek, and H. Balakrishnan. Chord: A scalable peer-to-peer lookup service for internet applications. In *SIGCOMM'01: Proceedings of the 2001 Conference on Applications, Technologies, Architectures, and Protocols for Computer Communications*, pages 149–160, 2001.

[22] D. Stutzbach and R. Rejaie. Understanding churn in peer-to-peer networks. In *IMC'06: Proceedings of the 6th ACM SIGCOMM Conference on Internet Measurement*, pages 189–202, 2006.

[23] C. Tang, Z. Xu, and M. Mahalingam. pSearch: Information retrieval in structured overlays. *ACM SIGCOMM Computer Communication Review*, 33(1):89–94, 2003.

[24] UDDI.org. UDDI Technical White Paper, 2000.

[25] Z. Xu and Z. Zhang. Building low-maintenance expressways for P2P systems. HP Laboratories, 2002.

[26] C. Zheng, G. Shen, S. Li, and S. Shenker. Distributed segment tree: Support of range query and cover query over DHT. In *IPTPS'06: Electronic Publications of the 5th International Workshop on Peer-to-Peer Systems*, 2006.

PART 3

VIRTUALIZATION

10

Computation Mobility and Virtual Worlds – Not Just Where You Work, But How You Work

Shivani Sud[1] and Cynthia Pickering[2]

[1]*Intel Labs, Santa Clara, CA 95054, USA; e-mail: shivani.a.sud@intel.com*
[2]*Intel IT Research, Chandler, AZ 85226, USA;*
e-mail: cynthia.k.pickering@intel.com

Abstract

As computing becomes a more integral important part of our lives and embedded in more things we use, people find themselves dealing with many heterogeneous interfaces, discrete experiences, and disruptions at device boundaries. Mobile phones, mobile internet devices (MIDs), tablets, PCs and TVs are elements of a diverse computing spectrum where disruption becomes more evident to the user, as they transition through mobile and fixed usages of these technologies. For the mobile enterprise user, the need for instant access and smooth transitions among on-the-go and desktop usages are more crucial since they impact productivity.

In addition, global team members who are remote need specialized tools to retain their sense of connectedness. Virtual collaboration tools such as virtual worlds help remote team members stay connected to each other and coordinate their work. We address these two emerging technology trends and discuss a solution for seamless computation mobility based on virtualization of the hardware platform for mobile users. The proposed solution transitions their active compute state among their many compute devices during a work day.

Anand R. Prasad et al. (Eds.), Future Internet Services and Service Architectures, 207–226.

Keywords: MID, virtualization, VM, Intel®, Atom™, mobile, live migration, virtual worlds, collaboration, compute continuum, LPIA.

10.1 Introduction

With the advent of better computation in small form factors, e.g. smart phones, it is becoming easier for users to access the same or similar applications on these devices as on their fully featured PCs. Additionally more devices that users own and interact with, are beginning to have more embedded intelligence. As a result, users have increased expectations of their devices beyond sharing content and media. They want their devices to work together to provide a more integrated, seamless experience that dynamically adapts to their changing needs and environment as they transit between different settings such as home, car, work, or hotel. Current technologies provide users with varied experience across different consumer electronics segments. Users must learn how to interact with different devices through new applications customized for that platform. Fortunately, some of these platform specific issues are being abstracted by the browser based delivery of content, applications, and cloud services. However these approaches are relevant for only newer applications that are written with these new (and still evolving) technology landscapes in mind.

For enterprise users, immersive virtual worlds that allow remote users to interact in ways more reflective of their interactions with the physical world are at the forefront of research. These immersive worlds provide users an experience that is richer than traditional digital computer interfaces and scale the experience beyond co-located objects and people in the physical world. Virtual interactions that closely imitate real world place increased computational demand on platforms. As small form factor and embedded device capabilities increase and they have more processing power, sensor integration, and connectivity to each other and the cloud, they become viable devices to enable more immersive interactions for their users.

Today's mobility and collaboration technologies are frequently used in combination by consumers and business users, e.g. email, calendaring, and social networking on smart phones and tablets. This trend will become even more pervasive over the next decade. In this chapter, we start by discussing the emerging heterogeneity in user devices, trends in convergence of mobile and desktop computing and the various technologies that provide users with seamless access to applications, state, and data across device boundaries. We then discuss another emerging trend of immersive virtual worlds and present

Figure 10.1 Users own and interact with many compute devices everyday.

a "day in the life" case study involving a mobile enterprise user using an immersive virtual environment while on the go. Next we present a solution based on virtualization and live migration techniques that provides seamless transition of this immersive collaboration environment across multiple devices. Finally we wrap up with security implications and a summary of emerging usage and technology trends that are making device boundaries more fluid.

10.2 Heterogeneity in Devices

There is a rise in the number and diversity of intelligent devices that a user encounters during the course of a day, from the desktops and laptops of 1990s and early 2000s. Computation devices targeted for market verticals are becoming a force de jour. Apple Inc's iPad* and iPhone*, and Amazon's Kindle* are changing the way people are getting their news and infotainment. Smart meters and other attempts at adding intelligent control to the home environment, buildings and smart spaces will soon add another dimension of diversity of the compute elements and devices that people will interact with in the future.

Figure 10.2 Users transition through many modes during a day, using their various devices as their portals to the connected digital world.

Since the dawn of the computing era, paradigms have changed dramatically in three major stages. Initially there were many users to a single computer. Then we transitioned to the personal computing era when everyone had their own computer. Today people have multiple computing devices that they carry with them and even more that they interact with during the day. The number of devices per user, complexity, and effort to keep up with these evolutions is increasing.

What users want is access to their information and services from whatever device is most convenient to them at a given time and situation – without being bothered by the limitations of the physical device boundaries. On-the-go users prefer access to their digital world through their smart phones. Those watching television may prefer to treat that as their portal of choice and those in the presence of a desktop or gaming console, may prefer to use those devices.

Current usage paradigms make 24/7 access to online information a set of discrete user experiences that is disrupted at the device boundary. Users need to install and use different applications that are customized for each market vertical. For example, to access media over internet there are different players

a) Different applications for smart phones and PCs

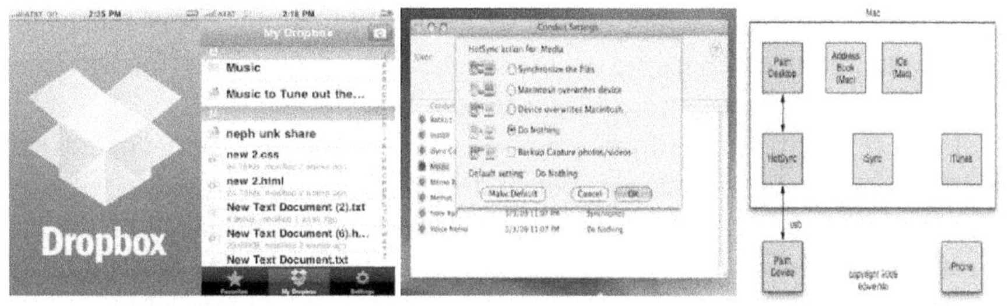

b) File and data sync applications to synchronize user information across devices

Figure 10.3 Discrete user experiences on different category of devices.

for smart phones, desktop and television widgets. When users are running an application on one device and want to just "migrate" their experience on another device that is more convenient, they need to quit their application on their current device and launch the appropriate application on the other device. This not only causes a disruption to the user experience but also to the user's current state information. For example, let us consider an enterprise user who is on a teleconference with remote team members and it is time for him/her to head home. The user needs to hang up the desk phone and dial back into the teleconference from his/her mobile phone. This would cause a disruption not only to him/her but also to other people who will hear tones as the caller disconnects and then rejoins from another phone. It is not only an annoyance but also makes technology intrusive and less user friendly. Ideally these technologies would span the compute continuum, without disruption to the user experience at device boundaries.

10.3 Software Trends in Convergence of Mobile and Desktop Computing

Web browsers have for some time recognized the diversity of the client devices and tried to provide some sort of uniform experience to the users by abstracting the platform differences. It was soon realized they were not able to provide native client performance and richness of client device diversity through this generalized paradigm. Thus, there was an evolution in the plug ins that provided users with better video and graphics in the browser framework (e.g. Adobe flash player*, Microsoft Silverlight*). These Rich Internet Application frameworks and others like Java, JavaScript, and Apple's proprietary framework based on HTML5 now provide faster OS independent development of dynamic web content and applications for different platforms, while providing native client performance.

However, there are still other considerations. The display area sizes are widely different for mobile devices and PCs, as are the available memory and processing power. This affects the latency and performance of the applications and hence the user experience. Additionally the modalities of interacting with the devices are now quite varied, from traditional PC mouse and keyboard to multi touch on mobile devices to gesture interfaces on gaming consoles. The diversity of the peripherals and interaction styles on these devices make writing a general one-size-fits-all application difficult. Additionally, application developers will still need to comprehend requirements for interacting with devices with varying resources.

These limitations of current application frameworks prevent wide deployment of applications written across all client devices, in a build once, deploy everywhere model, where native client platform performance and customized user experience can be achieved. This has led to an increased fragmentation with SDKs proliferating along with device types and categories.

Research in Always Best Connected networks [2] addresses the issues of providing devices with uniform connectivity to heterogeneous infrastructure with various wired and wireless systems such that users enjoy ubiquitous access to applications. It addresses access of hosted services and applications from the perspective of seamless broadband connectivity and how users can be provided with the optimal combination of connection technologies available from the perspective of the mobile device. However, it does not encompass the scope of heterogeneity of user's devices.

10.4 Computation Mobility – Compute Access across a User's Many Devices

Mobility is one of the key emerging technology trends. The desire for access to compute from any place and independent of the client device has manifested itself in various research projects and products in traditional PC based clients. Thus we looked for the techniques that have been explored both in research and products that could be extended to address emerging reality that users have multiple devices that they use as a portal to access information.

Much of this existing work has been based on the thin client and server based computing architectures, where user applications and their state are stored on some centralized servers. The networked thin client device needs minimal processing power to be able run a client program that provides user with access to the centralized server (e.g. X Client, VNC, remote desktop).

The response time and bandwidth requirements of these desktop virtualization solutions vary based on the protocol and compression algorithms they use and whether they are implemented in the user space or as kernel modules. Some of the protocols in this space are Virtual Network Computing (VNC) [4] (from Olivetti Research), X Windows System [3] (originally from MIT), Remote Desktop Protocol (Microsoft), Remote Frame Buffer Protocol [5] (Olivetti Research), Apple Remote Desktop Protocol, and Independent Computing architecture (by Citrix).

Client devices need network connectivity to access these services and user experience suffers if there are latencies in the network connection, and no functionality is available to the user in absence of connectivity.

Web desktops or webtops are another category of online desktop environments based on web servers. They provide access to the desktop environment from inside a web browser that provides access to web services, client server applications and rich internet applications. This is similar to the thin client described above in terms of remote computation and requirement for network connectivity to access the user data, configuration and compute state from the servers hosting these web desktops. They have not been as widely used or popular as the traditional thin clients, because they require applications to be re-developed for use inside browsers or similar client applications. This approach is also not extensible to legacy applications.

Alternative solutions based on virtualization of the desktop environment made it possible for enterprise clients to save their virtual desktops to servers and get access to them from another client device with access to the network.

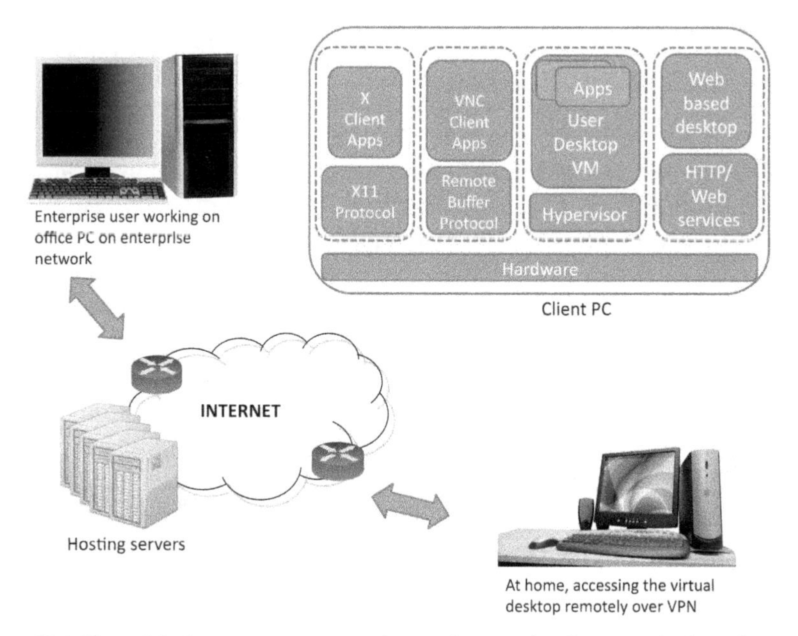

Figure 10.4 Hosted desktops: server computing modes, ranging from terminal services using remote buffers to hosted virtual machines and web based desktops.

VMware, Citrix and Parallels are some of the solution providers in the virtual desktop infrastructure space.

Process migration is another area of research interest in the operating system community [3, 7–9]. There has been less success in their wide deployment as these solutions tend to be either point solutions, require rewriting of applications using some programming language or framework, or are restrictive in other ways.

10.4.1 Virtualization on Smart Phones

Virtualization is not widely supported in processor hardware that powers the majority of today's smart phones. This makes it challenging to provide virtualization based computation mobility solutions on MIDs and smart phones.

However, virtualization brings several benefits to this market. The foremost is that of secure isolation of service provider space on the smartphone providing protection to the provisioned data and also for other trusted services and digital rights management. This secure isolation allows creation of an open space for users to install another operating system and applications of choice on the platform without being restricted only to the applications and services available from their service provider and its supported app-stores.

For enterprises, the diversity in the hardware of mobile devices that the end users want to use to access the enterprise network has exploded from conventional laptops to smart phones and now to netbooks and tablets. Users also do not use just one mobile device; they use and want to have the flexibility to use multiple devices, depending on what is convenient for their mode of activity. In office or home a laptop or desktop may be more suitable. When travelling for sales or customer visits, a tablet or netbook may be more appropriate (because of its form factor), and while going out for dinner, smartphone may be the most convenient portal to access information. IT departments end up having to train their staff to keep up with the new and legacy hardware that they need to support for their users. This either limits the IT policies and support to a restricted subset of devices, or requires additional IT staff and hence bigger budgets. Users get frustrated because they cannot use the device of their choice. With virtualization, enterprises can provision an enterprise client on the devices the users bring in, and not have to deal with the maintenance of the hardware diversity. Users have the freedom to use the hardware of their choice that enables access to personal and enterprise services in isolated containers on the same device.

Thus, hardware support for virtualization can be a huge benefit, as it overcomes the cost of software emulation on these mobile phone processors that are constrained for compute resources compared to their desktop counterparts. This makes hosting multiple VMs on these devices a reality and enables multiple usages on the same device.

10.4.2 Virtualization for Computation Mobility

We described a unique solution that leverages hardware virtualization extensions in Intel Atom powered MIDs for a seamless transfer of user state with other IA based PCs [1]. Atom is a Low Power Intel Architecture (LPIA) System-on-chip (SoC) microprocessor from Intel that supports the x86 instruction set designed for a target market of low power devices like MIDs. Live migration techniques have been used typically in the server environment for purposes of server consolidation and reliability [11] and are typically not used in client devices.

VM live migration works by abstracting away the differences of the underlying hardware to present an abstract notion of devices to the guest OS

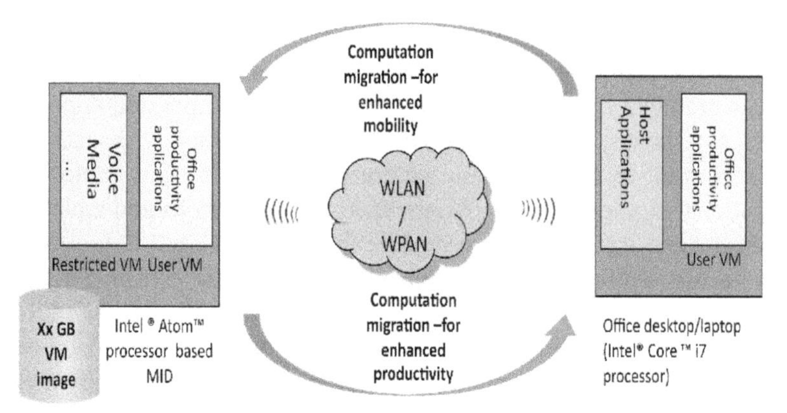

Figure 10.5 Using virtualization as a container for isolation of user and ISP space and also for migration.

running inside the VM. This makes it possible for it to map the state of the virtual devices in the guest OS to the hardware resources on the host.

The process of migration re-establishes the abstraction mapping between the actual hardware resources on the target and those required by the guest VM. It also re-establishes the connections to the hardware resources on the target and the network simulating the continuity of resource access to the end user.

Our proposed solution, based on VM live migration techniques, does not require a centralized server for hosting the virtual image and uses a local (W)LAN connection for migrating the users' active state between a MID and a PC without causing a disruption to the user state. MIDs provide the user with stand alone access to applications (if so natively supported) on their device, unlike cloud based or hosted services that depend on broadband connectivity.

So, users can enjoy the ease of mobility with complete access to their desired applications on the MID. They can also enjoy enhances productivity from opportunistically available devices with better processing capabilities by migrating executing binaries to them. And then, when they need to be mobile again, they can seamlessly migrate their active state back to their MIDs and continue.

We used hardware virtualization feature support on the MIDs to create a container for isolating the user OS and applications. The host OS of the MID provides the basic MID operation and functionality. The VM is where the user can run their favorite OS with its native applications which may not be ported to the host OS. With this solution, the user is not limited to the set of applications available on the host OS which may be controlled by the service provider or manufacturer of the device.

We proposed using the VM as a container for isolation of user application state, and using live migration techniques as discussed above, to migrate the applications from the MID to the PC with minimal disruption to the user.

We used the VM image over an NFS shared drive from the MID device. This reduces the amount of data that needs to be migrated from the source to the target device to get the migrated process active and running. The only information that needs to be transferred is the memory allocated to the guest OS. With the high bandwidth available in LAN technologies this time is almost seamless to the end user. Additionally, the live migration algorithm uses a cyclic process of copying memory pages from the source to the destination host iterating through only the dirty (write accessed) pages in subsequent iterations. The iterative process stops at a low fixed threshold, and the remaining dirty pages are copied en masse to the target, where the process resumes. This ensures a short down time such that the transition feels relatively seamless to the user.

Later in Section 10.7 we provide an example case study using a 3D virtual collaboration environment.

10.5 Immersive Virtual Worlds and Their Ubiquity

In this section, we discuss the other emerging trend of collaboration applications, specifically immersive virtual world based collaboration tools.

Immersive Virtual Worlds in the consumer gaming and social computing arena have user bases on the order of tens of millions to hundreds of millions and are estimated to be a $40 billion industry [13]. These applications typically require high performance compute and graphics, broadband connectivity, and sizeable memory footprint.

Meanwhile 8.75 million iPhones users [13] currently enjoy over 200,000 apps [13] including games, entertainment, social networking, and productivity applications.

Social computing applications such as Facebook with approximately 500 million users [14] and Twitter with over 75 million [14] are increasingly available on small form factor devices out of the box.

As technology for small form factor devices evolves, virtual worlds that previously ran only on more powerful PCs will become viable offerings on these platforms.

10.5.1 Virtual Worlds and the Mobile User

We conducted a case study to explore virtual world usage by knowledge workers in a business context, and implemented a research prototype for mobile usage of virtual worlds. Our technical approach could be equally applied to consumer scenarios.

In an enterprise with an inherent culture of global teaming, innovation, and meetings, information visualization and rich interactivity in virtual workspaces can improve the speed of making the right decisions. Our approach began by studying early adopter usage of today's immersive virtual worlds systems to increase understanding, readiness and acceptance of the potential business usages. We then looked at what would be needed to architect these systems to scale up/out to pave the way for next stage technology transfer and adoption by business users.

Based on prior research conducted within Intel we identified several unique requirements for global team collaboration [15]. Our findings included that two-thirds of Intel employees work on distributed teams and the same percentage also work on multiple teams that may or may not have overlapping interests [17]. These and other data led us to seek a solution that would seamlessly support the complexities discovered while enhancing team and personal productivity. To adequately represent the complexities, we needed an object-oriented information workspace where the use of a 3D user interface allows users to retain multiple contexts and share across them. We conducted an environmental scan of commercial and open source virtual worlds products oriented to enterprise collaboration scenarios and evaluated them against the following requirements.

The business oriented virtual teaming environment needs to support expressive interaction with team members via multi-media "social" channels such as voice, video, text and avatar gestures, team content creation and manipulation. It should persist the work context of the team between meetings, and allow work to continue outside of meetings. As most people participate on more than one team, it should be easy to move between multiple team work spaces and share information across them. Team members should be able to enter the team space at any time and notice if other members are present, and optionally interact with them. Many of the things that a team would do together if they are physically co-located should also be supported in the environment: whiteboard sketching, approach a team member and talk to them, show presentations or documents on a shared screen, brainstorm ideas on a wall of post-it notes, look at a video in real time, or interact with a simulation,

Table 10.1 Business usages for virtual worlds.

Business Virtual World Usage	*Description*
1. Knowledge Worker Project Team Room (see also Figure 10.6 and Section 10.7)	Persistent workspace for distributed cross organizational team members working together on a project
2. Knowledge Worker Staff Meeting	Persistent meeting place for distributed members in the same organization who meet regularly as a staff to coordinate
3. Training Scenarios	1. Simulation based training 2. Training curriculum/content development
4. Factory Operations & Equipment Bay Monitoring (Baywatch)	A semi-conductor manufacturing scenario that uses a 3D layout of manufacturing equipment bays, status boards and visuals used to monitor equipment operation status.
5. Product Design & Engineering "war room" that included Visualization for Collaborative Decision making	A special case of the knowledge worker project team room that also incorporates use of visual analytics to aid in dynamic allocation of data center compute resources to different design projects.

a model of a physical object or an application. With increasing mobility in the workforce, the requirement to support virtual world interactions from small form factor devices is also highly desirable.

To better understand these business requirements, we looked at five different types of teams representing common business usages that are described in Table 10.1. The first three usages include general capabilities for running team projects, organizational staff meetings, and conducting training sessions while the last two usages are more specific to Intel's business. We engaged with real teams who showed interest in trying our 3D virtual world application for real collaboration, and constructed a dedicated persistent virtual workspace for each team, using pre-constructed templates that could be easily tailored to the teams' needs via drag and drop.

We used an iterative approach for testing the capability with our volunteer teams in the above usage categories over three successive releases of a virtual world product selected from an environmental scan of vendors suited to use in a business setting. User feedback from the prior release was incorporated into each subsequent release. In some cases we used the same team across releases and in other situations, a new team volunteered. All teams were observed by a Human Factors Engineer (HFE), in the virtual world. Some HFE lab testing was completed as well. Each team had an initial orientation session to install the client software on their PCs, basics training, and hands on experimentation.

Teams also appointed a "champion" to organize the team workspace and lead meetings. In most cases this was the existing team lead or facilitator who already performed a similar role, but using different tools. The HFE continued to observe the teams over several meetings. During testing, we also used processor, graphics, audio and network performance metrics and indicators built into the client and logs on the server. On roughly a quarterly basis, findings were summarized and communicated to the vendor using the technology itself. A new release from the vendor triggered the beginning of a new test cycle.

It was no surprise that, during our testing, team members were often on travel or working from home especially if they were located in a different geographical region from the rest of the team. In these situations, access to the virtual world from a range of mobile devices would have been more convenient. For highly mobile workers, such as the sales force or executives live migration would have reduced disruption when switching to a different device during a meeting. We describe this in greater detail next.

10.6 Case Study: A Virtual World On-the-Go

To further explore the scenario of an on-the-go "Knowledge worker" collaborating remotely with a global team, we extended the proposed virtualization and live migration solution [1] with the virtual world collaboration environment in a research prototyping activity.

The "Knowledge Worker Project Team Room" use case described in Table 10.1 and shown in Figure 10.6 provided the basis to explore mobile usage and seamless session transfer across multiple devices. The Virtual Project Team Room shown in Figure 10.6 portrays a virtual office setting with desks, meeting rooms, and avatars representing knowledge workers on the same project. Each person usually works on more than one project with different people and different deliverables. In the Virtual world environment these projects would be represented as different scenes with unique work in progress visible on the walls. Team members go to a project's virtual room when they wish to work or meet with others about the project. There they can talk, text chat, edit and create deliverables, brainstorm, track progress and have meetings. They can easily resume work in progress because it is readily available and easy to see what has transpired since the last work session. In the real world, people do not stay in a single location or use a single compute device to do all of their project work and meetings. Throughout their day, the mobile knowledge worker moves from home to office and other locations.

Figure 10.6 Knowledge worker project team room.

While at the office they may take meetings and work on projects at their desk, in conference rooms, or in the cafeteria. Ideally they could transit among different locations and continue to collaborate while on the go, and then take advantage of the best device(s) at the destination, without disrupting the ongoing meeting. For example, the US employee may take an early meeting with Europe in the virtual world on their home PC, then migrate the active meeting session to their Smart Phone or MID for hands free transit to the office in time to make their next meeting.

When they get to the office they can supersize the meeting experience by migrating it to their more powerful Office PC with its large High Resolution Monitor. Later they may need to move to a conference room before the current meeting is over in time to make their next meeting; transferring to the MID or laptop helps achieve this goal. In the conference room, a large full resolution display panel pairs with the MID to provide greater visibility.

Our research prototype enables knowledge worker mobility in the real world while the team uses a virtual world to conduct business. The research prototype, consisted of a MID, PC/Laptop, wireless network, and displays. It included virtualization on the MID to host the 3D virtual world, and wirelessly transferred the virtualized session from the MID to execute on larger desktop computers and displays (see Figure 10.7). Using an Atom based MID allowed us to leverage hardware virtualization acceleration, and significantly improve application runtime performance on the mobile device. The cap-

Figure 10.7 Virtual world to go for a mobile knowledge worker.

ability is secure, convenient and enables mobile activities to move between physical locations.

10.6.1 Test Setup and Observations

We extended our test setup[1] as described in [1] to run a 3D Virtual Worlds collaboration application, the Teleplace* client [12], which relied on OpenGL for rendering the virtual world scenes. Initially we used the default client configuration for hardware accelerated 3D rendering.

When we ran our migration with this setup, we found that live migration failed. This is because the graphics adapters' capabilities of the two devices were widely different. The MID was using the integrated graphics module on the Atom SoC, whereas the PC had a discrete graphics processor with high end display capabilities.

When an application in the VM uses hardware assisted 3D rendering (like ours), it is dependent on the actual hardware and that state information needs to be translated into state information for the OpenGL library of the other adapter in the process of migration. Since we were using OpenGL based 3D application, one approach to do this was to collect the history of all OpenGL

[1] Test setup used an Intel Atom Z530 based Netbook (Fujitsu Lifebook U820*) as the mobile device, with a 1.6 GHz processor, 512 K cache and 1 GB of memory. The PC was a ThinkPad T60* notebook computer based on an Intel T2400 Core Duo 1.83 GHz processor with 2 MB cache and 2 GB of memory. Both devices are connected to a DLink DGL 4500* 802.11 Draft N WLAN AP. The host OS on both the T60* and Lifebook* is the Ubuntu 8.10 desktop edition. The guest VM is also created using the Ubuntu 8.10 desktop edition on the Lifebook. We set up the NFS server for hosting the VM image on the Lifebook*. The T60* has an NFS client installed and can access the VM image from the NFS share on the Lifebook*.
*Other names and brands may be claimed as the property of others.

commands (since the booting of the guest OS) ship them to the remote migration target, and let the OpenGL library on the guest OS of the target iterate through those commands to recreate the same state on the remote target's graphics adapter.

However, this approach is a point solution that would work only for OpenGL applications. Also, passing the history of OpenGL commands collected in the guest VM adds to the amount of data that needs to be migrated to the target device, and hence, to the live migration time. A large volume history could cause the migration to fail.

Consequently, we chose another alternative. We achieved the flexibility required for the migration process across disparate platforms by using software rendering instead of hardware assisted rendering. Even though the performance dropped from 35 fps to 20 fps on the PC with software rendering of the OpenGL commands, we were able to perform a seamless migration of the Teleplace* application between the MID and the PC.

In addition, this experiment also re-iterated our observations in [1] about the need for adapting to different hardware resources for an application that a user will use across client devices with varying capabilities.

10.6.2 Security – The Need to Dynamically Adapt Trust to Fluid Boundaries of Compute

As device boundaries blur and users' experiences extend to include various processing resources that may or may not be owned by the user, there are additional privacy and security concerns beyond those that need to be addressed from the single device. This creates the need for a conscious effort to categorize information and device types in order to provide layered security. These layers can span the spectrum from complete anonymous sharing (e.g. location based services based on proximity) to completely regulated opt-in models via third party mediated and/or temporal and spatial context based security. Mobility adds another dimension to the security challenges because of remote access, possible device loss, etc.

Devices can be categorized as the devices that are owned or trusted by a user, and the devices that are unknown and used on an ad-hoc basis. Then, for enterprise users we can apply policies based on the type of information and business rules for sharing.

A User's Many Devices Ownership of the devices reduces threats and intent of malicious attack introduced by multi-tenancy. It is easier to establish security among single owners' devices through introduction – e.g. shared tokens, to protect channels of communication and migration. Thus one can build on existing trust relationships.

Ad-Hoc Connections with Unknown Devices In ad-hoc collection of devices that include shared resources, the surface area of attack is increased because of possible conflict of interests or malicious intent. Hardware and software attacks are possible (e.g. man-in-the-middle, replay, video buffer scanners, sniffers) and privacy may be compromised.

In these cases there is the need for establishing trust through either an established third party (e.g. using a Private Key Infrastructure (PKI) mechanism) or through other means of introduction. It should be possible to form these trust relationships dynamically because of the diversity and number of processing elements that the user may encounter and these mechanisms will need to be easy to use to ensure adoption by users.

Virtualization and Security We discussed in a previous section how virtualization provides a mechanism for securing the content and application provisioned by a service provider from the user downloaded applications. In addition, when we use virtualization and live migration techniques for migration of user computation across to, say for example, a kiosk computer, it allows the user to leverage compute resources in infrastructure without leaving a disk footprint on the loaned computer, as the technique can be configured to only share the disk image over an NFS share. This reduces the possibility of inadvertently left behind VM images being misappropriated and misused.

10.7 Conclusion

Increases in complexity, number, and heterogeneity of user client devices along with users' desire for being connected at anytime from anywhere and from any device they choose as their portal to the connected digital world, drive the need for technologies that enable these usages without disrupting user experience.

There are fragmented software approaches to address the increasing heterogeneity in client devices, with some approaches trying to provide anywhere anytime access. These approaches provide access across devices but disrupt the user experience and require centralized services.

We show a method that uses VMs as a container for user applications to provide not only secure isolation between the user and service provider execution spaces, but also a mechanism to provide computation mobility across the user's many devices using live migration techniques, over a local wireless connection, with no additional broadband access required. It also ensures that the user has access to the standalone applications on the mobile device in the absence of connectivity.

In another technology dimension, virtual worlds are useful, versatile collaboration tools, which can significantly lower travel costs by reducing the need for face-to-face meetings. Mobile small form factor devices are on the rise in both business and consumer settings and are nearly capable of supporting the immersive interactive user experience in virtual worlds while on the go.

In our chapter we discussed the emerging trends in the computing industry that are making device boundaries fluid. The industry will need to adapt and evolve the proximal and remote infrastructure compute elements to deliver the best integrated user experience, while making the boundaries transparent to the user. Our research prototype demonstrated this in the context of another emerging usage trend by enabling the seamless transitioning of a 3D collaboration application across devices. The prototype leveraged unique hardware virtualization extensions in the MID category of devices and the commonality of industry standard bus architecture.

References

[1] S. Sud et al. Dynamic migration of computation through virtualization of the mobile platform. In *Proceedings of ICST MobiCASE 2009*, San Diego, October 2009.

[2] V. Gazis, N. Alonistioti, and L. Merakos. A generic architecture for 'always best connected' UTMS/WLAN mobile networks. *International Journal of Wireless and Mobile Computing*, 2(4), 2007.

[3] Standard X windows releases hosted by freedesktop.org (www.x.org).

[4] T. Richardson, Q. Stafford-Fraser, K.R. Wood, and A. Hopper. Virtual network computing. *IEEE Internet Computing*, 2(1), January/February 1998.

[5] T. Richardson and J. Levine. Remote Frame Buffer protocol, IETF Draft, http://datatracker.ietf.org/doc/draft-levine-rfb/.

[6] S. Osman, D. Shubhraveti, G. Su, and J. Nieh. The design and implementation of Zap: A system for migrating computing environments. OSDI, 2002.

[7] T. Boyd and P. Dasgupta. Process migration: A generalized approach using a virtualized operating system. In *Proceedings of the 22nd International Conference on Distributed Computing Systems (ICDCS 2002)*, Vienna, Austria, July 2002.

[8] J. Casas, D.L. Clark, R. Conuru, S.W. Otto, R.M. Prouty, and J. Walpole. MPVM: A migration transparent version of PVM. Technical Report, CSE-95-002, 1995.

[9] D. Milojicic, F. Douglis, and R. Wheeler. *Mobility: Processes, Computers, and Agents*. Addison Wesley Longman, February 1999.

[10] J. Smith. A survey of process migration mechanisms. *ACM SIGOPS Operating Systems Review*, 22(3), July 1988.

[11] C. Clark, K. Fraser, S. Hand, J.G. Hansen, E. Jul, C. Limpach, I. Pratt, and A. Warfield. Live migration of virtual machines. In *Proceedings of the 2nd Conference on Symposium on Networked Systems Design & Implementation*, Volume 2, 2005.

[12] Teleplace* – Virtual places for real work, www.teleplace.com.

[13] B. Reeves and J. Leighton Read. Total engagement – Using games and virtual worlds to change the way people work and businesses compete, Harvard Business Press. Nov 2009Apple by the Numbers: iPhone sales more than double, Mac holds up; Rusli E., TechCrunch, http://techcrunch.com/2010/04/20/apple-by-the-numbers-iphone-sales-more-than-double-mac-holds-up/, 20 April 2010.

[14] iPhone Apps Store, http://www.apple.com/iphone/apps-for-iphone/.

[15] M.G. Siegler. Facebook movie poster announces 500 million Facebook users before Facebook does. TechCrunch, http://techcrunch.com/2010/06/19/facebook-movie-poster/, 19 June 2010.

[16] S. Gaudin. Twitter now has 75M users; most asleep at the mouse. *Computer World*, http://www.computerworld.com/s/article/9148878/Twitter_now_has_75M_users_most_asleep_at_the_mouse, 20 January 2010.

[17] C. Pickering and E. Wynn. An architecture and business process framework for global team collaboration. *Intel Technology Journal*, 8(4), Toward the Proactive Enterprise, November 2004.

[18] K. Chudoba, E. Wynn, M. Lu, and M.B. Watson-Manheim. How virtual are we? Measuring virtuality in a global organization. *Information Systems Journal*, 15(4), 2005.

11

Evolution of Peer-to-Peer and Cloud Architectures to Support Next-Generation Services

Luca Caviglione[1] and Roberto Podestà[2]

[1]*Institute of Intelligent Systems for Automation, Genoa Branch, National Research Council of Italy (CNR), Via Opera Pia 13, 16145 Genova, Italy;*
e-mail: luca.caviglione@ge.issia.cnr.it
[2]*Nice-software Inc., Via Milliavacca 9, 14100 Asti, Italy;*
e-mail: ropode@gmail.com

Abstract

Nowadays, the Internet is widely populated with new services and architectures. Specifically, those relying upon the peer-to-peer (p2p) communication paradigm allow to exploit overlay networks. The latter are an important technological component to achieve network independency, and they realized capabilities in binding resources spanning over the Internet. Besides, cloud architectures are emerging as another way to manage services. New scenarios are continuously evolving both in terms of platforms and infrastructures. Such advancements impose the challenge of harmonizing the considerable amount of existing "clouds" into a global one. Therefore, a convergence between the two "worlds" is not only desirable but the most promising direction towards a new generation of a global computing infrastructures. In this perspective, this chapter discusses p2p architectures to provide next-generation services, such as resource reservation to support large-scale data streaming. Also, it will show new advancements in clouds in terms of Application Program Interfaces (APIs), service abstraction features and scalability properties.

Anand R. Prasad et al. (Eds.), Future Internet Services and Service Architectures, 227–246.

Keywords: Service Oriented Architectures (SOA), peer-to-peer, overlay, mash-up.

11.1 Introduction

The Internet is becoming ever more the preferred playground to develop and deploy new services and architectures. For instance, *peer-to-peer* (p2p) allows to create *overlay networks* and to bring scalability to the next step, also relying upon end-users' resources. Nevertheless, they guarantee *network independency*, which enables new paradigms to effectively reflect in services, as well as to overcome in a transparent way limitations imposed by the underlying physical deployment (e.g., the lack of multicast support). In this perspective, numerous techniques exploiting the usage of an overlay, to superimpose a service capable to bypass such constraint, have been under investigation. One of the most relevant efforts concerns Scalable Adaptive Multicast (SAM) techniques [1]. SAM has been mainly developed within the SAM Research Group (SAMRG) of the Internet Research Task Force (IRTF). Its core activity aims at investigating several multicast protocols, such as: Application Layer Multicast (ALM), Overlay Multicast (OM), and hybrid techniques (i.e., mixing overlay-based techniques with standard IP multicast if available).

However, multicast is only one of the numerous features needed in the modern Internet to enable the development of next-generation services, which are often introduced to accomplish requests of more demanding users. In fact, nowadays Internet adopters demand a rich set of "additional" requirements such as: (i) the possibility of accessing an application (or a functionality) also while moving; (ii) having an infrastructure to ensure the proper behavior when in presence of intermittent connectivity, for instance via caching; (iii) support some data delivery in a real-time fashion, for instance, for A/V communications or for receiving multimedia contents. In this perspective, one of the major issues concerns the need of providing some kind of Quality of Service (QoS) guarantees when using network applications.

Besides, cloud architectures are emerging as one of the most interesting novelties of recent years. More than a decade of research on grid computing and its intersection with the Service Oriented Architecture (SOA) style resulted in providing computational power as an *utility on demand*. The scenario continuously evolves with new *platforms*, *infrastructures* and *services*, which are appearing and growing within the Internet. The next challenge will be how to harmonize the considerable amount of existing clouds into a single global one. The p2p technologies provide a proven capability in binding resources spanning over the Internet, as demonstrated by many successful applications. Apart from the ubiquitous file-sharing networks, we can mention the distributed look-up infrastructure exploited by popular Voice over IP (VoIP) applications (see, e.g., [2] and references therein).

Therefore, a convergence between p2p and cloud is not only desirable but it appears as one of the most promising directions towards a new generation of a global computing infrastructure. In fact, p2p platforms and overlays can provide a *backbone* to "mash-up" the services deployed in the various clouds. Furthermore, p2p also exhibits a high resistance against *churn*, which is the continuous process of nodes' arrival and departure, as well as high scalability properties. Even if we can rely on a variety of "legacy" p2p frameworks, the complex and heterogeneous nature of the Internet, both in terms of infrastructure and amount of layered/overlapped services, accounts for additional design efforts. In this vein, p2p architectures to support the creation of next-generation services can be empowered with optimization techniques.

The remainder of the chapter is structured as follows: Section 11.2 portraits next-generation services deployable through p2p and discusses how-to perform optimization through an example. Section 11.3 discusses the integration of cloud architectures by the mash-up of existing services on top of p2p platforms and overlays, while Section 11.4 concludes the chapter and proposes future directions and key research to be done.

11.2 Using p2p to Deploy Next-Generation Services

As hinted in the introduction, the p2p communication paradigm has been widely adopted for the following reasons: (1) it allows to develop an overlay network to be "juxtaposed" over the physical one; and (2) it offers a systematic approach to also exploit end users' resources to keep the resulting architecture running.

Concerning (1), in the past the main usage of overlay-based technologies was aimed at recovering constraints and limitations imposed by the underlying network architecture. The major problems encountered by modern deployments, which are "classically" mitigated by relying upon overlays networks and the p2p paradigm, are:

- the presence of Network Address Translation (NAT) devices reduces the end-to-end transparency of transport level connections. Overlays can be adopted to offer an isolated playground to develop traversal techniques, especially by offering built-in mechanisms, e.g., rendezvous nodes and connections relayers;

- the growing demand of multimedia and real-time constrained contents dramatically increases the need of having IP multicast available at the network level. Even if both IPv4 and IPv6 already offer multicast support, it is seldom employed. For the case of IPv4, network operators do not enable it to their customers for security concerns. Instead, for the case of IPv6, its slow diffusion does not account for a definitive solution. In this perspective, overlay can be adopted to exploit the aforementioned ALM-based techniques;
- modern network deployments are highly heterogeneous, also joining wired and wireless parts, which exhibit different characteristics, e.g., in terms of bandwidth and delays. Overlays can be used to provide a unified manner to offer basic network-centric services, such as QoS support. They can also be used to compensate transparently intrinsic behaviors of the underlying data bearer, as well as to provide applications in a "steady" environment;
- today mobility is very common among users, reflecting in nodes entering/leaving the system continuously, due to intermittent connectivity issues affecting wireless transmission (especially in urban areas). Since p2p architectures have an intrinsic resistance against churn, they are very effective against such behavior.

Regarding (2), the main advantages of adopting the p2p approach are as follows:

- modern services do rely upon sophisticated software and network infrastructures rising costs due to hardware and bandwidth requirements. Moreover, calibrating the needed amount of resources is a complex task, which increases the risk in the start-up phase of a new next-generation service. Conversely, in p2p, users do not solely request a service, but they also contribute, distributing the overall system infrastructure; this allows to flatten costs by reducing hardware and bandwidth requirements at the service provider's side. In fact, the p2p approach intrinsically encourages self-organization, thus adjusting the needed amount of resources on-the-fly, and it may avoid over/under provisioning of resources. We point out that a user can be everyone interested in joining a p2p-enabled service, e.g., an enterprise or a service provider (SP);
- p2p-based services do not have a single point of failure. It is possible to augment, via intrinsic replication of functionalities, the robustness of

the service infrastructure. This reflects an increased resistance against attacks such as Denial of Service (DoS).

With such desirable properties, overlays are fundamental building blocks for developing applications and services for the future Internet. In addition, they are naturally placed at the application layer, thus they can be easily joined (or unified) with a pre-existent SOA. Summing up, by adopting the p2p paradigm, applications can "see" a *service* rather than a *network*. For instance, in the case of ubiquitous p2p file-sharing applications, users can simply "perceive" a service to retrieve a desired content. Nevertheless, the user is not able to reach the blueprint of the system, i.e., centralized or distributed. This allows "specialization" of the system, to cope with specific requirements, without exposing resources to the users, or more important, to developers.

11.2.1 Overlays to Support Data Streaming in Interactive Grids

As a case study we present how to use *optimization* and overlay techniques to perform data streaming over next-generation interactive grids. Also we propose some numerical results, partially taken from [5], where a detailed mathematical framework is also discussed, to prove the effectiveness of the approach.

The introduction of new services has been also a hot topic in the field of next-generation grids. One important new service concerns the support of instrumentations and it implies issues about real-time communication. In fact, the real-time access to instrumentation was already envisaged in the original grid vision [6]. Advancements in the integration of devices, together with the availability of standardized signaling solutions [7], concretely raise the problem of supporting such a "vision" with proper network requirements. This requires a non-trivial support of QoS from the underlying network architecture, and a careful planning to avoid the under/over provisioning of resources. Therefore, this "enriched" grid vision needs sophisticated and strict QoS guarantees. However, the grid being a highly distributed framework, it can be also built on top heterogeneous network deployments. Some practical issues preventing the guarantee of QoS for such applications are: (i) the grid could span different autonomous systems (ASs) or ISPs, having non-uniform service level agreements (SLAs), thus impeding a global traffic engineering strategy (e.g., by using standard signaling solutions such as the MPLS-TE or RSVP-TE); (ii) the underlying infrastructure (or well-defined "isles") could not support QoS, being straight best-effort technologies; (iii) grid nodes are often deployed to the border of the network, thus preventing intervention on capacity allocations within the data bearer (e.g., adjusting RSVP-based reservation disciplines); (iv) the heterogeneity of the infrastructure in terms of transmission technology (e.g., mixed wired and wireless communications) account for difficulties in exploiting traffic management in a unified manner; (v) for security or management reasons, a higher management layer could be required, e.g., in order to uniform the access to resources, or to provide a unique Application Program Interface (API) or infrastructure.

To cope with such issues, we exploit an optimized overlay to merge instrumentation and grid nodes. Its aim is to deliver the data flow produced by an instrument in a real-time fashion to a given set of grid nodes and end-users, having the available resources optimized. All the involved components are connected through an IP network, i.e., the Internet. The reference scenario is depicted in Figure 11.1, where multiple domains with different QoS requirements could be "unified" by exploiting the isolation properties of the overlay acting as a kind of "specialized" Content Distribution Network (CDN).

The core architectural components are:

- Grid Node (GN): this is responsible both of forwarding the stream through the distribution overlay tree, which is composed of the interconnection of different GNs (depicted in Figure 11.1 with bold lines), and of serving end users. We point out that a GN could be deployed for: providing an entry point for the source; solely forwarding the stream to remote GNs (i.e., it is a simple data transit point); serving end users or a mix of the previous points.
- Decision Maker (DM): this is the tracker-like entity that could be merged within a GN, in charge of optimizing the usage of resources, by exploiting different performance costs (e.g., to maximize the allotted user population). It exchanges status information with GNs and issues back to them the optimized strategies to be executed. Relying upon a tracker-like entity has the following advantages: it enables the easy establishment of overlays *à la* BitTorrent; the decision is taken in a unique point at the border of the network allowing to change service policies quickly. Lastly, the content provider can be co-located within a GN.
- Users: they consume the stream on their host or appliances. They could be collaborative environments adopted for displaying results, or machineries running services for data processing purposes. We suppose that the produced data flow has significance only if received in a well-defined moment in time. For instance, this could be the case of an instrument

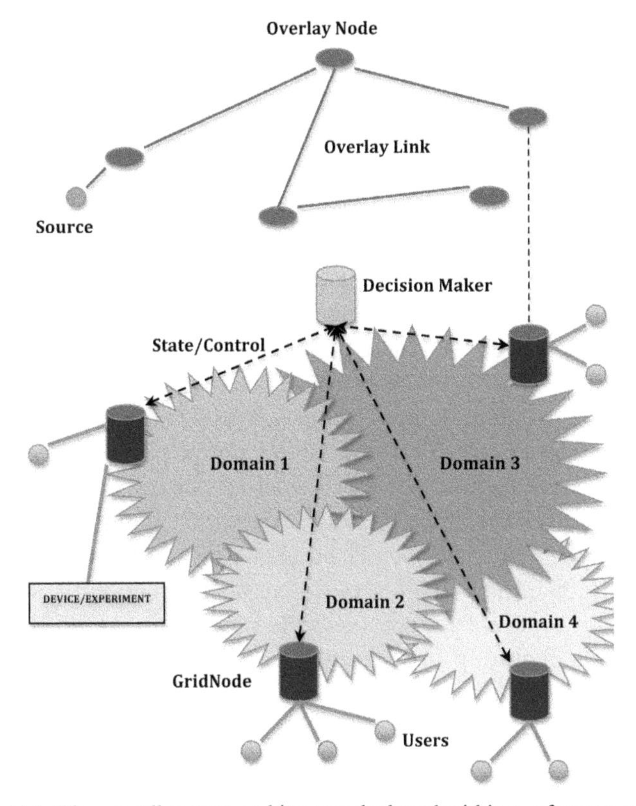

Figure 11.1 The overall system architecture deployed within a reference scenario.

sending measurements to be processed "live" while observing impacts over the surrounding environment.

- Source: this provides the data for the stream. This could also be the resulting composition of different instruments (e.g., a synthetic aperture radar) or the final step of a complex measurement chain. From an architectural viewpoint, this is perceived as a monolithic entity (in Figure 11.1 it is depicted as a device or an experiment).
- Overlay Node (ON): this is the specialized entity responsible to distribute the stream of data through the overlay. The resulting overlay can be merged within the grid architecture, thus resulting in another service layer that can be accessed. We emphasize that in this work GN and ON are "collapsed" in a unique entity. Thus, they are interchangeable.
- Overlay Link: this is a transport-layer connection joining two ONs.

The aim of the DM is to control the evolution of the process by optimizing the bandwidth resources dynamically by employing a discrete-time dynamic model (see [5] for the mathematical framework). Put briefly, each ON is equipped with a buffer of fixed length and manages its bandwidths to accomplish the following (competing) tasks: (1) allocating a portion of resources to receive the incoming stream; (2) allocating a portion of the bandwidth to propagate through the overlay the amount of data received; (3) using the remaining resources to serve the attached user population. In addition, to increase the flexibility of the architecture, a given ON can also: perform a local compression of the stream, e.g., by employing proper coding schemes. We constrain the compressed replica of the stream to be only delivered to the attached user population, so as to avoid the propagation of a "poor" quality stream; implement a local Call Admission Control (CAC) scheme for admitting/removing end-users, to absorb fluctuations in the bandwidth allocation and to not endanger the delivery of the stream. We point out that strategies employed for the CAC are locally computed, i.e., they are not performed by the DM. In fact, the evolution of the number of end users served by GN follows its own dynamic system, controlled by the bandwidth share devoted to them and the quality of the stream. The optimization of these parameters reflects an optimal CAC strategy. To absorb brief disconnections, end-users are equipped with a local buffer, and they are properly uniquely identified to be re-admitted before their local buffers exhaust. The DM, upon receiving state information (i.e., bandwidths, the quality of the stream and the user population) computes the optimal control vectors and sends back the optimal strategy. Performing optimization over the whole range would be unfeasible in a real-time context, then to keep computational times manageable, we adopt a model predictive control scheme that minimizes the cost over K stages. In this work we successfully employed a number of stages $K = 3$.

To prove the effectiveness of the proposed solution, we showcase a test scenario composed by a source producing a data stream at a constant bit-rate of 10 Mbit/s. The duration of the test has been set equal to two hours, to reflect a quite long experimental analysis, and with a sampling time $\Delta T = 5$ s. Data can be locally encoded at the 100, 75 or 50% of the original rate. The buffer length L of the ONs has been selected taking into account IETF recommendations on transmitting real-time multimedia streams [8], i.e., it can store 10 seconds of "playback". The overlay topology is a tree: without loss of generality, if loops are present, we can always assume the presence of a proper spanning tree algorithm. It is composed by seven ONs equipped with OC-1 or OC-3 links, which correspond to 155.52 and 51.84 Mbit/s, respectively.

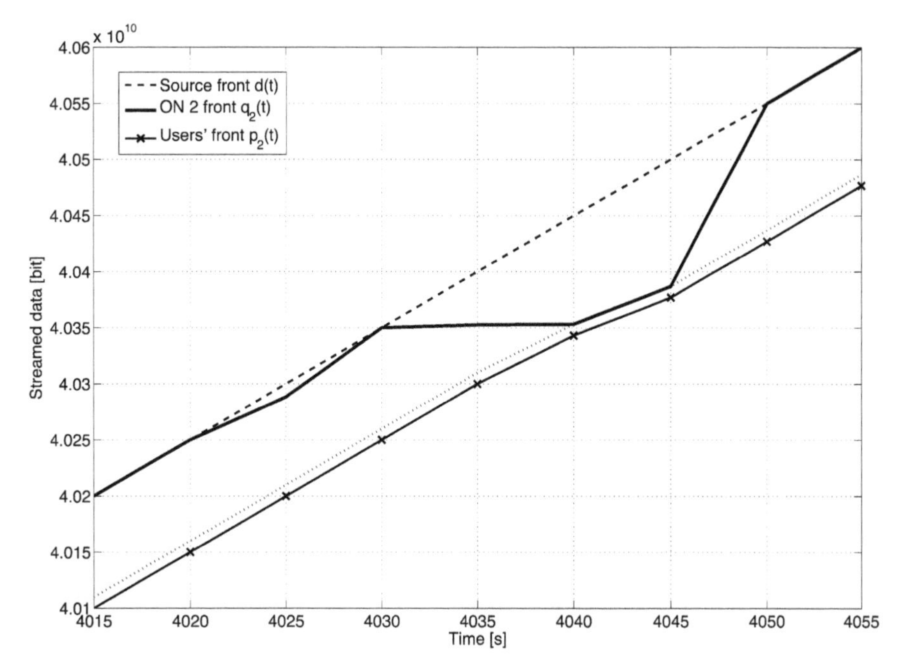

Figure 11.2 Close-up of a critical situation of an ON responsible both of sending data to other ONs and to end-users.

Figure 11.2 depicts a close-up of the behavior of an ON delivering the stream by using the proposed optimized approach. Specifically, we investigate an ON responsible of delivering data both to local users and to other two ONs. Informally, we named as "fronts" the amount of data sent or received up to a given time. Therefore, we define as *source front* the amount of data produced by the source, while *ON front* and *users's front* are the amount of data received by a given ON and by attached users, respectively. As long as all the fronts have the same slope (i.e., the same rate), the stream is correctly delivered to all the components. However, at the time step $T = 4030$ an increase in the user population forces the DM to allocate more bandwidth to server users. Thus, the amount of data received by other ONs within the overlay decreases. At $T = 4045$ the optimizer reacts and reallocates bandwidth. Also it performs momentary end-users disconnections, which are not perceived owing to the presence of local buffers. Then the stream is correctly received and propagated through the overlay again.

11.3 Cloud Architectures

Recently, cloud computing emerged as one of the most important innovation in the field of distributed computing. It harmonizes different technologies in a single view, and binds them to a robust business model by providing a new generation of services never appeared on the Internet before. A transparent view of a vast number of grouped machines is obtained by a web-based interface masking a distributed back-end, which is typically composed by a single or multiple super-computers. Even if innovative, cloud computing spawns from the convergence of pre-existent technologies, such as: SOA, grid computing concepts, web technologies, and hardware virtualization. Three different levels of software abstraction are addressed by cloud computing: the application cloud, the platform cloud and the infrastructure cloud.

Before going into detail, we further investigate the origin of the cloud vision, which appears not an absolute novelty. In fact, back in 1961, John McCarthy of MIT roughly defined the cloud computing vision:

> if computers of the kind I have advocated become the computers of the future, then computing may someday be organized as a public utility just as the telephone system is a public utility ... The computer utility could become the basis of a new and important industry.

Then, cloud computing is an "enrichment" of such vision. It provides a variety of services (i.e., applications, platforms and infrastructures) through simple access interfaces available on the Internet. Details on the provider are not known, as well as the source of services, which is referred as the *cloud.*

Similar concepts were at the basis of grid computing in the second part of 1990s. The famous symmetry between computing and the electrical power grid introduced by Carl Kesselmann and Ian Foster [6] contains the cloud vision. In fact, the initial grid computing aim was a virtual organization collecting CPU cycles, memory and storage resources disregarding details on *how* and *where*. However, being mainly "project-based", grid computing efforts never produced a successful business model able to get over their own boundaries. Among others, we mention: (i) the integration of High Performance Computing (HPC) applications running on supercomputers belonging to different administrative domains as the Enabling Grid for E-SciencE (EGEE)[1]; (ii) mixing virtualization technologies with parallel computing as Grid5000 [10]; (iii) the usage of web-based technologies to simplify

[1] http://egee2.eu-egee.org/.

the access to parallel applications for E-Science as Globus[2]; (iv) industrial applications as in the case of Entropia [11].

The cloud computing vision is more precise as it underlines the availability of clean interfaces to access the cloud, and generally speaking it is "one step beyond". It brings together the large part of technical advancements achieved by grids and associates them to the mass-utilities business model thus fitting completely grid initial vision. The great emphasis given by cloud computing to the adoption of standardized interfaces eases the accounting of the provided services, and finally, makes it as the most promising utility of the future [12].

The first cloud level is the *application cloud* and actually is not completely new. In fact, it refers to Software as a Service (SaaS) applications, which appeared before the cloud advent. SaaS exposes a web interface leveraging distributed resources to the users, in a totally transparent way. This is typically done by interacting with distributed resources through the Simple Object Access Protocol (SOAP) or the HTTP-REST mechanism. As possible examples, we mention Gmail and Google Maps. The underlying computing facility can scale the access structure of the application for a potentially huge amount of concurrent users, as well as to manipulate in parallel many requests to a large amount of not-correlated data. The interaction with application clouds also takes advantage from Web 2.0 technologies allowing the presentation of complete interfaces without any additional requirements at the client-side. All issues related to the application management are delegated to the cloud provider, which owns both the infrastructure and the application provided as a service.

The *platform cloud* level is actually a real novelty and is referred to as Platform as a Service (PaaS). It provides a set of APIs, which can be used to develop service oriented applications. Also, there is a web interface for installing components on the platform, and to set Service Level Agreements (SLAs). The concept behind platform cloud reminds frameworks such as Java EE or Java Spring. In fact they offer automatic and transparent data persistence to ease the development of complex enterprise applications. Similarly, a platform cloud allows as applications the automatic provisioning of an adequate number of computational resources and storage capabilities, according to the needed SLA. We mention here, as possible examples of platform cloud

[2] http://www.globus.org/.

those based on the Google AppEngine[3] and Microsoft Azure.[4] Obviously, each platform implies technological constraints to the development style (for instance, Python for AppEngine and .Net for Azure). The distinguishing feature of this type of cloud combines two aspects: the complete support of the application development by providing specific programming interfaces and the ownership of the underlying computing infrastructure. Although the cloud provider does not own the application code which belongs to the cloud user. The application profits go to the cloud user who in turn, pays to the cloud provider the platform use.

Finally, the third level is the *infrastructure cloud*, commonly known as Infrastructure as a Service (IaaS). In this case, the cloud provider owns just the computing infrastructure and does not constrain the user regarding the type of application or the technology used to develop such an application. In IaaS, virtualization has two main roles: (i) it allows a quick configurability and deployment of resources fitting user wishes; (ii) it provides the necessary isolation preserving the infrastructure from malicious or unwanted behaviors possibly arising from user applications. This is guaranteed by the virtual machine wrapping all the applications running on top. Virtualization also allows avoiding any conflicts among different user needs, e.g., different OS or library versions.

As an example of IaaS solutions, we mention the Amazon Elastic Compute Cloud (EC2) that is the current market leader in IaaS. Its back-end software is provided as a proprietary technology, which is not commercialized in any form. However, some open-source IaaS frameworks are available as Nimbus,[5] Eucalyptus[6] and OpenNebula.[7] These projects were conceived in scientific communities with strong ties with the grid computing world. Rather, Amazon offers a set Web Services Description Languages interfaces (EC2 WSDLs) to integrate directly an application with the infrastructure cloud. Given the Amazon leadership, all other infrastructure clouds also offer support to the EC2 WSDLs, which is a de-facto standard. Apart from providing access to clouds running their own software, the main relevance of these projects stays in the distribution of cloud software. In fact, they have driven the deployment of private clouds addressing specific industrial needs as well as the advent of new players in the cloud computing providers landscape.

[3] http://code.google.com/appengine/.

[4] http://www.microsoft.com/azure/windowsazure.mspx.

[5] http://workspace.globus.org/.

[6] http://eucalyptus.cs.ucsb.edu/.

[7] http://www.opennebula.org/.

Furthermore, some clouds are partially open to the public and they are called hybrid. In this rapidly evolving scenario, which involves IaaS, PaaS and SaaS types, it is impossible to identify how many clouds are publicly available or hybrid. However, it poses the challenge on how to realize an inter-cloud federation towards a single exploitation of all (or a sub-set) available clouds.

As will be discussed in the next section, it is our opinion that a key role can be played by p2p technologies. The cloud interfaces mash-up will be the way to federate transparently clouds into a single global one.

11.3.1 Cloud Mash-up

The cloud scenario is rapidly evolving and various open-source implementations are favoring the advent of new cloud providers. IaaS software is used to set-up cloud on top of legacy data-centers and not only to expose infrastructure services, rather as base-line to deploy application as SaaS. Also public administration adopted this model and several clouds are appearing with the direct involvement of Governs (see, among the others, https://www.apps.gov/ and http://data.gov.uk/apps). The inter-cloud concept aims at exploiting in a single view all publicly available services offered by the different clouds. It may appear an impossible effort trying to "tame" this kind of spontaneous phenomenon. It is continuously growing, driven by an interest involving potentially many millions of users. This means the next years will see the birth of thousands of different clouds. It is our opinion that a harmonization of the whole process will take place, and finally, mechanisms *to federate* different clouds will definitely be used in the next few years. Many developments support this perspective. It will be convenient for providers to agree with others the cloud sharing thus optimizing strategically the computing resources exploitation. This kind of partnership will be the way to contrast big market players for small/medium players. For governments and public administration, it will be an opportunity to increase cooperation with allies and to improve the synergies among different departments. We call this kind of cloud sharing or inter-cloud concept as *cloud mash-up*. We used the term mash-up to identify the heterogeneous clouds integration exactly as currently, web mash-up is used to refer web pages composed by different content sources. Many providers (e.g., Google Maps, Yahoo Maps, etc.) allow invoking their contents through clear SOAP or HTTP APIs. Thus, the web mash-up exposes the contents retrieved by those APIs within a single view.

The cloud mash-up adopts the same concept of web mash-up to cloud resources to obtain the inter-cloud exploitation among different providers.

Practically, cloud mash-up is an entity acting as a mediator; in a nutshell, it brokers resources from possibly different clouds. Compared to the web mash-up targeted to web content, it involves different types of resources which basically reflect the different natures of cloud architectures, namely IaaS, PaaS and SaaS. However, it can take advantage from the technical basis of cloud computing. As we have shown, cloud architectures basically leverage the same standards to expose services independently on the physical cloud technology. While the latter may vary for hardware/OS type, capacity and configuration, the first is strongly bounded to web services technologies, which naturally facilitates mutual integration. It could be noted that web services already provide a brokering technology, namely the Universal Description Discovery and Integration (UDDI). However, it does not seem appropriate for cloud services. Its lack of flexibility makes it hard to envisage its adoption. The main effort to reduce its monolithic structure was the UDDI Business Registry (UBR) project aimed at federating UDDI registries belonging to IBM, Microsoft, HP and SAP. However, it did not provide really applicable results. Further investigations [13] tried to focus on the adoption of p2p technologies to get more flexibility from UDDI.

All solutions provided require changes in the UDDI specifications. Therefore, it is reasonable to disregard a general purpose, complex technology as UDDI, and to focus on a different brokering technology possibly closer to what cloud computing provides. Basically, we discuss the cloud mash-up at the application level to provide a way to automatically cope resources (e.g. Virtual Machines) belonging to heterogeneous clouds. We will simply call the entity allowing the mash-up *broker* and it involves the use of p2p technology. The p2p infrastructure provides a natural backbone able to support the needed requirements. As shown in the previous sections, overlay networks provide a way to overcome current Internet limitations by providing multicast, barriers bypass, QoS in interactive grids, etc. This approach makes overlays a SOA-oriented communication level, which was proposed in an ontology as a further type of cloud, the Communication as a Service (CaaS) [14]. This perspective shows that the integration of overlay services with other SOA-based architecture is technically feasible.

The designed cloud mash-up aims at connecting pieces of software running on resources available from different cloud providers. A common use case is when a user books a set of virtual machines from different IaaS providers. Afterwards, the user may have to run a distributed application on top of those machines and a mean to seamlessly bind them is not provided by anyone. A possible issue may arise from the fact that some resources have a private IP address as shown by a test of the Information Brokering Service Provider (IBSP) [15], where computational resources coming from an IaaS cloud are joined to other ones (ADSL connected machines and LAN computers) to serve seamlessly an overlaying parallel program. Another possible reason is the use of specific name-spaces over the cloud resources. At this level, cloud mash-up should provide a *cloud-neutral* brokering service. Basically, such service allows the resources look-up according to a pluggable identifiers space. Furthermore, this model can be adopted to merge scenarios involving SaaS enterprise application. Specific plug-ins could support the integration of an ERP and a CRM running as SaaS on different clouds, respectively. Equally, PaaS integration could also be useful by setting up an overlay name space for a specific platform type. By following many concepts traced by IBSP, we can draw a broker for cloud mash-up. We will use the term resource to identify a unit booked independently on the cloud architectures it comes from (i.e. IaaS, PaaS or SaaS). The possible useful functionalities are the following ones: (1) cloud resources name-space and lookup, (2) resources connectivity enabler, and (3) resources information system.

Point (1) regards the registration and the look-up of the resources into the mash-up according to a chosen name-space. This functionality implies the exposition of an enrollment function for every type of resource requested by a mash-up. The broker provides an identifier system which is used to identify resources. A unique identifier is assigned to a resource. Identifiers can be ranges of private IP addresses, namespaces or any other identification system required by an application running on IaaS, PaaS-provided machines and SaaS applications. The assignment of identifiers can be done by the broker (in a pseudo-DHCP way) or statically configured at the cloud side. In both cases, the broker maintains the identifier consistence. Different identifier systems can coexist within the broker, thus serving concurrently multiple mash-ups. Once a resource is enrolled, the broker stores an association between a unique identifier and a resource representation including, at least, its physical location. Naturally, the broker exposes a simple look-up method to retrieve a target endpoint (e.g. physical location as an URL or an IP) when a resource is looking for another one given a target identifier.

Concerning point (2), the resources connectivity enabler refers to the actions performed by the broker to set-up the connectivity between two enrolled resources. It is an inter-mediation functionality, it supports the start-up of the connection between two resources, which otherwise would not be able to communicate. After this set-up phase, the resources communicate directly without any other broker intervention. The connectivity enabler functionality relies on schemes of messages relayed between the requester resource and the requested resource through the broker. Those schemes may comprise a simple request/reply messages exchange as well as complex inter-mediations composed by messages loops involving the requester, the callee and even third-part resources. Forwarded messages can be enriched and modified by the broker according to specified pattern depending on the client platform. This functionality can be useful for a variety of situation that can span from network information (e.g., bypass of NATs or firewalls, overlay routing rules, etc.) to public security information exchange.

Finally, with regard to point (3), the broker has a flexible information system where resources information is stored. Basically, a resource is registered with an association between a unique identifier and a representation of its actual location. However, distributed applications may require more than just location (e.g., system information, integration mapping rules, etc.). Therefore, it is possible to add additional information to a given entry. Such information can be registered at the enrollment stage, or updated via a method exposed to modify the entries content. Furthermore, the information system also provides a resource discovery not based on a precise identifier. It can be useful for those cases requiring a resource search on the base of some feature without an a-priori knowledge of the resource endpoint. This kind of search can be useful for a vast variety of cases (e.g., search for specific computational capabilities, specific software configuration, etc.). This capability is supported by a publish/subscribe-based system. Every mash-up can implement an application grammar used by resources to publish topics useful for a not identifier-based resource discovery. Resources can then subscribe to topics according to the module grammar. Such a grammar can be composed by a few keywords or based on some ontology.

A desirable feature of the broker is a flexible and failure-resistant structure to get scalability and robustness at the same time, mainly to avoid single point of failures and to provide multiple access points. This infrastructural requirement leads naturally to p2p technologies. Specifically, the approach based on the structured overlay network exploiting Distributed Hash Tables [16] and [17] (DHTs) seems quite appropriate. Put briefly, a DHT is a hash table distributed among a set of cooperating nodes. As a normal hash table it allows to place and get items as key-value pairs. The broker stores into a DHT the identifier generated for an enrolled cloud resource as key. The value associated to a key has to provide information helping to locate the resource as its public IP address. We prefer the DHT adoption to the unstructured overlay network because DHTs provide better scalability. Moreover, the usage scenario of the broker should not affect the main known drawback of DHTs. Their main capability is to provide a scalable logarithmic keys lookup time. The scalability is not a trivial issue in unstructured overlay networks being based on queries broadcasting. The solution consists in the layering of nodes into a hierarchical classification. However, it introduces critical points of failure in the structure and does not nullify completely the lack of scalability. Unstructured overlay networks actually have a good possibility to perform range queries which are not possible with DHTs. To overcome this limitation, DHT implementations provide basic APIs to perform multicast communication based on the publish/subscribe pattern. This can be particularly useful for the functionalities exposing further functionalities more than the simple look-up. They can be useful for searches which are not strictly based on an identifiers system, but rather on resource-, application- or platform-based features depending on predefined semantic. In this case, the information stored as value in the DHT has to fit the chosen semantic to public the topics needed for the multicast communication.

We implemented the cloud mash-up by wrapping an existing DHT API, namely FreePastry.[8] It provides access to the insert/lookup operations and the multicast communication based on the publish/subscribe pattern. On top of the DTH wrapper, it is possible to customize the three functionalities previously described according to the specific mash-up needs. As shown in Figure 11.3, the resulting architecture is equivalent to the one of IBSP. A distributed set of cooperating nodes host a wrapped DHT portion and a plug-in for a certain mash-up. Different cloud resources implementing the desired mash-up contact the broker to get in touch with others part of the same integration scenario.

Finally, further possible types of cloud mash-up can be addressed. We have seen how interactive grids can leverage p2p technologies to get QoS in data streaming. This approach can be taken into consideration for cloud mash-up at network level. In fact, when cloud resources coming from different providers or when different data centres have to communicate, they use the public Internet which obviously involves QoS issues. The methodology described in Section 11.2.1 may be adopted "as is" to achieve this goal. Approaches as described in Section 11.1 may be used in cooperation with cloud grammars based on grid components description languages [18]. In this sense, a cloud mash-up at access level may allow the merge of a cloud claim to what a global cloud registry can offer.

---

[8] http://www.freepastry.org/.

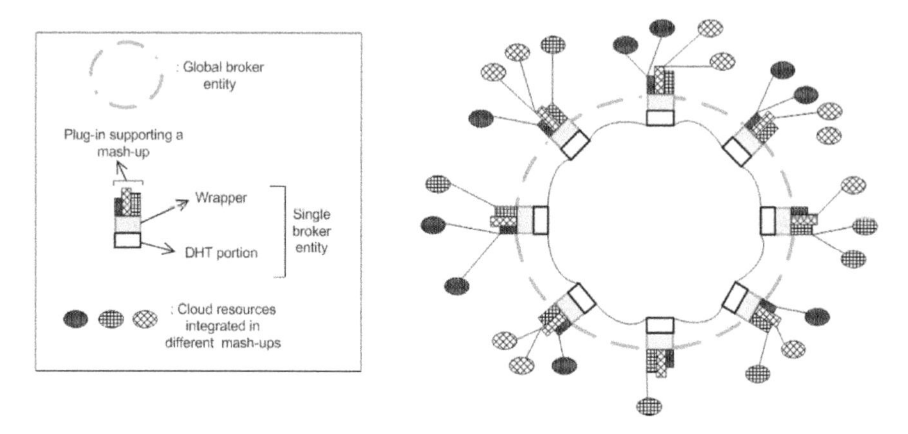

Figure 11.3 An IBSP-equivalent architecture.

11.4 Conclusions

In this chapter, we investigated how to use p2p and cloud technologies to support next-generation services. As discussed, the ability of p2p of creating overlay networks can be used to deploy what we defined as "network independent" services. The latter can be merged also with those offered by cloud architectures. Specifically, we investigated a specialized content distribution architecture to support the real-time data delivery in interactive grids. Then, we discussed the key role of p2p in "joining" mash-ups within clouds. In fact, as proposed, the cloud to mash-up interfaces can be operated in a p2p fashion in order to federate transparently clouds into a single global one.

There is still much work to be done, especially in the field of prototypal implementation, but the subject is very promising. Nevertheless, such solutions being located at the application layer, the efforts needed can be justified by their "technological" independency. Thus, the underlying network deployment and computing infrastructure can evolve, without affecting the design of the p2p and cloud frameworks used to spawn new next-generation services. Finally, in order to avoid "balkanization" of the technological space (e.g., in terms of APIs and requirements) a standardization process must be meticulously addressed, especially by starting in the very early stages of the development of this disruptive mix.

References

[1] J. Buford and M. Kolberg. Hybrid overlay multicast simulation and evaluation. In *Proceedings of the 6th IEEE Consumer Communications and Networking Conference (CCNC2009)*, pages 1–2, January 2009.

[2] L. Caviglione. Enabling cooperation of consumer devices through peer-to-peer overlays. *IEEE Transactions on Consumer Electronics, IEEE*, 55(2):414–421, May 2009.

[3] K. Fujii and T. Suda. Semantics-based dynamic service composition. *IEEE Journal on Selected Areas in Communications*, 23(12):2361–2372, December 2005.

[4] J. Buford, E. Celebi, and P. Frankl. An integrated peer-to-peer data and service dissemination system. In *Proceedings of 4th IEEE Consumer Communications and Networking Conference (CCNC2007)*, pp. 546–549, January 2007.

[5] L. Caviglione and C. Cervellera. Design, optimization and performance evaluation of a content distribution overlay for streaming. *Computer Communications Journal.* Available online at http://dx.doi.org/10.1016/j.comcom.2010.04.047.

[6] I. Foster, C. Kesselman, and S. Tuecke, Anatomy of the grid. *International Journal of Supercomputer Applications*, 15(3), 2001.

[7] L. Caviglione and L. Veltri. A p2p framework for distributed and cooperative laboratories. In F. Davoli, S. Palazzo, and S. Zappatore (Eds.), *Distributed Cooperative Laboratories – Networking, Instrumentation and Measurements, Proceedings of 2005 Tyrrhenian International Workshop on Digital Communications*, Sorrento, Italy, July 2005. Springer, Norwell, MA, pages 309–319, 2006.

[8] J. van der Meer, D. Mackie, V. Swaminathan, D. Singer, and P. Gentric. RTP payload format for transport of MPEG-4 elementary streams. Network Working Group, IETF, RFC 3640, November 2003.

[9] G. Lin, G. Dasmalchi, and J. Zhu. Cloud computing and IT as a service: Opportunities and challenges. In *Proceedings of the IEEE International Conference on Web Services*, September 2008.

[10] F. Cappello et al. Grid5000: A large scale, reconfigurable, controllable and monitorable grid platform. In *Proceedings of the 6th IEEE/ACM International Workshop on Grid Computing (GRID2005)*, Seattle, WA, USA, November 2005.

[11] A. Chien, B. Calder, S. Elbert, and K. Bhatia. Entropia: Architecture and performance of an enterprise desktop grid system. *Journal of Parallel and Distributed Computing*, 63(5):597–610, May 2003.

[12] R. Buyya, C.S. Yeo, S. Venugopal, J. Broberg, and I. Brandic. Cloud computing and emerging IT platforms: Vision, hype, and reality for delivering computing as the 5th utility. *Future Generation Computer Systems*, 25:599–616, 2009.

[13] R. Podestà. A lightweight inter-node operation for UDDI cloud. In *Proceedings of the International Workshop on Middleware for Web Services (MWS) part of the 12th IEEE Enterprise Computing Conference (EDOC2008)*, Munich (Germany), pages 397–400, September 2008.

[14] L. Youseff, M. Butrico, and D. Da Silva. Towards a unified ontology of cloud computing. In *Proceedings of Grid Computing Environments Workshop (GCE08)*, held in conjunction with SC08, November 2008.

[15] R. Podestà, V. Iniesta, A. Rezmerita, and F. Cappello. An information brokering service provider for virtual clusters. In Proceedings of 17th International Conference on Co-

operative Information Systems (CoopIS 2009), Lecture Notes in Computer Science, Vol. 5870, pages 165–182, Springer-Verlag, 2009,

[16] F. Dabek, E. Brunskill, M.F. Kaashoek, D. Karger, R. Morris, I. Stoica, and Balakrishnan. Building peer-to-peer systems with chord, a distributed lookup service. In *Proceedings of the 8th Workshop on Hot Topics in Operating Systems (HotOS-VIII)*, Elmau/Oberbayern, Germany, May 2001.

[17] A. Rowstron and P. Druschel. Pastry: Scalable, distributed object location and routing for large-scale peer-to-peer systems. In *Proceedings of the IFIP/ACM International Conference on Distributed Systems Platforms*, Heidelberg, Germany, November 2001.

[18] M. Migliardi and R. Podestà. Parallel component descriptor language: XML based deployment descriptors for grid application components. In *Proceedings of the 2007 International Conference on Parallel and Distributed Processing Techniques and Applications*, Las Vegas, Nevada (USA), June 2007.

12

End-to-End Service Provisioning in Network Virtualization for the Next Generation Internet

Qiang Duan

Department of Information Science and Technology, The Pennsylvanian State University Abington College, 1600 Woodland Road, Abington, PA 19001, USA; e-mail: qduan@psu.edu

Abstract

Although the Internet has become a global information infrastructure in a relatively short period of time, it is facing many new challenges. Heterogeneous networking technologies are expected to coexist in the Internet for supporting a wide spectrum of applications with different networking requirements, which requires fundamental changes in the Internet architecture and service model. However, the current Internet lacks the ability to adopt disruptive innovations in network architectures and service delivery models. To fend off this ossification, network virtualization has been proposed as a key diversifying attribute of the next generation Internet. A key technical issue for network virtualization in the Internet lies in interaction between the service and infrastructure providers to enable collaboration across heterogeneous network infrastructures for end-to-end service provisioning. The Service-Oriented Architecture (SOA) provides an effective approach to coordinating heterogeneous systems to meet diverse application requirements, which is essentially the same challenge faced by network virtualization. Therefore applying the SOA principles in network virtualization may greatly facilitate end-to-end service provisioning in the future Internet. This chapter will introduce the notion of network virtualization in the next generation Internet,

Anand R. Prasad et al. (Eds.), Future Internet Services and Service Architectures, 247–269.

review the SOA and its basic principles, discuss the application of SOA in network virtualization to facility end-to-end service provisioning in the Internet, and investigate some key technologies for service delivery in a SOA-based network virtualization environment.

Keywords: service provisioning, network virtualization, Service-Oriented Architecture, next generation Internet.

12.1 Introduction

The Internet has become a critical information infrastructure for global commerce, media, and defense in a relatively short period of time. Numerous applications have been developed and deployed on top of the Internet, and these applications have a wide spectrum of requirements on the networking services provided by the Internet. The highly diverse application demands require research on a variety of networking technologies and alternative network architectures to offer new Internet services. It is unlikely that any single network architecture or technology can offer the Internet services that meet all the application requirements; therefore coexistence of heterogeneous networking systems with different architectures and implementation technologies will be an essential feature of the next generation Internet. Due to the diversity of the coexisting network architectures and technologies and the wide spectrum of supported applications, end-to-end network service provisioning will become a challenging issue in the future Internet. How to coordinate heterogeneous networking systems to offer Internet services that meet the various application requirements become a significant research problem.

A key to solving this problem lies in flexible collaborations among heterogeneous networking systems and effective interactions between applications and underlying networks. The main barrier to such collaborations and interactions is the relative opaqueness of the current Internet architecture, which is mainly caused by the lack of cross-domain communications of operation information and insufficient information exchange between networks and the applications utilizing them. The lack of inter-domain collaborations in the current Internet architecture also limits Quality of Service (QoS) provisioning within the boundaries of Internet Service Providers (ISPs). However, a wide variety of applications, including VoIP, video telephony, online gaming, e-commerce, and multimedia streaming applications, all require a level of end-to-end QoS guarantee across multiple network domains. The inter-domain QoS mechanism available in the current Internet relies on bilateral cooperation between service providers [15], which is limited to a small set of service classes and not flexible enough to support the wide spectrum of application requirements. Therefore, fundamental changes in network architecture are required for the next generation Internet.

However, the current Internet is relatively resistant to fundamental changes. The significant capital investment in current network infrastructures and competing interests of its major stakeholders create a barrier to any disruptive innovation in Internet architecture and service delivery. The end-to-end design principle of IP requires global agreement and coordination across various ISPs to deploy any fundamental change in the Internet architecture and service model. To fend off ossification of the current Internet, network virtualization has been proposed as a key diversifying attribute of the future inter-networking paradigm and is expected to play a crucial role in the next generation Internet.

Essentially, network virtualization decouples network service provisioning from data transportation infrastructures. Through network virtualization, the role of the traditional ISPs is separated into two independent entities: Infrastructure Providers (InPs) which manage the physical network infrastructures, and Service Providers (SPs) which create virtual networks to offer end-to-end network services by aggregating networking resources from multiple InPs. Network virtualization enables multiple network architectures to cohabit on a shared physical substrate, and allows an SP to build virtual networks on top of heterogeneous network infrastructures to meet different application requirements. In network virtualization environments, the end-to-end services are offered by SPs by synthesizing networking resources from various infrastructures. Therefore, one of the key technical issues must be addressed in network virtualization is effective interaction between SPs and InPs to coordinate heterogeneous network infrastructures for end-to-end service provisioning.

The Service-Oriented Architecture (SOA), which initially emerged in Web services and Grid computing areas, provides an effective mechanism for coordinating heterogeneous systems to support different application requirements. This is essentially the same challenge faced by end-to-end service provisioning in network virtualization. Therefore, applying the SOA principles in network virtualization may greatly facilitate service provisioning in the next generation Internet, which is the topic that this chapter focuses on. The rest of this chapter first gives an overview about the notion of network virtualization, the benefits of this new networking paradigm, and the technical challenges to its implementation. Then the SOA is introduced, analyzing its key principles, and investigating the application of the SOA in network virtualization. The SOA-based network virtualization allows effective collaboration across heterogeneous network infrastructures; thus it may greatly facilitate end-to-end network service provisioning. This chapter will particularly discuss two key technologies for SOA-based network virtualization – capability modeling for infrastructure services and performance evaluation for end-to-end service delivery. A general approach to modeling networking capabilities of heterogeneous network infrastructures is described in this chapter; and then based on this model, analytical techniques for evaluating performance of end-to-end service delivery in SOA-based network virtualization environments are developed.

12.2 Network Virtualization for the Next Generation Internet

Network virtualization was first proposed as an approach to developing virtual testbeds for the evaluation of new network architectures [2]. For example, in PlanetLab [3] and GENI [13] projects, network virtualization was employed to build open experimental facilities for researchers to create customized virtual networks to evaluate new network architectures. Then the role of virtualization in the Internet shifted from a tool for architecture evaluation to a fundamental diversifying attribute of the inter-networking paradigm [11, 28]. By enabling diverse network architectures to coexist on a shared physical substrate, network virtualization can extenuate the ossifying forces in the current Internet and stimulates continual innovations on network architectures and technologies.

Essentially, network virtualization follows a well-tested principle – separation of policy from mechanism – in the network area. In this case, network service provisioning is separated from data transportation mechanisms; thus dividing the traditional role of ISPs into two entities: infrastructure providers which manage the physical infrastructures, and service providers which create virtual networks for offering end-to-end services by aggregating resources from multiple infrastructures [8].

2cInfrastructure providers (InPs) are in charge of operations and maintenance of physical network infrastructures and offer their resources through programmable interfaces to different service providers instead of providing direct services to end users. Service providers (SPs) lease networking re-

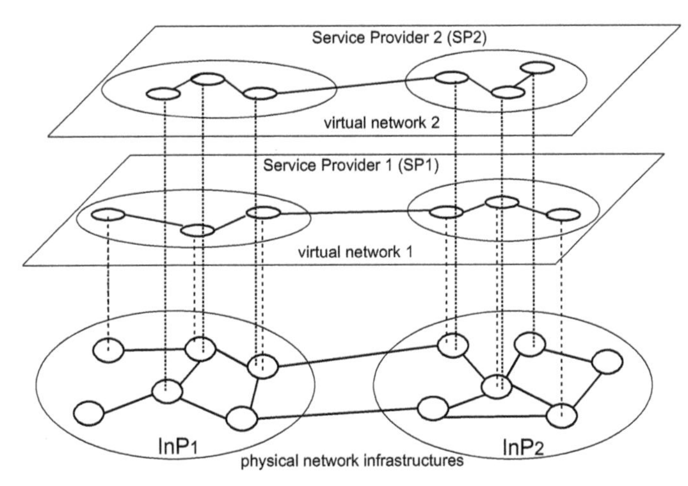

Figure 12.1 An illustration of a network virtualization environment.

sources from multiple InPs to create virtual networks and deploy customized protocols in the virtual networks by programming the resources in multiple infrastructures. Each virtual network is composed and managed by a single SP, which synthesizes the networking resources allocated in underlying infrastructures to offer end-to-end services to the end users. A virtual network is a collection of virtual nodes connected by a set of virtual links to form a virtual topology, which is essentially a subset of the underlying physical topology. Each virtual node could be hosted on a particular physical node or could be a logical abstraction of a networking system. A virtual link spans a path in the physical network and includes a portion of the networking resources along the path. Figure 12.1 illustrates a network virtualization environment, in which the service providers SP1 and SP2 construct two virtual networks by using resources from the infrastructure providers InP1 and InP2.

Network virtualization will bring a significant impact on service provisioning in the future Internet. By allowing multiple virtual networks to cohabit on a shared physical substrate, network virtualization provides flexibility, promotes diversity, and promises increased manageability. The best-effort Internet today is basically a commodity service that gives network service providers limited opportunities to distinguish themselves from competitors. A diversified Internet enabled by network virtualization offers a rich environment for innovations, thus stimulating the development and deployment of new Internet services. In such an environment, SPs are released from the requirement of purchasing, deploying, and maintaining physical network equipment, which significantly lowers the barrier to entry of the Internet service market. Network virtualization enables a single SP obtain control over the entire *end-to-end* service delivery path across network infrastructures that belong to different domains, which will greatly facilitate the end-to-end QoS provisioning.

Recently, network virtualization has attracted extensive research interest from both academia and industry. A new network architecture was proposed in [28] for diversifying the Internet, which enables various meta-networks built on top of a physical substrate. The CABO Internet architecture proposed in [11] decouples network service providers and infrastructure providers to support multiple network architectures over shared infrastructures. The EU FP7 4WARD project has also adopted network virtualization as a key technology to allow the future Internet to run virtual networks in parallel [18]. A network virtualization architecture was proposed in [26] and analyzed from both technical and business perspectives. Some standard organizations are also embracing the notion of network virtualization into their standard specifications. For example the Next Generation Network (NGN) architecture defined by ITU-T follows a key principle of separating service-related functions from underlying transport-related technologies [27]. The Open Mobile Alliance (OMA) recently developed an Open Service Environment (OSE) that delivers network services by composing standard service enablers, which are virtualization components of networking resources [23]. An overview of more research efforts and progress in the area of network virtualization can be found from the survey given in [8].

Although much progress has been made toward network virtualization, this research field is still in its early stages. There are several technical challenges which must be addressed to realize the vision of virtualization-based Internet. One of the primary challenges centers on the design of mechanisms and protocols needed to enable automated creation of virtual networks for service provisioning over a global, multi-domain substrate comprising heterogeneous network infrastructures. Since an SP utilizes the networking resources provided by multiple network infrastructures to offer end-to-end services, the key to solving this problem lies in effective interactions between the SP and its InPs that enable flexible collaborations across heterogeneous network infrastructures.

The nature of the interaction between an SP and the multi-domain infrastructure substrate is intrinsically more complex than the relationship between a single user and a network providing a reserved bandwidth flow. An SP needs to be able to determine what resources are available within an infrastructure in order to formulate virtual networks for end-to-end service provisioning. InPs need to make certain information available in a machine-readable form. Such information must be specific enough to allow SPs to construct virtual networks that meet application requirements. On the other hand, InPs may not want to expose detailed information about their network infrastructures. Automated mechanisms are also needed to respond to SP's request for infrastructure services and to configure and orchestrate network infrastructures.

12.3 The Service-Oriented Architecture (SOA)

The SOA is a system architecture initially developed by the distributed computing community, especially in the area of Web services [24] and Grid computing [12], as an effective solution to coordinating the computational resources in multiple heterogeneous systems to support various application requirements. The SOA is described in [7] as "an architecture within which all functions are defined as independent services with invocable interfaces that can be called in defined sequences to form business processes." The SOA can be considered as a philosophy or paradigm to organize and utilize services and capabilities that may be under the control of different ownership domains [20]. Essentially, the SOA enables virtualization of various computing resources in form of services and provides a flexible interaction mechanism among services.

A service in the SOA is a computing module that is self-contained (i.e., the service maintains its own states) and platform-independent (i.e., the interface to the service is independent with its implementation platform). Services can be described, published, located, orchestrated, and programmed through standard interfaces and messaging protocols. All services in the SOA are independent of each other and service operation is perceived as opaque by external services, which guarantees that external components neither know nor care how services perform their functions. The technologies providing the desired functionality of the service are hidden behind the service interface.

The key elements of SOA and their interactions are shown in Figure 12.2. A service provider publishes a machine-readable document called *service description* at a service registry. The service description gives descriptive information about the functions provided by the service and the interfaces for utilizing such functions. When a service customer, either an application or another service, needs to utilize computing resources to perform a certain function, it starts a service discovery process to locate an available service

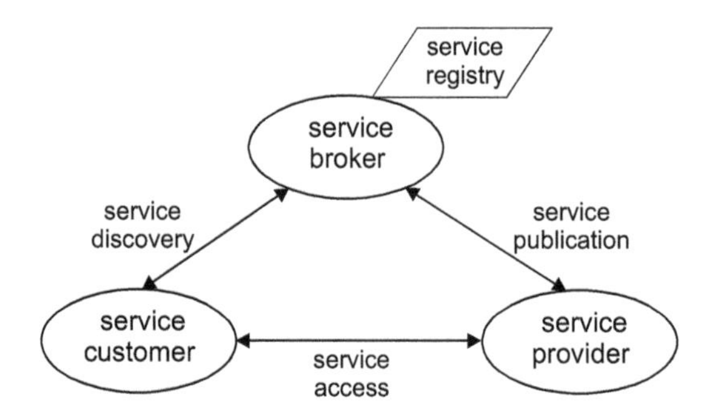

Figure 12.2 Key elements of the SOA and their interactions.

that meets its requirement. Typically a service broker handles service discovery for service customers by searching the service descriptions published at the registry and selecting a service that matches the criteria specified by the customer. The broker may select and consolidate multiple services into a composite service to meet the customer's requirements. After discovering a service, the service customer contacts the service provider and invokes the service by following the interface defined in the service description.

A key feature of SOA is the "loose-coupling" interaction among heterogeneous systems in the architecture, including service providers, service customers, and the service broker and registry. The term "coupling" indicates the degree of dependency any two systems have on each other. In a loosely coupled exchange, systems need not know how their partner systems behave or are implemented, which allows systems to connect and interact more freely. Therefore, loose coupling of heterogeneous systems provides a level of flexibility and interoperability that cannot be matched using traditional approaches for building highly integrated, cross-platform, inter-domain communication environments. It is this feature that makes the SOA a very effective architecture for coordinating heterogeneous systems to support various application requirements.

Although the SOA can be implemented with different technologies, Web services are the preferred environment for realizing the SOA promise of maximum service sharing, reuse, interoperability. Key technologies for realizing SOA include the technologies for service description, service publication, service discovery, service composition, and message delivery. The standard for web service description is Web Service Description Language (WSDL) [31], which defines the XML grammar for describing services as a collection of communicating endpoints capable of exchanging messages. Web service publication is achieved by Universal Description Discovery and Integration (UDDI) [19], which is a public directory with standard interfaces for publishing and searching service descriptions. Simple Object Access Protocol (SOAP) [30] is an XML-based messaging protocol on which web services rely to exchange information among them. Web service composition describes the execution logic of service-based functions by defining their control flows. The Business Process Execution Language for Web Services (BPEL4WS) [21] provides a standard for Web service composition.

12.4 Applying the SOA in Network Virtualization

The SOA is a very effective architecture for coordinating computational resources in heterogeneous systems to support various application requirements. Essentially the same challenge, namely coordinating the heterogeneous networking systems for offering end-to-end services to meet highly diverse application requirements, is faced by network virtualization for the next generation Internet. Therefore, applying the SOA principles in network virtualization may greatly facilitate end-to-end service provisioning in the Internet of the future. In such a SOA-based network virtualization environment, the networking resources and capabilities of each network infrastructure are abstracted into an *infrastructure service* that can be offered by the InP to SPs. The end-to-end network services provided to end users are constructed by SPs through discovering, composing, and accessing infrastructure services.

The SOA-based network virtualization paradigm is shown in Figure 12.3. In this paradigm a network service provisioning layer is deployed between the end users (typically network-based computing applications) and the underlying shared infrastructure platform. The infrastructure platform consists of heterogeneous networking systems that are encapsulated into infrastructure services. Each InP compiles a description of its infrastructure service and publishes it at the service registry. By publishing an infrastructure service description, an InP can advertise the networking functions and capabilities of its infrastructure without exposing internal implementation details. The service provisioning layer consists of SPs which offer network services directly to end users, a service registry where infrastructure service descriptions are published and maintained, and a service broker which searches and discovers appropriate infrastructure services for different SPs.

One key function of the service provisioning layer is to assemble the infrastructure services provided by InPs into end-to-end network services to meet the requirements of different end users. When an SP needs to util-

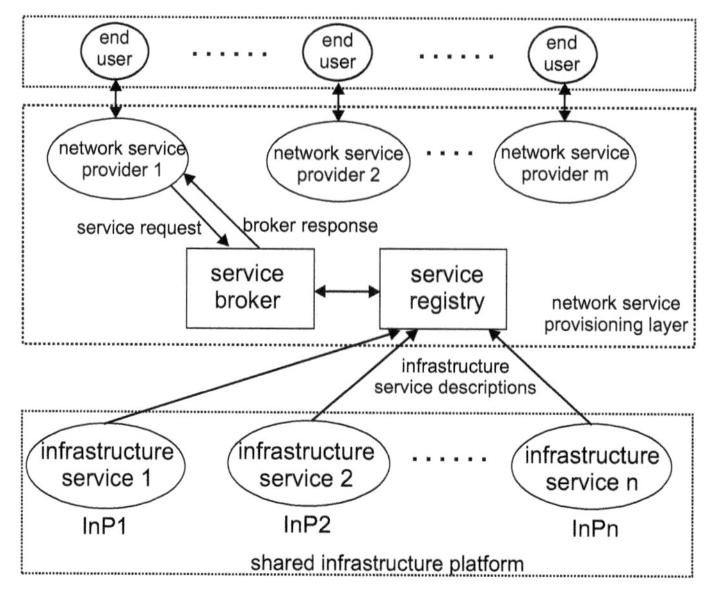

Figure 12.3 The SOA-based network virtualization paradigm.

ize networking resources in infrastructures to offer an end-to-end service, it sends a request to the service broker, which will search the infrastructure service descriptions published at the service registry and select the appropriate infrastructure service (or a set of infrastructure services) for the SP. Then the SP can synthesize the discovered infrastructure services to construct a virtual network for end-to-end service provisioning. The service broker acts as a mediator between infrastructure providers and service providers in network virtualization environments. The broker simplifies the process of matching service provider's requirements to available networking resources by aggregating offers from multiple infrastructure providers.

In a SOA-based network virtualization environment, end-to-end network service provisioning is offered by an SP through orchestrating infrastructure services provided by InPs. Therefore, the end-to-end delivery system for a network service is composed of a series of tandem service components, each of which represents an infrastructure service. Figure 12.4 shows such a service delivery system constructed on top of n network infrastructures, denoted as I_i, $i = 1, \ldots, n$. The service component S_i represents the infrastructure service provided to this service delivery system by the InP of I_i.

Application of the SOA principles in network virtualization makes loose-coupling a key feature of the interaction between SPs and InPs and the

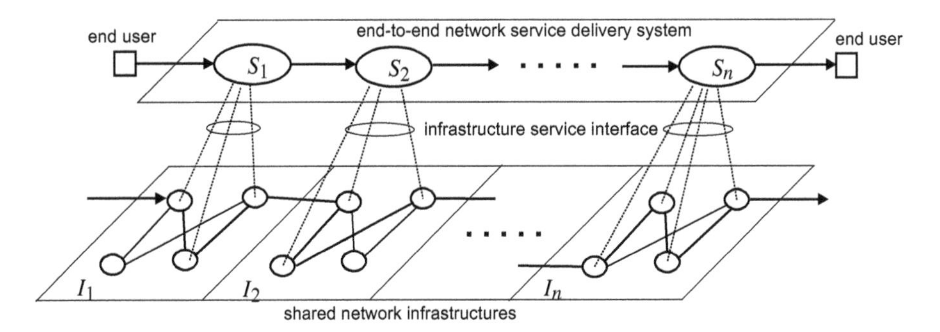

Figure 12.4 An end-to-end service delivery system in SOA-based network virtualization.

collaboration among heterogeneous infrastructures. Therefore, such a network virtualization paradigm inherits the merit of a SOA that enables flexible and effective collaborations across heterogeneous systems for providing services that meet diverse application requirements. SOA-based network virtualization also gives Internet service providers the ability to view their underlying infrastructures more as commodities and allows infrastructure development to become more consistent. This enables a faster time to market as new service initiatives can reuse existing services and components, thus reducing design, development, testing, and deployment time in addition to the cost and risk of new service development.

12.5 Service Description and Discovery for SOA-Based Network Virtualization

With service description and discovery being the key technologies in SOA, the description and discovery of infrastructure services are the keys to realizing the SOA-based network virtualization. A distinguishing feature of infrastructure services is their networking capabilities of guaranteeing a certain level of QoS to an SP. The description of an infrastructure service should provide sufficient information about the networking capability of the infrastructure. Therefore, capability modeling for infrastructure services is a key technical issue must be addressed in SOA-based network virtualization. In addition, a key to infrastructure service discovery and composition is to select and synthesize the appropriate infrastructure services for meeting the end-to-end service performance requirements. Therefore, evaluating end-to-end service performance in SOA-based network virtualization environments also becomes a significant research problem.

The currently available service description and discovery technologies in SOA must be enhanced to support network virtualization. The WSDL specification focuses on providing functional information about services and UDDI lacks effective mechanisms to publish and search non-functional features such as service provisioning capabilities. Recently, World Wide Web Consortium (W3C) developed WS-Policy [33] and WS-PolicyAttachment [32] specifications for describing non-functional characteristics of Web services. WS-Policy aims to provide a general-purpose framework and model for expressing service characteristics such as requirements, preference, and capabilities as policies. WS-PolicyAttachment defines a mechanism to associate the policy expressions with the existing WSDL standard. The WS-Agreement specification [22] developed by the Open Grid Forum (OGF) defines a protocol between service providers and users for establishing and managing service level agreements. These specifications have made significant progress toward QoS-enable Web/Grid service description and discovery. However, modeling and describing network service capabilities are left open as domain-specific issues in these specifications.

Service composition is also an important aspect of SOA. Currently the standard BPEL4WS provides a workflow-like definition language that describes sophisticated business processes that can orchestrate Web services. The technologies specified in WS-Coordination [10] and WS-Transaction [9] complement BPEL4WS to provide mechanisms and protocols for transaction processing systems, workflow systems, and other applications to coordinate multiple Web services. Web Service Choreography Description Language (WS-CDL) [29] is a W3C candidate recommendation that describes how peer-to-peer Web service participants collaborate. The above-mentioned specifications were mainly developed in the context of business process integration and distributed computing. Their applications to networking and network virtualization need further investigation. Recently, research efforts toward network service composition included the Ambient network composition system reported in [4] and a service composition mechanism for optical Internet that was developed in [1].

Recent research results on network modeling and description have also been reported in the literature. For example, Lacour and colleagues developed a network description model using the directed acyclic graph (DAG) to describe network topology [17]. This description model focuses on a functional view of network topology instead of the networking capabilities needed by infrastructure service description in network virtualization. A Network Description Language (NDL) was proposed in [14] as a semantic schema for describing networks. The NDL serves more as a vocabulary to present network topologies than an approach to modeling network service capabilities. The applications of NDL reported in [14] mainly focus on optical networks, but the next generation Internet may consist of a wide variety of network infrastructures with different implementations.

Heterogeneity of the network infrastructures coexisting in the future Internet and the resource abstraction feature of the SOA-based network virtualization bring new challenges to service modeling and performance evaluation. Traditional network modeling and analysis approaches are typical for a particular network architecture, for example either for circuit switching networks or packet switching networks. Currently available performance analysis on network services is based on assumptions about the technologies employed in the analyzed networks, such as the data forwarding mechanisms, routing protocols, traffic control schemes, and packet scheduling algorithms. Diversity of network infrastructures in network virtualization environments requires a general model that is applicable to various heterogeneous networking systems. Such a model should also support the composition of multiple infrastructure services into an end-to-end network service. SOA-based network virtualization allows an SP access to underlying networking resources through an infrastructure service interface without knowing the infrastructure implementation details. Therefore, performance analysis techniques for end-to-end service delivery in such a networking paradigm must be agnostic to the implementations of underlying infrastructures.

In the next section we will introduce a general approach to modeling and describing the networking capabilities of heterogeneous infrastructure services. Based on this service capability model, techniques for evaluating the end-to-end service performance in SOA-based network virtualization will be developed in Section 12.7. The performance evaluation techniques form a basis for discovering and selecting the appropriate infrastructure services that meet the performance requirements of end-to-end service provisioning, which will be discussed in the same section as well.

12.6 Capability Modeling for Infrastructure Services in Network Virtualization

The core function provided by an infrastructure service is data transportation across the network infrastructure. The data transport capability of an infrastructure service can be described by two features: (a) the *connectivity* supported by the infrastructure, which can be described by enumerating the pairs of sources and destinations between which the infrastructure can transport data; and (b) the *capacity* of data transportation between each pair of source-destination. Therefore, in order to describe the networking capability of the infrastructure service $\mathcal{S}$ provided by an infrastructure with m ingress ports and n egress ports, a *capability profile* can be defined as the following $m \times n$ matrix:

$$\mathbf{C} = \begin{pmatrix} c_{1,1} & c_{1,2} & \cdots & c_{1,n} \\ c_{2,1} & c_{2,2} & \cdots & c_{2,n} \\ \cdots & \cdots & \cdots & \cdots \\ c_{m,1} & c_{m,2} & \cdots & c_{m,n} \end{pmatrix} \tag{12.1}$$

where the matrix element $c_{i,j}$ is defined as

$$c_{i,j} = \begin{cases} 0 & \text{no route from ingress } i \text{ to egress } j \text{ exists in } \mathcal{S} \\ Q_{i,j} & \text{a route } R_{i,j} \text{ from ingress } i \text{ to egress } j \text{ exists in } \mathcal{S} \end{cases} \tag{12.2}$$

and $Q_{i,j}$ is called the *capacity descriptor* for data transportation from ingress i to egress j through this infrastructure.

In the above-defined capability profile, element $c_{i,j} = 0$ when the network infrastructure does not support data transportation from the ingress i to the egress j. Therefore, all non-zero elements in this profile describe the connectivity feature of the infrastructure service. If the infrastructure provides a route from i to j, then the data transport capacity is described by the descriptor $Q_{i,j}$, which is defined as follows.

The concept of *service curve* from *network calculus* theory [5] is adopted here for designing a general capacity descriptor. Let $T^{\text{in}}(t)$ and $T^{\text{out}}(t)$ respectively be the accumulated amount of traffic of a flow that arrives at and departs from a server by time t. Given a non-negative, non-decreasing function, $S(\cdot)$, we say that the server guarantees a *service curve* $S(\cdot)$ for the flow, if for any $t \geq 0$ in the busy period of the server,

$$T^{\text{out}}(t) \geq T^{\text{in}}(t) \otimes S(t) \tag{12.3}$$

where $\otimes$ denotes the convolution operation in min-plus algebra defined as

$$h(t) \otimes x(t) = \inf_{s:0 \leq s \leq t} \{h(t-s) + x(s)\}$$

Essentially, a service curve gives the minimum amount of service capacity that a server offers to a customer, which is independent of the server implementation.

A Latency-Rate (LR) server gives a more tractable capacity descriptor for data transportation in infrastructure services. If a server guarantees a service curve

$$S(t) = \max\{0, r(t - \theta)\}$$

to each flow, then the server is called an LR server, denoted as $P[r, \theta]$, where the θ and r are respectively called the latency and rate parameters of the server.

In the capability profile for an infrastructure service, the service curve guaranteed by the network route $R_{i,j}$ is employed as the capacity descriptor $Q_{i,j}$. Since a service curve is a general data structure independent of infrastructure implementations, it is flexible enough to describe the transport capacities of heterogeneous infrastructures with different network architectures and implementation technologies. In a network infrastructure where a route $R_{i,j}$ can be modeled by an LR server $P[r_{i,j}, \theta_{i,j}]$, the capacity descriptor for the route can be represented by a data structure with two parameters $[r_{i,j}, \theta_{i,j}]$, which is called an LR descriptor.

An LR server gives a capacity model for network connections provided by typical network infrastructures. Typically, an SP requires each InP to provide a certain amount of data transport capacity (the minimum bandwidth), and such a transport capacity guarantee is described by the rate parameter r in an LR descriptor. Data transportation in a network infrastructure also experiences a fixed part of delay that is independent with traffic queuing behavior in the infrastructure; for example signal propagation delay, link transmission delay, router/switch process delay, etc. The latency parameter θ of an LR descriptor is to characterize this fixed part of delay, which may be seen as the worst-case delay experienced by the first traffic bit in a busy period of a networking session through this infrastructure.

Currently there are various mechanisms available for measuring and managing network state information, for example the technologies reported in [16, 25], which could be used to obtain the data for constructing the capacity descriptors and the capability profile for a network infrastructure. The methods of collecting network state information are implementation dependent and may vary in different infrastructures, but the capability profile with its service curve-based capacity descriptors provides all infrastructure services with a general and standard approach to describing their networking capabilities.

12.7 Evaluation of Service Performance Bounds in Network Virtualization

12.7.1 End-to-End Service Capability Modeling

In a large scale networking environment such as the Internet, a path for end-to-end service delivery typically traverses multiple network infrastructures. In the SOA-based network virtualization, the data transportation provided by each underlying infrastructure is virtualized as an infrastructure service, and the transport capacity offered by each infrastructure can be described by a capacity descriptor in the service capability profile. The end-to-end network service is offered by an SP through synthesizing a set of infrastructure services; therefore the end-to-end service delivery system consists of a series of tandem infrastructure service components. The first step toward analyzing end-to-end service performance is to compose the capacity descriptors of all the infrastructures in this service delivery system into one descriptor that models the data transport capacity of the end-to-end path. The service curve-based capacity description presented in Section 12.6 supports such composition of capacity descriptors.

Assume that a network service delivery system consists of service components $S_1, S_2, \ldots, S_n$, which respectively guarantees the capacity descriptors $S_1(t), S_2(t), \ldots, S_n(t)$, then it is known from network calculus that the capacity descriptor for the end-to-end service delivery system, denoted as $S_e(t)$, can be obtained as

$$S_e(t) = S_1(t) \otimes S_2(t) \otimes \cdots \otimes S_n(t) \tag{12.4}$$

Suppose each component S_i guarantees an LR descriptor; that is, $S_i(t) = P[r_i, \theta_i] = \max\{0, r_i(t - \theta_i)\}$, it can be proved that the capacity descriptor of the end-to-end service delivery system is

$$P[r_e, \theta_\Sigma] = P[r_1, \theta_1] \otimes \cdots \otimes P[r_n, \theta_n] \tag{12.5}$$

where

$$r_e = \min\{r_1, r_2, \ldots, r_n\} \quad \text{and} \quad \theta_\Sigma = \sum_{i=1}^{n} \theta_i \tag{12.6}$$

Equations (12.5) and (12.6) imply that if each service component in an end-to-end service delivery system can be modeled by an LR server, then the transport capacity of the entire service system can be described by an LR descriptor. The latency parameter of the end-to-end system is the summation of the latency parameters of all service components in the system, and the end-to-end service rate parameter is the minimum service rate of all the service components.

12.7.2 Load Profile for Service Delivery System

Service performance evaluation also requires an approach to characterizing the traffic load on a service delivery system. Let $T^{in}(t)$ denote the accumulated amount of traffic that arrives at the entry of a service delivery system by the time instant t. Given a non-negative, non-decreasing function, $A(\cdot)$, the service delivery system is said to have a *load profile* $A(\cdot)$ if for any time instant $s : 0 < s < t$,

$$T^{\text{in}}(t) - T^{\text{in}}(s) \leq A(t - s) \tag{12.7}$$

The load profile defined above is equivalent to an arrival curve in network calculus. The load profile gives an upper bound for the amount of traffic that the end user can load on a service delivery system. Defined as a general function of time, this profile is applicable to describe the traffic loaded by various applications onto a service delivery system.

Currently, most QoS-capable networking systems apply traffic regulation mechanisms at network boundaries to shape arrival traffic from end users. The traffic regulators most commonly used in practice are leaky buckets. A service delivery system under a networking session constrained by a leaky bucket has a load profile

$$A(t) = \min\{pt, \sigma + \rho t\}$$

where p, ρ, and σ are respectively the peak rate, sustained rate, and maximum burst size of the traffic load.

12.7.3 Bandwidth and End-to-End Delay Bounds

Among various network service performance parameters, the evaluation techniques presented in this section focus on the minimum bandwidth and the maximum service delay, which are two important performances required by most QoS-sensitive applications. Network calculus provides an effective approach to analyzing the minimum bandwidth and maximum delay performances guaranteed by an end-to-end network service. Given the end-to-end capacity descriptor $S_e(t)$ for a service delivery system, the minimum bandwidth guaranteed by this system can be determined as

$$b_{\min} = \lim_{t \to \infty} [S_e(t)/t] \tag{12.8}$$

Suppose the traffic load on the system is described by the profile $A(t)$, then the maximum delay performance of the system is

$$d_{max} = \max_{t:t\geq 0}\left\{\min\{\delta : \delta \geq 0\ \ A(t) \leq S(t+\delta)\}\right\} \tag{12.9}$$

Since the LR server models the data transport capacities provided by typical infrastructure services and the leaky bucket traffic regulator is widely deployed in network infrastructures, we are particularly interested in evaluating end-to-end bandwidth and delay performances of service delivery systems with an LR capacity descriptor and a leaky bucket load profile. Suppose an end-to-end service delivery system has an LR capacity descriptor $P[r_e, \theta_\Sigma]$ and a load profile $A(t) = \min\{pt, \sigma + \rho t\}$, then from equations (12.8) and (12.9) it can be shown that the minimum bandwidth and the maximum delay guaranteed by the system are respectively

$$b_{\min} = \lim_{t\to\infty} \frac{r_e(t-\theta_\Sigma)}{t} = r_e \tag{12.10}$$

$$d_{\max} = \begin{cases} \theta_\Sigma + \left(\frac{p}{r_e} - 1\right)\frac{\sigma}{p-\rho} & (p > \rho,\ r \geq \rho) \\ \theta_\Sigma & (p = \rho,\ r \geq \rho) \end{cases} \tag{12.11}$$

12.7.4 A Simple Numerical Example

A simple numerical example is given here to illustrate the application of the performance evaluation technique. Suppose a service delivery system crossing three network infrastructures to provide a packet stream delivery. The network infrastructures include a local area network with 100 Mb/s link capacity (e.g., a switching Ethernet); a wireless access network with 50 Mb/s link capacity (e.g., a WiMAX network); and a long distance backbone network with 1 Gb/s link capacity. The data transportation capability offered by each network infrastructure to the service delivery system is given by an LR descriptor. Assume that the packet stream traffic is shaped by a leaky bucket regulator when entering the network infrastructure, and the load profile parameters are peak rate $p_1 = 3.2$ Mb/s, sustained rate $\rho_1 = 1.12$ Mb/s, and the maximum burst size $\sigma_1 = 1.13$ Mbits [6]. The relation between the end-to-end delay upper bound achieved by the service delivery system and the available transport capacity that the service system obtains from underlying infrastructures is given in Figure 12.5. This figure shows that the maximum end-to-end service delay is a decreasing function of the available infrastructure service capacity. This implies that given a service delay requirement,

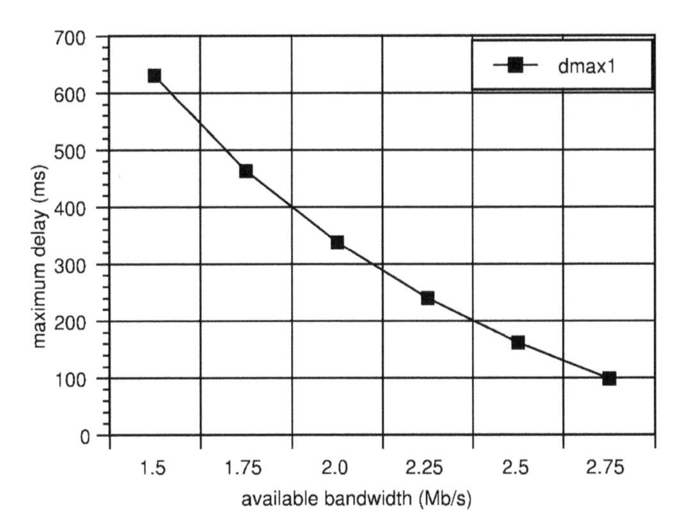

Figure 12.5 The maximum end-to-end service delay and the available infrastructure transport capacity.

the service broker must discover and select the infrastructure service(s) that provide sufficient amount of transport capacity for the SP.

12.7.5 Performance-Based Infrastructure Service Discovery

The above developed techniques for evaluating service performance bounds can be employed by a service broker in a SOA-based network virtualization environment to discover, select, and compose the appropriate infrastructure services in order to provide end-to-end network services that meet application performance requirements. Three aspects of information are needed by a service broker for performance-based infrastructure service discovery: (a) the networking capabilities of available infrastructure services; (b) the performance requirements of the end-to-end service; and (c) the characteristic of network traffic load of the service. Information (a) can be obtained from the capability profiles published by InPs as a part of their service descriptions. Information (b) and (c), which specify the demand of a network service, should be provided by the SP to the service broker.

In order to describe the highly diverse networking demands of the service requests from various SPs, a general *demand profile* should consist of at least three elements: an address set that specifies the source and destination addresses for data transportation; a load profile that characterizes the traffic of a network service; and a performance requirement set that gives the required performance parameters for the service. Different performance parameters may be required by various services but the minimum bandwidth and the maximum delay for data transportation are typical requirements. Such a demand profile can be submitted by an SP to the service broker as a part of its infrastructure service request.

After receiving a service request from an SP, the broker searches the infrastructure service descriptions published at the service registry and discover one or (more typically) a sequence of infrastructure services that can provide data transportation between the source and destination specified in the demand profile. The broker can employ the developed performance evaluation techniques to predict the end-to-end service performance bounds that can be achieved by the discovered infrastructure services. The capability profiles of the infrastructure services, which are included in the infrastructure service descriptions published at the service registry, provide the information needed by the performance evaluation process. The broker then compares the predicted achievable performance bounds with the performance requirements given in the demand profile to decide if the discovered infrastructure services can be selected for offering the end-to-end network service.

The above-mentioned technique focuses on describing and discovering network infrastructure services to meet performance requirements for end-to-end service provisioning in SOA-based network virtualization environments. Other functions of an infrastructure service broker system include publishing and updating infrastructure service descriptions, searching a service registry for available infrastructure services, and composing infrastructure services into virtual network services. SOA and Web services standards provide a framework for service description, registration, discovery, and composition; however their applications in the field of networking in general and network virtualization in particular are still open for research. Though this area is still in its early stages, it has attracted extensive research interest. The cross-fertilization between SOA and network virtualization, which traditionally are two independent fields, could come up with innovative ideas and solutions that may significantly contribute to the construction of the future Internet.

12.8 Conclusions

Various heterogeneous networking systems with different architectures and implementation technologies will coexist in the next generation Internet to provide services for meeting a wide spectrum of application requirements.

To face this challenge, network virtualization has been proposed as a key diversifying feature of the future inter-networking paradigm and is expected to play a crucial role in the next generation Internet. One of the key technical issues for realizing network virtualization centers on effective interactions between network service providers and infrastructure providers to enable flexible collaborations across heterogeneous infrastructures for end-to-end service provisioning. The SOA emerging in Web services and Grid computing areas provides an effective architecture for coordinating heterogeneous systems to meet diverse application requirements, which is essentially the same challenge faced by network virtualization in the Internet. This chapter discusses application of the SOA principles in network virtualization and develops a SOA-based network virtualization paradigm. In this new networking paradigm, the networking capabilities of infrastructures are virtualized as infrastructure services, which are composed by service providers into end-to-end network services. The SOA-based network virtualization enables loose coupling interactions between service delivery systems and underlying network infrastructures and allows flexible collaboration across heterogeneous infrastructures, which may greatly facilitate end-to-end service provisioning in the future Internet. This chapter particularly discusses two key technologies for the SOA-based network virtualization – capability modeling of infrastructure services and performance evaluation for end-to-end service delivery. A general approach to modeling and describing networking capabilities of heterogeneous infrastructures is presented, and analytical techniques for evaluating end-to-end service performance in SOA-based network virtualization are developed. The service capability model and performance evaluation methods presented in this chapter are applicable to various heterogeneous network infrastructures in the future Internet.

References

[1] C.E. Abosi, R. Nejabati, and D. Simeonidou. A novel service composition mechanism for the future optical internet. *Journal of Optical Communication Networks*, 1(2):106–120, 2009.

[2] T. Anderson, L. Peterson, S. Shenker, and J. Turner. Overcoming the Internet impasse through virtualization. *IEEE Computer*, 38(4):34–41, 2005.

[3] A. Bavier, M. Bowman, B. Chun, D. Culler, S. Karlin, S. Muir, L. Peterson, T. Roscoe, T. Spalink, and M. Wawrzoniak. Operating system support for planetary-scale network services. In *Proceedings of the 1st Symposium on Network System Design and Implementation (NSDI'04)*, 2004.

[4] F. Belqasmi, R. Glitho, and R. Dssouli. Ambient network composition. *IEEE Networks*, 21(1):47–52, January 2008.

[5] J.L. Boudec and P. Thiran. *Network Calculus: A Theory of Deterministic Queueing Systems for the Internet*. Springer Verlag, 2001.

[6] M. Butto, E. Cavallero, and A. Tonietti. Effectiveness of the leaky bucket policing mechanisms in ATM networks. *IEEE Journal of Selected Areas of Communications*, 9(4):335–342, 1991.

[7] K. Channabasavaiah, K. Holley, and E. Tuggle. Migrating to a Service-Oriented Architecture. *IMB DeveloperWorks*, 2003.

[8] N.M.M.K. Chowdhury and R. Boutaba. Network virtualization: State of the art and research challenges. *IEEE Communications*, 47(7):20–26, 2009.

[9] L.F. Cabrera et al. Web Services Atomic Transaction (WS-Atomic Transaction) version 1.0 available at http://www.ibm.com/developerworks/library/specification/ws-tx/, 2005.

[10] L.F. Cabrera et al. Web Services Coordination (WS-Coordination) version 1.0 available at http://www.ibm.com/developerworks/library/specification/ws-tx/, 2005.

[11] N. Feamster, L. Gao, and J. Rexford. How to lease the Internet in your spare time. *ACM SIGCOMM Computer Communications Review*, 37(1):61–64, 2007.

[12] I. Foster and C. Kesselman. *The Grid: Blueprint for a New Computing Infrastructure*. Elsevier, 2004.

[13] GENI Planning Group. GENI design principles. *IEEE Computer*, 39(9):102–105, 2006.

[14] J. Ham, P. Grosso, R. Pol, A. Toonk, and C. Laat. Using the network description lanaguage in optical networks. In *Proceedings of the 10th IFIP/IEEE International Symposium on Integrated Network Management*, May 2007.

[15] P. Jacobs and B. Davie. Technical challenges in the delivery of interprovider QoS. *IEEE Communications*, 43(6):112–118, 2005.

[16] A. Kind, X. Dimitropoulos, S. Denazis, and B. Claise. Advanced network monitoring brings life to the wareness plane. *IEEE Communicatins*, 46(10):140–146, 2008.

[17] S. Lacour, C. Perez, and T. PriolA. Network topology description model for Grid application deployment. In *Proceedings of the 5th IEEE/ACM International Workshop on Grid Computing*, 2004.

[18] N. Niebert, S. Baucke, I. El-Khayat, M. Johnsson, B. Ohlman, H. Abramowica, K. Wuenstel, H. Woesner, J. Ouittek, and L.M. Correia. The way 4WARD to the creation of a future Internet. In *Proceedings of the IEEE 19th International Symposium on Personal, Indoor and Mobile Radio Communications*, 2008.

[19] OASIS. Universal Description, Discovery and Integration (UDDI) version 3.0.2, 2005.

[20] OASIS. Reference Model for the Service-Oriented Architecture version 1.0, 2006.

[21] OASIS. Business Process Execution Language for Web Services (BPEL-WS) version 1.1, 2007.

[22] Open Grid Forum (OGF). Web Services Agreement Specification (WS-Agreement), 2007.

[23] The Open Mobile Alliance (OMA). OMA Service Environment Architecture, 2007.

[24] M.P. Papazoglou. *Web Services: Principles and Technology*. Pearson, 2008.

[25] R. Prasad, M. Murray, C. Dovrolis, and K. Claffy. Bandwidth estimation: metrics, measurement techniques, and tools. *IEEE Network*, 17(6):27–35, 2003.

[26] G. Schaffrath, C. Werle, P. Papadimitriou, A. Feldmann, R. Bless, A. Greenhalgh, A. Wundsam, M. Kind, O. Maennel, and L. Mathy. Network virtualization architecture:

proposal and inital prototype. In *Proceedings of the 2009 ACM Workshop on Virtualized Infrastructure Systems and Architectures*, 2009.

[27] J. Song, M.Y. Chang, S.S. Lee, and J. Joung. Overview of ITU-T NGN QoS control. *IEEE Communications*, 45(9):116–123, 2007.

[28] J. Turner and D.E. Taylor. Diversifying the Internet. In *Proc. of IEEE Global Communication Conference*, 2005.

[29] World Wide Web Consortium (W3C). Web Services Choreography Description Language version 1.0, 2004.

[30] World Wide Web Consortium (W3C). Simple Object Access Protocol (SOAP) version 1.2, 2007.

[31] World Wide Web Consortium (W3C). Web Service Description Language (WSDL) version 2.0, 2007.

[32] World Wide Web Consortium (W3C). Web Service Policy Attachment (WS-PolicyAttachment) version 1.5, 2007.

[33] World Wide Web Consortium (W3C). Web Service Policy Framwork (WS-Policy) version 1.5, 2007.

13

Enabling Wireless Sensor Network within Virtual Organizations

Salvatore F. Pileggi, Carlos E. Palau and Manuel Esteve

Departamento de Comunicaciones, Universidad Politécnica de Valencia (UPV), Camino de Vera S/N, 46022, Valencia, Spain; e-mail: salpi@upvnet.upv.es, {cpalau, mesteve}@dcom.upv.es

Abstract

During the last few years sensor systems and networks have been the subject of an intense research activity that has determined relevant advances for both basic technologies (e.g. wireless technology, smart processors and memories) and architectures. Current advanced platforms potentially allow several large-scale business scenarios in which complex services concretely work in the context of virtual organizations. In the future, the key issue for the massive dissemination of services based on WSN will be, probably, the result of the consolidation of the convergence among several outstanding issues: the improvement of the reliability of WSN architectures due also to a general costs reduction; the standardization of infrastructures will be able to provide an interoperable and flexible interaction model useful in the context of both common Service Oriented Architectures and the Internet of Things; finally, semantic technologies (according standardized ontologies) could enable innovative human-service and service-service interaction models.

Keywords: sensor networks/systems, virtualization, virtual organizations, Service-Oriented Architecture, sensor web, semantic technologies, semantic sensor web.

Anand R. Prasad et al. (Eds.), Future Internet Services and Service Architectures, 271–293.

13.1 Introduction

During the last years, sensors have been increasingly adopted in the context of several disciplines and applications (military, industrial, medical, homeland security, etc.) with the aim of collecting and distributing observations of our world in everyday life [9]. Sensors progressively assumed the critical role of technological bridges between the real world and information systems, through increasingly more consolidated and efficient solutions that enable advanced heterogeneous sensor grids.

Sensors are currently located everywhere and the relevance of their role is growing in everyday life. They can work as independent stand-alone objects or as part of complex networks, performing cooperative tasks in order to reach common goals. Current sensor networks are able to detect and identify simple phenomena or measurements as well as complex events and situations.

Regardless of domain specific purposes, or concrete sensor technologies, physical sensors are commonly embedded within electronic systems in which generated information is available for final applications. During the last few years sensors were progressively adopted in the context of more complex applications: also due to advances in low power wireless technology, sensors often act as remote objects and active part of large-scale heterogeneous networks. The infrastructure (commonly referred to as middleware) for accessing information, as well as for managing, controlling and monitoring sensors resources, assumes the key role of virtualizing physical resource [18]: virtual resources could be embedded within real systems according to several models and perspectives.

On the other hand, business and exploitation models are also quickly evolving: applications and/or resources sharing, both with an increasing complexity of interaction models, among heterogeneous actors not only determines the enabling of services on a large scale (Internet) but also a computation model intrinsically more complex that enables services in the context of Virtual Organizations [15]. The concept of Virtual Organization (VO) refers to a dynamic set of individual and/or institutions defined around a set of resource-sharing rules and conditions. All these virtual organizations share some commonality among them, including common concerns and requirements, but may vary in size, scope, duration, sociology, and structure.

The following are examples of VOs: "the application service providers, storage service providers, cycle providers, and consultants engaged by a car manufacturer to perform a scenario evaluation while planning a new factory; members of an industrial consortium bidding on a new aircraft; a crisis management team and the databases and simulation systems that they use to plan a response to an emergency situation; and members of a large, international, long-term high-energy physics collaboration. Each of these examples represents an approach to computing and problem solving based on collaboration in computation- and data-rich environments" [15].

At the moment, the massive exploitation of sensor services, despite a constantly growing demand, appears to be limited by several factors characterized by the subtle coexistence of economic and technical aspects.

First of all, large-scale sensor systems are intrinsically expensive solutions [16]: even low-cost advanced devices are currently available and wireless technologies allow a cost reduction for deployments, maintenance implies human intervention; this could result as especially hard in hostile environments; periodic maintenance is required for battery-powered devices; furthermore, in order to assure high-performance (normally in terms of lifetime) and a certain robustness (e.g. fault tolerance) and reliability, architectures have to include several software layers that implement an extended number of functionalities for communication, self-managing and configuration of the networks; the greater part of sensor architectures are ad-hoc solutions due to the great extension and articulation of application domain and the lack of standards of protocols and mechanisms.

Furthermore, the lack of standardization also impacts data models and access interfaces; this is a great limitation for suitable integration among heterogeneous systems as well as for the building of web-centric applications interacting with several data sources or sensor systems. The evolution of a web to a semantic-based model (Semantic Web) implicitly determines a convergence with new models for knowledge representation and building. Semantic Web is a group of methods and technologies to allow machines to understand the meaning or "semantics" of information on the World Wide Web [12].

Innovative strategies in the development of large-scale Sensor Systems according a general cost reduction both with important efforts for the standardization of protocols, mechanisms, data models and access interfaces are expected in the future. At the same time, convergence with emerging technologies (as semantics) could provide an additional challenge: machine interpretable semantics annotation could improve the interaction models [19] among intelligent actors (e.g. Agents) that could be able to automatize the resource discovery and other tasks commonly performed by humans or with human supervision.

This chapter is structured in two main sections that have, respectively, the aim of providing a short but exhaustive overview on Sensor Systems and Networks (including a brief state of the art of sensor technologies and related current limitations) and on overall Sensor Service Architectures in the context of several logic environments (Virtual Organization [15], Sensor Web [7] and Semantic Sensor Web [27]). These environments are progressive concepts, each one characterized by its own issues and challenges.

13.2 Sensor Systems and Networks

In this section the generic concept of Sensor System/Network is restricted to Wireless Sensor Network (WSN [14]); this is due to two main reasons: WSN is considered the most flexible and complex sensor architecture as well as its protocol stack being the most extended and structured; other sensor systems could be considered as a simplification (or sub case) of WSN in one or more aspects.

WSN technology is the result of the convergence of wireless communication mechanisms, limited resource computation and sensor technologies as in the common means; moreover, wireless communication capabilities allow random deployment of nodes (hard and more expensive using wired sensors), as well as mobile environments similar to Mobile Ad-Hoc Networks (MANETs [10]).

Modern wireless sensor platforms [14] are advanced smart nodes that provide relatively limited computation, storage, communication and sensing resources. These platforms, even if autonomous nodes, are designed ad-hoc for advanced networking guaranteeing high performance in event-driven environment for their ability of "sleeping" (low-power consumption state) during inactive periods. WSN architectures implicitly assume the presence of one or more high-resource devices (called Sink or Base Station) with the main aim of collecting data generated by sensor nodes.

As shown in Figure 13.1, the application domain of WSN potentially covers several disciplines that include (but are not limited to) environmental monitoring, saving resources, enabling new knowledge, enhancing safety and security, improving food, high confidence transport, preventing failures, military use, improving productivity, process control and health; this really addresses several theoretical points of view on the definition and structuring of main domain: the greater part of approaches assumes a unique domain and the structured definition of a number of subdomains in the function of concrete applications; other approaches prefer structuring the main domain

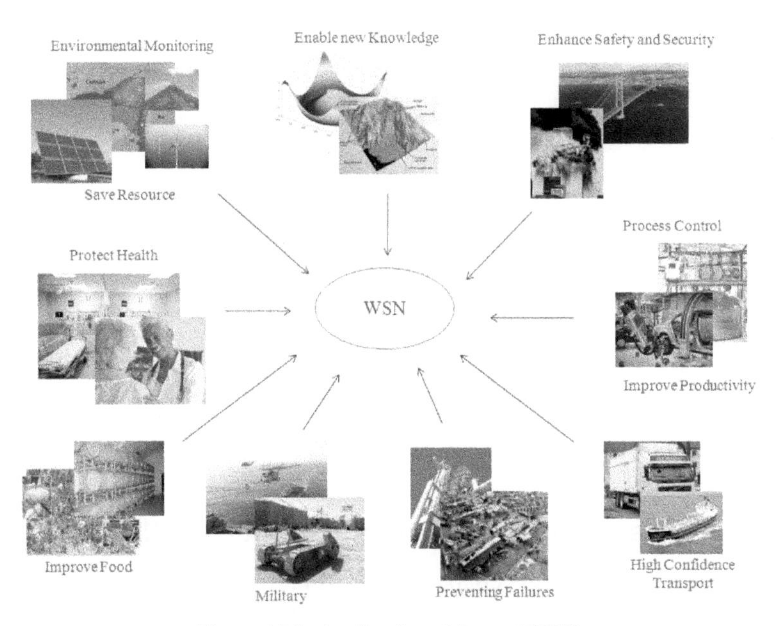

Figure 13.1 Application visions of WSN.

in the function of host behaviors (mobile, nomadic or static hosts) as well as others propose a perspective according to topological/network features (large-scale, clustered, etc.); alternative views allow the structuring of the domain considering the Quality of Service (as in [8]).

Regardless of the theoretical approach for the structuring of sensor systems domain, the most direct consequence to an extended and potentially structured domain is the objective difficulty for the engineering of generalized mechanisms and models; even several multi-domain approaches can be adopted, they cannot satisfactorily cover all sensor domains; the greater part of WSN solutions are ad-hoc networks.

Several generalized solutions [11] were proposed during the last years according to a multi-domain approach in which concrete solution are built on the top of a common core framework. Extensions or particularizations of the proposed mechanisms could be required. The performance and the effectiveness of these solutions should be carefully evaluated within the different application domains.

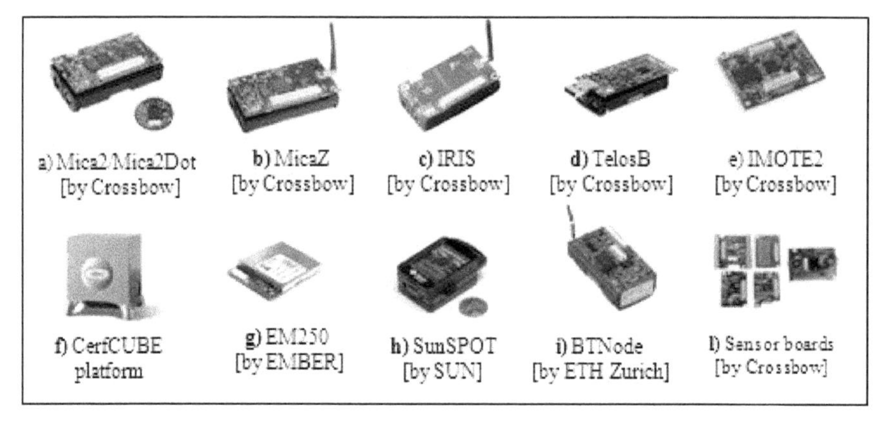

Figure 13.2 Some of the most common Wireless Sensor Node platforms.

13.2.1 Wireless Sensor Network: A Brief State of the Art

As mentioned before in this chapter, WSN is really a multi-disciplinary topic resulting from the convergence of several technologies and fields, each one featured with its own key issues, limitations and challenges. A vertical analysis of each topic, as well as of the relationships and trade-offs among them, although really interesting, is outside the scope of this chapter. In order to provide an exhaustive overview on WSN both with a realistic vision of real architectures, five different points are identified as follows:

- *Basic Technologies.* Current wireless sensor platforms are designed according to a development philosophy oriented to resource optimization under the realistic hypothesis of limited resource devices. Most devices are designed as an event-driven model: they spend the inactive time (the major part of their lifetime) "sleeping" (power consumption next to zero) and they react to internal (timeout expiration) or external (message reception) events by switching their state to active (significant power consumption); the state returns to sleeping at the end of scheduled operations. The major advances of the last few years in the field of low-power and low-rate wireless technology represent the real key issue: last generation wireless technology makes available transmitters designed ad-hoc for robust communication of little amount of impulsive information [6]. Last generation Wireless Technology, both with the miniaturization of low-power processors and memories, allows the concrete deployment of really advanced commercial solutions. Some of the most relevant platform prototypes are represented in Figure 13.2.

- *Software Platforms.* Software platform includes both the operating system for resource managing/accessing and the programming environment. Current solutions can be compact versions of generic-purpose operating systems (e.g. Linux), solutions directly resulted by embedded systems or ad-hoc platformas designed for wireless sensor networks. The most relevant ad-hoc operative system is TinyOS [1] that proposes an extremely efficient managing models both with a programming environment based on NesC (a module-oriented C-like language optimized for limited resource and event-driven models). During last few years alternative environments based on JAVA and .NET technologies were proposed; on the one hand they address an interesting convergence point with related general purpose technologies, on the other hand their performances appear, at the moment, inferior to C-like platforms.
- *Protocol Stack.* This is the software implemented and executed by sensor nodes. Due to the extended application domain, protocol stack can be featured by a variable number of layers that can propose a different complexity. This same reason also determines an objective difficulty for defining an unique reference model for the protocol stack as in other networks or domains. A large part of the available models, even if not explicitly related to concrete applications, is built according to a global approach (e.g. e-SENSE protocol stack architecture [26]) that includes functionalities characterized by a different level of abstraction (e.g. wireless sensor network and service infrastructures).
  In the context of this chapter, any high level infrastructure for enabling physical resources within final systems is considered out of protocol stack architecture and is referred to as middleware infrastructure.
  A possible simplified approach (Figure 13.3) includes several interdependent layers, each one composed by one or more protocols or mechanisms. This model does not match with any concrete implementation.
  A core set of protocols defines basic functionalities (communication); due to high density that characterizes the majority of real topologies and the need of efficient communication with Base Station(s), communication protocols are designed according a multilayer approach that includes, at least, two hierarchical organized levels that define extended MAC functionalities and routing strategy [3] (shortest path, energy-aware, location-aware, etc.). Additional mechanisms for the improving of reliability and robustness (e.g. replicated information sent through several paths) can be integrated at this level. The CORE functionalities

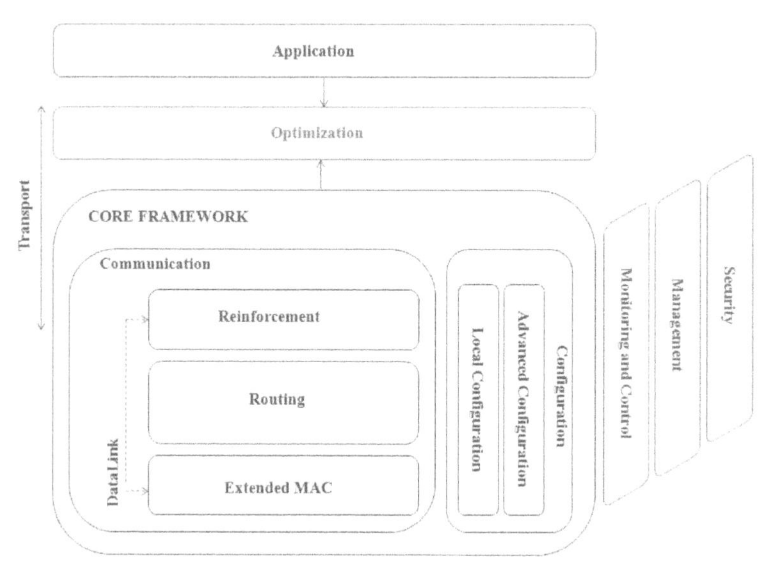

Figure 13.3 A reference model for Protocol Stack.

also include protocols for network configuration on which communication mechanisms logically work; also this set of functionalities is modeled on several layers that include basic mechanisms (e.g. neighbors discovery, local configuration and routing table performing) as well as overlay configuration (e.g. clustering [2], Figure 13.4, or other topology control schema). Under the assumption of several base stations, that are simultaneously working, clustering techniques (Figure 13.4) allow a n-size network to work as m sub-networks of n/m average size. This increases, in determinant way, the scalability of WSN solutions in terms of performance (lower consumptions due to shortest paths) and robustness. All high-level mechanisms (management, monitoring, control, security) are logically built over CORE functionalities as well as application level mechanisms.

- *Middleware.* Middleware infrastructure progressively assumed a central role within WSN architectures. Its fundamental aim is enabling devices and resource within main systems allowing pervasive access and remote control. The reference model is, at the moment, SOA (Service oriented architecture); this is mainly for its modularity that allows suitable integration of embedded services within complex systems and for the functional interoperability assured by access interfaces defined according interoperable formats; furthermore, the concrete application of some

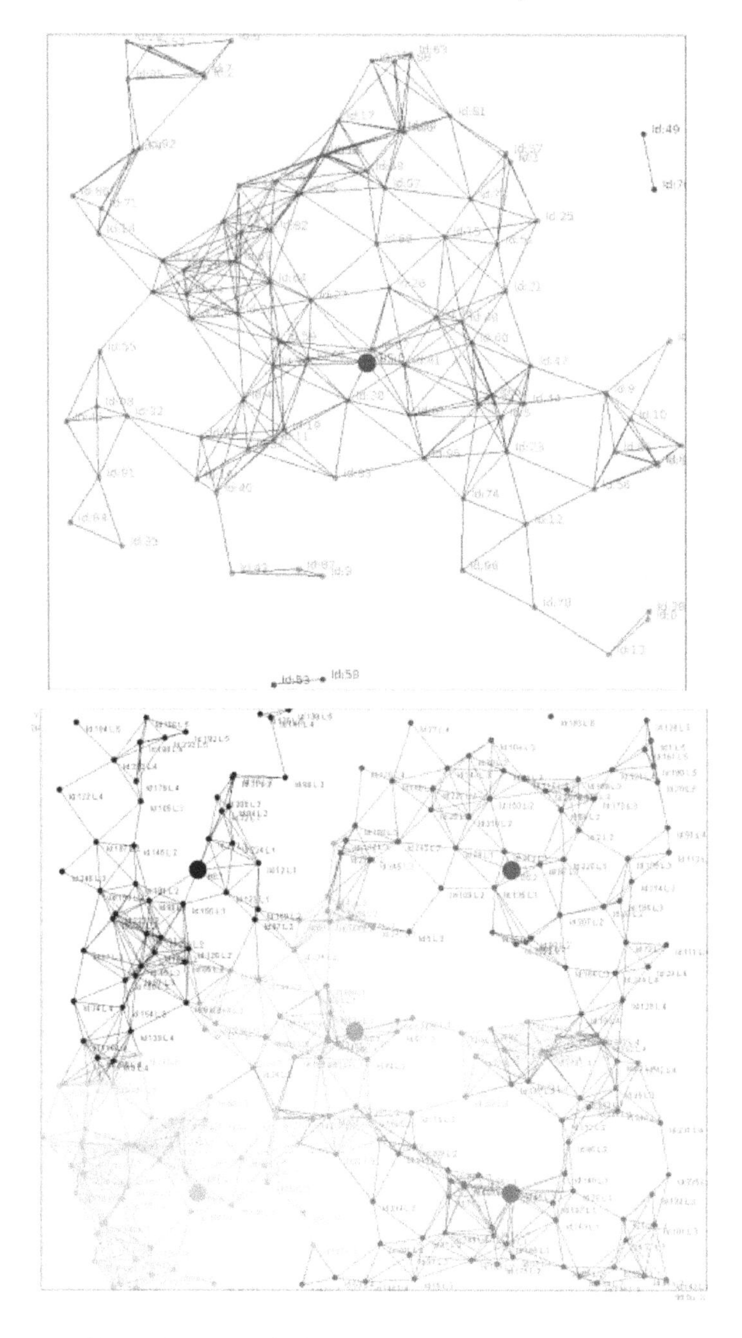

Figure 13.4 Network vision (up) and clustered vision (down).

of the most relevant principles of software engineering (as the clear differentiation between service interface and service implementation) makes SOA as a suitable and flexible solution for the implementation of complex approaches (e.g. content-aware, context-aware services). Advanced infrastructures were developed on improved service models as Grid Technologies [17, 22]. Grid computing is the combination of computer resources from multiple administrative domains applied to a common task, usually to a scientific, technical or business problem that requires a great number of computer processing cycles or the need to process large amounts of data [15]. Adopting flexible models also enables dynamic infrastructures to support emerging heterogeneous environments as Internet of Things [29] or Cooperative Objects.

- *Modeling, Simulation and Evaluation.* Theoretical models, as well as development methodologies, appear as the next to be consolidated. Solid theoretical models could enable more effective simulation techniques, as well as standardized evaluation models could provide a preliminary comparison among different architectures and complex mechanisms acting in the same application domain [23].

13.2.2 Sensor Networks within Virtual Organizations

Regardless of the internal composition, structure and complexity, all VOs, as defined in the previous sections, propose a common convergent point of sharing resources. VOs are spontaneous relationships among individuals and institutions that implicitly reflect the increasing dispersion of resources and the intrinsic complexity of modern society.

The flexible, secure and coordinated resource sharing among dynamic collections of individuals, institutions, and resources requires complex distributed computing models in order to assure dynamic authentication, authorization, resource access, resource discovery, and to meet other challenges.

An exhaustive discussion about the main issues and challenges related to dynamic sharing of resources as well as solutions to approach the enabling of generic resources within VOs is outside the scope of this chapter.

Enabling sensor resources within VOs allows the development of novel technical and business scenarios. As example, Social Sensors [25] will be considered. Social Sensors are understood as a technological bridge that allows sensor data and resource to be published and shared within social networks (Figure 13.5).

Figure 13.5 A schematic representation of Social Sensors.

Concrete examples of social networks potentially interested in sensor resources sharing are people that practice sports that need detailed information about environmental conditions (surf, windsurf, climbing, etc.), scientific communities interested in environmental or personal data monitoring, and so on. Each one of these examples has its own key issues (e.g. privacy) and challenges (e.g. cost).

Each social network has its own internal organization that commonly results in the definition of several roles on the base of members preferences. A dynamic sharing model provides a flexible, advanced (e.g. context-aware) and efficient infrastructure for the interaction with resources and data in relationship with user role(s).

At the same time, a resource is shared by more than one social network. Due to the different internal organization, interests and goals of considered social networks, also the interaction model is characteristic of a concrete social network. For instance, the complexity of the interaction model increases both with the complexity of the internal organization of social networks as well as in presence of complex business models and actors (outside a single social network) with special grants or roles (e.g. resource manager).

13.2.3 From Science to Reality: Current Limitations

The intense research activity of the last few years and the solid feedback provided by the constantly increasing commercial interest determine a relatively consolidated technology for Wireless Sensor Network. However, the massive dissemination of large-scale WSNs is not happening. Figure 13.7 resumes an analysis on the issue provided by onWorld WSN report (2005).

As shown, with the conviction that massive sensor dislocation is required only in certain environments, there is great satisfaction with respect to device size, bit-rate and architecture security; also consumption (in relation with available batteries) and cost appear satisfactory. Several doubts are related with interoperability and interferences and, finally, an absolute dissatisfaction is detected with respect to the reliability or robustness of overall architectures and on the ease of installation and maintenance.

This report is surely a good reference even if, probably, it just provides a simplified analysis of the current limitations; the main limitation of this analysis is considering each reported issue as independent with respect to the others. Where this is not the case, the lack of interoperability determines a strong impact on the overall cost of large-scale architectures mainly due to the lack of standards at several levels; novel business scenarios and models are hard to propose in reality. The real costs are also affected by the massive presence of wireless technologies in everyday life that make several environments (e.g. metropolitan areas) as critical and hostile from a communication point of view especially considering that the greater part of WSN works adopting low-power wireless technologies. Furthermore, real architectures require a higher level of reliability and robustness than supervised and/or laboratory experiments; the cost of improving the reliability/robustness as well as any other non-functional requirements normally increases with the scale of network or system. At the same time, the maintenance is a critical task respect to costs due to the cost of human intervention; it could be the determinant issue in hostile or hard to access environments. Even with the support of self-deploying and management mechanisms, random deployment is currently a critic issue due to the strong differences that can characterize different real environments.

At the moment there are several approaches targeted to realize the massive dissemination into the real world of WSNs:

- First of all, the constant improvement of technologies allows ever more advanced solutions; mayor efforts aim to assure reliable architectures also in the presence of variable and unpredictable conditions. Currently solutions based on clustering appear as the more concretely applicable: cluster headers are represented by Base Stations that allow the partitioning of a network of size n in m clusters of average size n/m [22]. Base Stations infrastructure require a more expensive hardware and mechanisms (according to hierarchical organization and routing) but allow high-scale WSN working as in low-scale solutions.
- An ad-hoc approach appears intrinsic in WSN solutions considering an extremely extended application domain. However, development methodologies could progressively evolve from the current application-specific approach to a domain-specific approach. Moreover, a well structured domain definition could allow multi-domain approaches. A substantial cost reduction could be provided by solution standardization at several levels (e.g. protocol stacks) that makes innovative business scenarios easier to be applied.
- Large-scale WSNs intrinsically imply a certain cost, especially in terms of maintenance. Competitive scenarios are required. Enabling WSN in a logically higher context (Virtual Organization) and on a higher scale (Internet) could and should allow innovative business models based on complex resource sharing. The idea of several interacting VOs that are sharing sensor resource needs (or advises) an evolving sensor web model, which will be subject of the next section.

13.3 Sensor Service Infrastructure

The idea of web-accessible shared sensor resources working on a large scale implies complex multilayer architectures that allow physical resource or

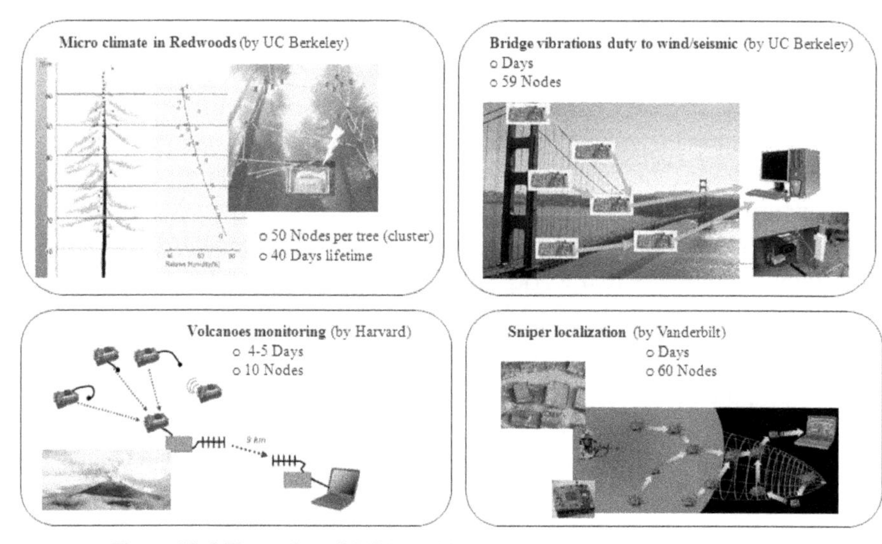

Figure 13.6 Examples of WSN architectures working into the real world.

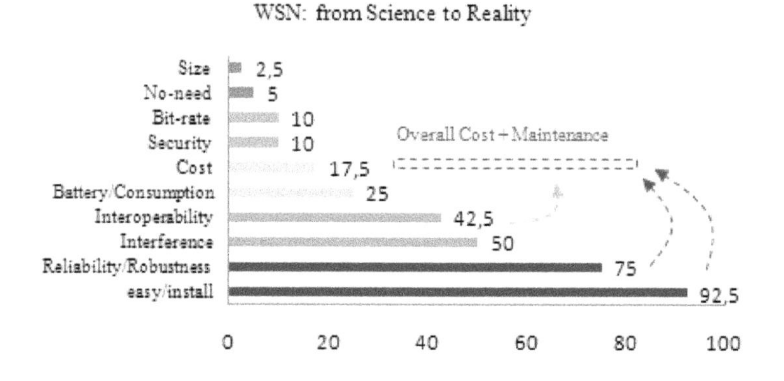

Figure 13.7 Analysis of the current limitations of WSN massive deployments.

data pervasively accessible by multiples actors in the context of complex heterogeneous VOs (Figure 13.8).

The structure of the architecture model represented in Figure 13.8 has three progressive interpretations that will be discussed in the following sections.

These architectures should assure both basic and advanced functionalities in a context of interoperability as well as a support infrastructure for the managing, monitoring and control of physical resources.

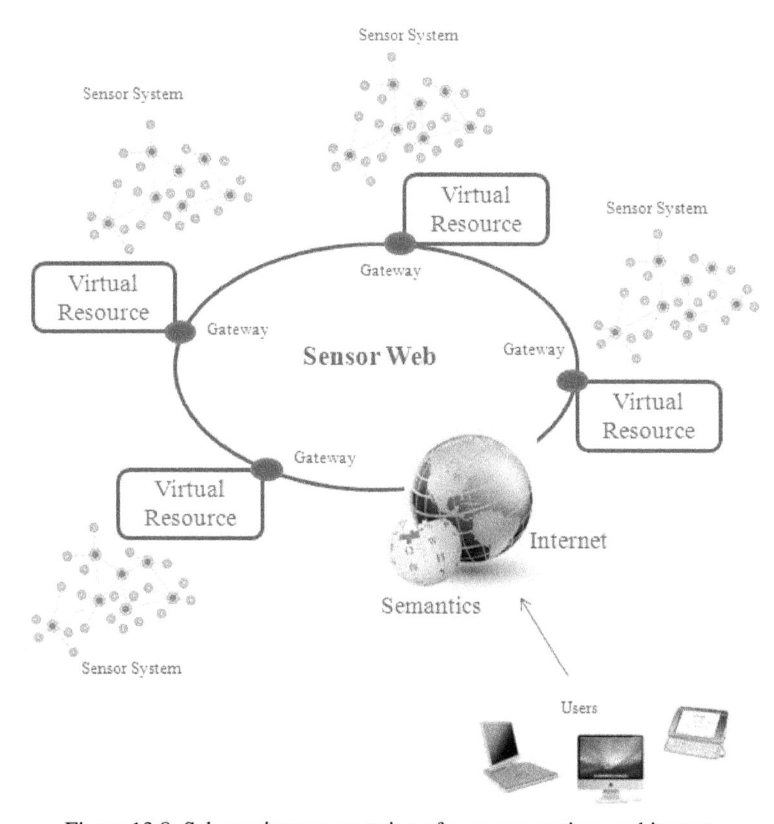

Figure 13.8 Schematic representation of a sensor services architecture.

The design of concrete solutions is determined by services purpose and application domain as well as by interaction model. In fact, sensor services can potentially work within several logic environments according different interaction models. Regardless of the VO's complexity, three progressive interaction models can be identified and they will be analyzed below: first of all Virtual Resource (it assumes sensor resources enabled within global systems), then Sensor Web (web-enabled interoperable environment) and, finally, Semantic Sensor Web (semantic models for interaction and interoperability).

The main aim of this analysis is to emphasize the impact of interaction model on the architectures design.

13.3.1 Virtual Resource

As shown in Figure 13.8, sensor systems and networks are available as Virtual Resources for end users and applications. A Virtual Resource is a highly abstracted set of functionalities that should fully represent systems and networks within global systems.

Service Oriented Architecture proposes a modular approach for embedded systems. Web Service technology allows suitable and interoperable web enabling as well as the capability of designing pervasive interfaces for services and mechanisms.

In practice, a Virtual Resource is a two side set of services (called Gateway or Enabler) designed in order to work within multi-role organizations:

- The front side interface implements all functionalities required by end-users and applications. Independent of concrete systems, resources or generated data can be accessed through this interface eventually according complex policies (e.g. context-aware). Furthermore infrastructures for on-demand data-acquisition and event-driven infrastructures for content pushing can be implemented.
- The backside interface includes all functionalities related to remote resource management, monitoring and control. Resource managing is commonly referred to both information system and network level mechanisms. This class of services is commonly accessed by administrative roles or by network/system owner.

A Virtual Resource is simply a set of interfaces for remote access and managing of sensor networks and systems. This abstraction level physically enables resources within VOs but it does not meet logic requirements: more complex services resulting by the interactions among Virtual Resources (e.g. social interaction [25]) really require a higher level of abstraction and interoperability (Sensor Web).

13.3.2 Sensor Web

As imaginable, the idea of sensor networks disseminated everywhere around the world implicitly assumes they are not connected among them as well as the associated information systems are not integrated. This scenario can be summarized as too much data and not enough knowledge.

The term Sensor Web was first used by Kevin Delin of NASA in 1997 [7] to describe a novel wireless sensor architecture where individual autonomous nodes could act and coordinate as a whole performing stand alone observations or cooperative tasks. In other contexts the term is simply used to refer to sensors connected to the Internet or World Wide Web.

In the context of the last generation systems and networks, Sensor Web is commonly defined as "Web-accessible sensor networks and archived sensor data that can be discovered and accessed using standard protocols and application interfaces."

Sensor web is a general purpose concept that is progressively assuming importance within several application domains such as large-scale geographic information system (GIS) that are especially well suited for environmental monitoring, social sensors and, for instance, for all sensor systems working in accordance with business models that assume reference services enabled within complex VOs.

Sensor Web is a progressive concept that can be suitably modeled on interoperable infrastructure (e.g. Service Oriented Architecture – SOA) but it is partially limited, at the moment, by the lack of standardization at several levels (access interfaces, data models).

Web Service proposes a technological realization of Sensor Web in which services implement Gateways as in the meaning of the previous section. It allows web-accessible virtual resources composition and orchestration, as well as advanced discovery and complex binding according several approaches (e.g. content-aware, context-aware). Computational Grids [15] propose an interesting technological environment for Sensor Web: sensor networks can off-load heavy processing activities to the Grid and Grid-based sensor applications can provide advance services for smart-sensing by deploying scenario-specific operators at runtime.

Sensor Web is a sufficiently general concept to efficiently allow suitable enablement of any sensor system within VOs; at the same time, it is a more specific logic environment in respect to generic Internet of Services [30] theoretically easier to be standardized.

An improved interoperability level could enable several innovative large-scale business scenarios: virtual resources are potentially available and shared in the context of complex VOs that can also interact.

13.3.3 Semantic Sensor Web

Semantic Sensor Web [27] would be an evolving extension of Sensor Web that introduces a semantic layer in which semantics, or meaning of information are formally defined.

Semantics should integrate web-centric standard information infrastructures improving the capabilities of collecting, retrieving, sharing, manipulating and analyzing sensor data (or associate phenomena) as well as potential interoperability between systems through semantic interactions [13].

The convergence of semantic technologies in the Sensor Web could mainly introduce the following advantages:

- Improved interoperability model that can positively affect any aspect directly or indirectly related to intelligent operations (e.g. data analysis).
- Extended rich data models that allow the representation of knowledge according to several perspectives and abstraction levels as well as the ability of inferring knowledge on the base of semantic relationships.

An analysis of current limitations and disadvantages introduced by semantic technologies is outside the scope of this chapter.

At the moment, semantic technologies are applied in several sensor architectures. Common applications have the aim of providing advanced support to information description and processing, data management, interoperable networking, dynamic representation of situations and system states, advanced analysis of data and classification.

Semantic interoperability would improve common interoperability models: basic interoperability assumes the interchange of messages among systems without any interpretation; functional interoperability integrates basic interoperability models with the ability of interpreting data context under the assumption of a shared scheme for data fields accessing; semantic interoperability introduces the interpretation of the meaning of data.

Semantic interoperability is a concretely applicable interaction model under the assumption of adopting rich data models (commonly called Ontologies) composed by concepts within a domain and the relationships between those concepts. So, each system has a semantic representation of interest aspects (Figure 13.8).

The availability of standard languages for Ontology definition and specification (e.g. Ontology Web Language, OWL), as well as extensible and standard reasoners (e.g. Jena, Pellet [28]) able to work on them, allows the engineering of intelligent ontology-driven actors [24]. Ontology-driven interaction takes advantage by semantic rules and relationships implemented by ontology: the actor receives as input the ontology, it is able to interpret it; so, it does not need to implement semantic rules and relationships (commonly dependent of reference data models) that, now, are explicitly expressed in the semantic schema according to an interoperable model.

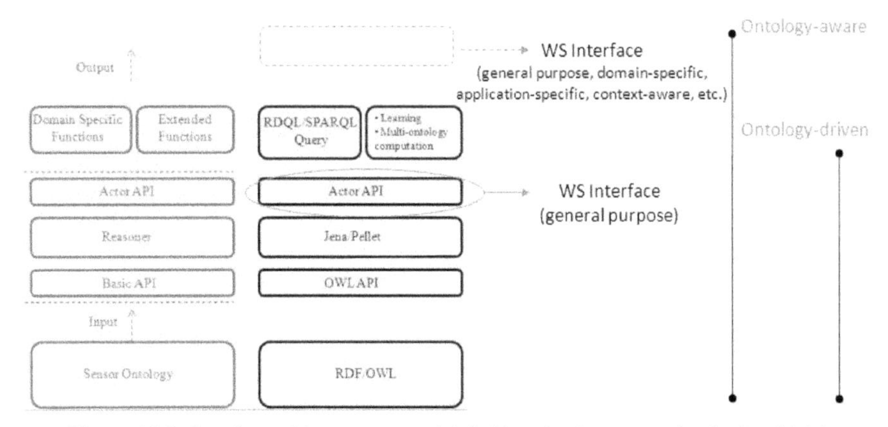

Figure 13.9 Ontology-driven actor model (left) and reference technologies (right).

Semantic technologies are partially reverting to the common view of actor intelligence: intelligence is not implemented by actors (that are understood as interpreters) but it is implicitly resident in the knowledge model.

As shown in Figure 13.9, ontologies defined according to standard and interoperable technologies (e.g. RDF or OWL) are inputs for actors designed on the top of standard reasoners able to automatically interpret ontology semantics; extended intelligence layers could be designed providing actors with domain-specific querying and/or added capabilities (e.g. learning or multi-ontology computation).

Several software layers can be built as extension of the model represented in Figure 13.9; these layers can be designed for reaching several goals; common solutions provide extended functions for supporting ontology-aware computation: actors are able to switch their behaviors and functionalities in function of the input ontology.

Sensor Ontology Engineering resulting from knowledge modeling of the sensor domain and other interest aspects (e.g. data) is, probably, the real key issue for the concrete realization of a Semantic Sensor Web.

Due to the fundamental lack of standardized methodologies for knowledge engineering (semantic knowledge in this case), mapping real knowledge on semantic schemas is, probably, the most creative task for the concrete engineering of a Semantic Sensor Web [20].

The sensor domain poses several peculiarities that allow simpler approaches than for generic knowledge building. Even considering that each sensor system could have its peculiarities that could be reflected in semantic

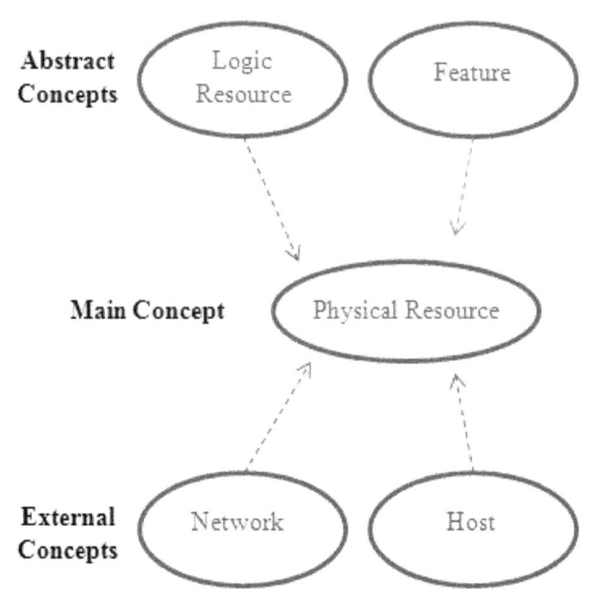

Figure 13.10 A simplified view of a possible approach for sensor knowledge engineering.

representations, the majority of systems could be represented according to two main semantic structures: the Domain Ontology and the Data Ontology. The first one should support resource-centric interaction; the second one data-centric interaction [24].

The main goal of the Domain Ontology is the model of the sensor domain [24]. A possible simplified semantic scheme for domain ontology could be the model represented in Figure 13.10 [21]; as shown the central concept (Physical Resource) is the result of the convergence of domain specific semantics and inferred external concepts (e.g. Network, Host); an abstract semantic layer includes high level features definition and logic resource; this last layer differs with respect to lower layers because it does not define a full sub-domain/domain: composing concepts have finite means only in the context of main domain (sensor domain). Concrete implementations depend on interaction scope. Typically, the classification of resources according to several perspectives (e.g. functional) is required [21]. The Domain Ontology has a key role for allowing search, discovery and all interactions that assume there is no previous information about the considered system [24].

A great number of systems just need to interchange information, such as sensor data [24]. In this last case, systems require ad-hoc structures in order to search, discovery or retrieve data. This class of semantic scheme (Data Ontology) is conceptually different with respect to Domain Ontology (resource-centric structure) because it is data-centric. The main purpose of a Data Ontology is defining the model for interchanging data as well as the meaning of data [24].

Knowledge engineering methodology is currently an open research issue: building knowledge on semantic ontologies could be an attractive option considering interoperable schemes and semantic annotation standards provided by current semantic technologies. However, there are several efforts aimed at promoting a convergence among multiple semantic domains (for example, the use of standardized vocabularies for ontology concepts [20]).

13.4 Conclusions

Enabling sensor systems within VOs, eventually on a large scale (Sensor Web), has great commercial potential supported by consolidated and flexible technologies (e.g. sensor technologies and web-services) that constantly propose innovative scenarios also for the suitable convergence with other emerging technologies (e.g. semantics).

The concrete realization of Sensor Web is evidently related to business and exploitation models; these models could be extended to innovative scenarios if current solutions are standardized and key issues (robustness, reliability) improved according to a general costs reduction.

Furthermore, there are great expectations for the concrete realization of the Semantic Sensor Web in which semantics should improve current interaction models.

Finally, also the progressive and definitive commercial availability of last generation sensor networks (e.g. large-scale Mobile Sensor Network and Multimedia Sensor Networks [4]) could enable further development of business scenarios.

References

[1] The TinyOS 2.x Working Group. Tinyos 2.0. In *Proceedings of the 3rd International Conference on Embedded Networked Sensor Systems*, 2005.

[2] Ameer Ahmed Abbasi and Mohamed Younis. A survey on clustering algorithms for wireless sensor networks. *Computer Communications*, 30(14/15):2826–2841, 2007.

[3] Kemal Akkaya and Mohamed Younis. A survey on routing protocols for wireless sensor networks. *Ad Hoc Networks*, 3(3):325–349, 2005.

[4] I.F. Akyildiz, T. Melodia, and K.R. Chowdury. Wireless multimedia sensor networks: A survey. *IEEE Wireless Communications*, 14(6):32–39, 2007.

[5] M. Baqer and A. Kamal. S-sensors: Integrating physical world inputs with social networks using wireless sensor networks. In *Proceedings of 5th International Conference on Intelligent Sensors, Sensor Networks and Information Processing (ISSNIP'09)*, 2009.

[6] Paolo Baronti, Prashant Pillai, Vince Chook, Stefano Chessa, Alberto Gotta, and Y. Fun Hu. Wireless sensor networks: A survey on the state of the art and the 802.15.4 and ZigBee standards. *Computer Communications*, 30(7):1655–1695, 2006.

[7] Mike Botts and Alex Robin. Bringing the sensor web together. *Geosciences*, 6:46–53, 1997.

[8] Dazhi Chen and Pramod K. Varshney. Qos support in wireless sensor networks: A survey. In *Proceedings of International Conference on Wireless Sensor Network 2004*, 2004.

[9] D.E. Culler. Wireless sensor networks – Where parallel and distributed processing meets the real world. In *Proceedings of 19th IEEE International Parallel and Distributed Processing Symposium*, 2005.

[10] S. Dhar. Manet: Applications, issues and challenges for the future. *International Journal of Business Data Communications and Networking*, 1(2):66–92, 2005.

[11] A.Z. Faza and S. Sedigh-Ali. A general purpose framework for wireless sensor network applications. In *Proceedings of 30th Annual International Computer Software and Applications Conference*, 2006.

[12] A. Gerber, A. Van der Merwe, and A. Barnard. A functional semantic web architecture. In *Proceedings of European Semantic Web Conference*, 2008.

[13] Vincent Huang and Muhammad Kashif Javed. Semantic sensor information description and processing. In *Proceedings of Second International Conference on Sensor Technologies and Applications*, 2008.

[14] Su Weilian, Y. Sankarasubramaniam, I.F. Akyildiz, and E. Cayirci. A survey on sensor network. *IEEE Communication Magazine*, 40(8):102–114, 2002.

[15] I. Foster, C. Kesselman, and S. Tuecke. The anatomy of the grid: Enabling scalable virtual organizations. *International Journal of Supercomputer Applications*, 15(3):200–222, 2001.

[16] K. Kulkarni, S. Tilak, K. Chiu, and T. Fountain. Engineering challenges in building sensor networks for real-world applications. In *Proceedings of 3rd International Conference on Intelligent Sensors, Sensor Networks and Information*, 2007.

[17] Hock Beng Lim, Yong Meng Teo, Mukherjee P., Vinh The Lam, Weng Fai Wong, and See S. Sensor grid: Integration of wireless sensor networks and the grid. In *Proceedings of the IEEE Conference on Local Computer Networks*, 2005.

[18] W. Masri and Z. Mammeri. Middleware for wireless sensor networks: A comparative analysis. In *Proceedings of IFIP International Conference on Network and Parallel Computing*, 2007.

[19] G. Niezen et al. From events to goals: Supporting semantic interaction in smart environments. In *Proceedings of 2010 IEEE Symposium on Computers and Communications*, 2010.

[20] Daniel Oberle et al. Dolce ergo sumo: On foundational and domain models in the smartweb integrated ontology (swinto). *Web Semantics: Science, Services and Agents on the World Wide Web*, 5(3):156–174, 2007.

[21] Salvatore F. Pileggi. A novel domain ontology for sensor networks. In *Proceedings of International Conference on Computational Intelligence, Modelling and Simulation (CimSim2010)*, 2010.

[22] Salvatore F. Pileggi, Carlos E. Palau, and Manuel Esteve. Grid sensor/actuator network architecture. In *Proceedings of International Conference on Wireless and Mobile Communications*, 2006.

[23] Salvatore F. Pileggi, Carlos E. Palau, and Manuel Esteve. Analysis techniques and models for resource optimization in wireless sensor/actuator network environment. In *Proceedings of IFIP International Federation for Information Processing, Wireless Sensor and Actor Networks*, Vol. 248, pages 23–34, 2007.

[24] Salvatore F. Pileggi, Carlos E. Palau, and Manuel Esteve. Building semantic sensor web: Knowledge and interoperability. In *Proceedings International Workshop on Semantic Sensor Web (SSW 2010)*, 2010.

[25] Takeshi Sakaki, Makoto Okazaki, and Yutaka Matsuo. Earthquake shakes twitter users: Real-time event detection by social sensors. In *Proceedings of 19th International Conference on World Wide Web*, 2010.

[26] W. Schott et al. E-sense protocol stack architecture for wireless sensor networks. In *Proceedings 16th IST Mobile and Wireless Communications Summit*, 2007.

[27] Amit Sheth, Cory Henson, and Satya S. Sahoo. Semantic sensor web. *IEEE Internet Computing*, 12(4):78–83, 2008.

[28] Evren Sirin et al. Pellet: A practical OWL-DL reasoner. *Web Semantics: Science, Services and Agents on the World Wide Web*, 5(2):51–53, 2007.

[29] Patrik Spiess, Stamatis Karnouskos, Dominique Guinard, Domnic Savio, Oliver Baecker, Luciana Moeira, and Vlad Trifa. Soa-based integration of the internet of things in enterprise services. In *Proceedings of IEEE International Conference on Web Services*, 2009.

[30] Matthias Winkler, Jorge Cardoso, and Gregor Scheithauer. Challenges of business service monitoring in the internet of services. In *Proceedings of International Conference on Information Integration and web-based Applications and Services*, 2008.

14

Virtual Terminals for IMS

Jean-Charles Grégoire and Salekul Islam

Centre Energie Matériaux Télécommunications, INRS, 800 De la Gauchetière Ouest Bureau 6900, Montréal, Québec H5A 1K6, Canada;
e-mail: {gregoire, islam}@emt.inrs.ca

Abstract

While the IMS core and IMS-based services have generated sustained interest, the IMS terminal has received less attention. Yet terminal restrictions, in user interaction capability, memory or compute power can severely constrain the deployment of new services. In this chapter, we explore how terminal virtualization can be integrated in the IMS architecture to alleviate some of these constraints. The benefits of the proposed model are to streamline the user's physical terminal and facilitate dynamic access to new applications while at the same time preserving the integrity of the core IMS architecture and taking full advantage of its features. We present the architecture supporting virtualization and analyze its impact on control and media communications and their integration with IMS. Furthermore, we present a Java Portlet-based prototype implementation. We discuss the key features of our implementation and explore two different models for media integration, each with benefits of their own.

Keywords: IMS client, virtual terminal, portlet, media services.

14.1 Introduction

The IP Multimedia Subsystem (IMS) [4] is an infrastructure created to allow the integration of all multimedia services in a single, unique framework,

Anand R. Prasad et al. (Eds.), Future Internet Services and Service Architectures, 295–314.

based on the IP protocol. Originally created for wireless communications, the standard has been adopted for other forms of access, namely DSL and cable. This vehicle for convergence has however encountered limited success so far, for a number of reasons. One key reason we focus on in this chapter is the difficulty of deploying applications on user equipment (UE).

Even though so-called "smart phones" have become quite pervasive and present the fastest growing segment in the industry, this market is still quite fragmented between several hardware/software manufacturers and no – or too few – agreed upon standards. Most software platforms, for reasons of resource management, still impose restrictions in the applications which can be executed on a phone and the degree of concurrency in which they can execute. Combined with different restrictions in terms of display, storage capacity or APIs supported, we can imagine how difficult it can be to create, deploy and support service-clients on such devices. These difficulties lead us to look into virtualization as an alternate vehicle for deployment.

Virtualization offers a new paradigm for service provisioning and opportunities for resource tradeoffs between compute resources and communications. While most of the recent efforts on virtualization have focused on moving the server-side computing resources into the cloud, there is also an increased interest in the simplification of the terminal functions. For the general public, what is key is to simplify the operations of the personal computing/communications device in order to access new services more easily with a simple, straightforward and pleasant experience.

In this chapter, we expose the benefits of using a virtualization model for the user-side implementation and deployment of IMS-based services. We begin in Section 14.2 with a presentation of the key features of IMS and follow in Section 14.3 with a sketch of the principles of virtualization, focusing on their application to terminals. Section 14.4 explains different aspects of our proposed architecture. Section 14.5 introduces a demonstration version of our service.

14.2 IMS, Core and Terminals

IMS provides a unique, highly scalable signalling infrastructure for services, based on the extensive use of the Session Initiation Protocol (SIP) [22], including support for security, quality of service and billing. Unlike the traditional end-to-end view of the Internet underlined by the predominance of the Web, IMS is really meant to increase value for services by providing support within the network, thereby helping network/infrastructure providers

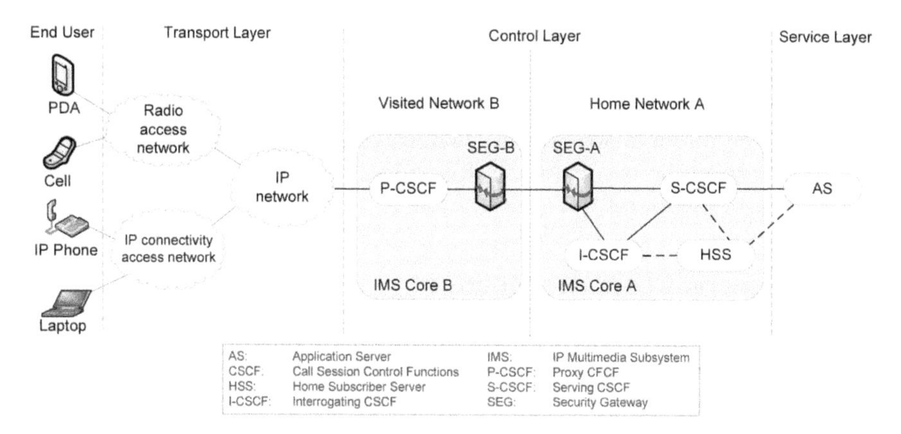

Figure 14.1 Functional inter-domain architecture of Next-Generation Networking (NGN) IMS.

to become part of the value chain. In traditional Web services, the network providers (e.g. Internet Service Providers, ISP) work as a bit-pipe and they have no share in the revenues earned through service delivery. On the other hand, the IMS core network provider (who might be the access network provider as well) receives revenue by deploying value added services.

The functional inter-domain architecture of Next-Generation Networking (NGN) IMS is shown in Figure 14.1, where only the key components of the IMS core are highlighted. This IMS model has been designed to work in an access network-agnostic fashion. Therefore, any SIP-enabled UE should be able to communicate with the IMS core network and, after successful authentication and authorization, be eligible to receive services from the IMS core.

14.2.1 IMS Core

The core functional components for call control include different Call Session Control Functions (CSCF), such as Proxy CSCF (P-CSCF), Interrogating CSCF (I-CSCF) and Serving CSCF (S-CSCF). The P-CSCF is a SIP proxy that sits on the path of all signalling messages. In the inter-domain architecture, the P-CSCF is located in the visited network while the service is provided from the home network. The S-CSCF is a SIP proxy and the signalling plane's central node; it registers users and provides services to them. The I-CSCF, a SIP proxy located at the edge of the home network, takes part in user roaming. The Home Subscriber Server (HSS) is the master user database and supports the IMS network entities to handle calls/sessions. It contains subscription-related information (i.e. user profiles) and performs authentication and authorization of the user. The visited and home networks are connected through two Security Gateways (SEG), which establish a trusted link between the two networks following the Network Domain Security (NDS)/IP specification [2].

14.2.2 Generic Client Architecture

Although there is no universal standard for IMS client architecture and functions, currently many activities are focusing in the specification of IMS client. The Eurescom's generic IMS client framework has been designed within the P1656 project (Definition of an open and extendible IMS Client Framework) [20]. Similarly, the Fraunhofer Institute for Open Communication Systems (FOKUS) has developed an extensible IMS client framework, myMONSTER (Multimedia Open InterNet Services and Telecommunication EnviRonment) [11]. Ericsson has also developed an IMS client platform [17], which is based on the Java Community Process JSR-281 [10].

As an example, we see that IMS stacks of myMONSTER are built on 3GPP IMS and JSR-281 specifications. By decoupling the service logic from the presentation layer, it supports multiple presentation layers (e.g. Swing, SWT, Widgets, JavaFX, etc.). myMONSTER is not only an IMS client but also a tool for creating and communicating with a wide range of rich terminal applications that are based on NGN, IPTV and WEB standards. Thus, it supports the convergence of applications from three domains, including Communication, Television and Web.

By studying different IMS frameworks, it could be concluded that they all result into a quite identical horizontal architecture for the IMS client framework [17], which is shown in Figure 14.2. This highly modularized and expandable architecture is composed of four layers:

1. An abstraction layer to encapsulate all hardware and OS-related dependencies.
2. A protocol layer to implement the underlying communications and media transmission protocols.
3. A control layer consisting of a core part and a service part.
4. An interface layer that might be a Graphical User Interface (GUI).

In addition to these generic client frameworks, the EUREKA Mobicome project proposes a widget-based IMS client architecture [25] that supports

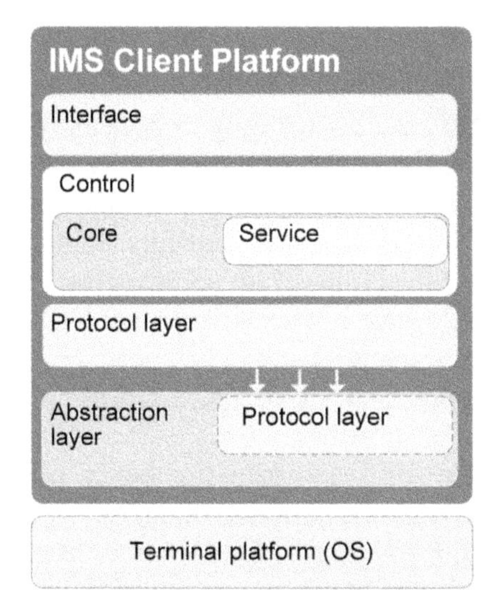

Figure 14.2 Horizontal architecture for IMS clients.

personalized and dynamic IMS client. A widget is a small application implemented using Web 2.0 technologies. A widget is a portable code fragment that can be installed and executed without additional compilation either inside a browser or in a standalone fashion. In a widget-based architecture, unlike with traditional IMS clients, the end user does not communicate directly with the IMS client framework. The user is exposed with a personalized user interface, which is composed of IMS widgets.

14.2.3 Limitations

The horizontal layering of the IMS client framework (shown in Figure 14.2) results in a highly modular and extensible IMS client architecture. Using the proper APIs (e.g. JSR-281 APIs [10]), an application could be assembled together using the underlying IMS services. Hence, the prevailing one-to-one mapping between an application and IMS service could be overcome. The operators could respond to market dynamics and quickly bring out new applications. Despite these advantages, the existing models still have several drawbacks, which are due to the implementation of the entire client framework inside the UE: most of the existing IMS clients' software is either shipped embedded with the device or is downloaded and installed from the Web. The entire client framework, including the user interface and supported

IMS applications, is bundled inside a single software package. This rigid architecture introduces several limitations:

1. Users access IMS services not only from heterogeneous networks (e.g. mobile, fixed broadband, WLAN) but also from a variety of devices (e.g. mobile phones, PCs, laptops, PDAs). A wide variety of alternatives is to be expected in terms of user authentication, identity management, protocols implemented, hardware supports, etc. Hence, an IMS client might not be deployable in a specific type of UE or access network.
2. If the IMS client framework has to be updated (e.g. due to implementation of a new application) every single UE using that IMS client needs to update the client software.
3. Implementing the full IMS client framework might overload the UE, especially for devices with limited resources.
4. No personalization or user-centric interface is possible in traditional IMS clients. Only the Mobicome project [25] supports such facilities. However, although provided with a widget-based user interface, UEs still must implement the underlying client framework. Moreover, communications between the user and the underlying client framework are not optimized as they must be relayed by the Mobicome proxy server.

14.3 Virtualization

Virtualization is a well-established concept in computer science, dating back to virtual systems in mainframes [19]. In a nutshell, this term designates the process of recreating a (virtual) hardware or software environment by emulation on top of a real system. It has become quite popular again in the more recent past as a tool for flexible support and deployment of server software and compose around issues of OS and application legacy. In other guises it has also been applied for storage, networks and databases.

14.3.1 Virtualization in Networking

Virtualization in networking is mostly observed on network architectural level for hiding the underlying infrastructure network. For example, *Cabo* [15] presents a high-level architecture for a flexible and extensible system that supports multiple simultaneous network architectures through network virtualization. Cabo identifies two different entities: infrastructure providers (e.g. the ISPs) who manage the substrate resources and service providers who operate their own customized network inside the allocated slices. The

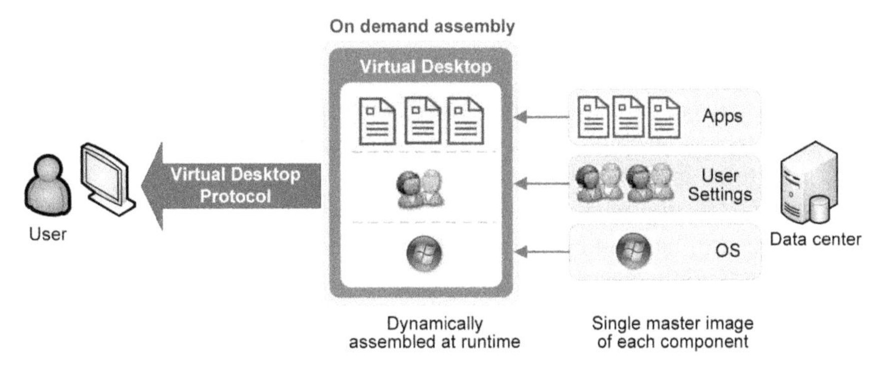

Figure 14.3 Architecture of XenDesktop virtual desktop computing.

two-layer virtualization of Cabo has been further refined in *Cabernet* (Connectivity Architecture for Better Network Services) [26], which presents a three-layer network architecture by introducing a connectivity layer between the infrastructure provider and the service provider. *VNet* (Virtual Network) [23] proposes to further split the role of the connectivity layer into Virtual Network Provider (VNP) and Virtual Network Operator (VNO). This provides more granular splitting of responsibilities with respect to network provisioning, network operation, and service specific operations.

Our usage of virtualization here is closer to an appliance or desktop, which is known as desktop virtualization or virtual network computing [21]. In the following we discuss desktop virtualization and its use in our design.

14.3.2 Benefits through Desktop Virtualization

Desktop virtualization can be achieved in a number of ways and with various technologies, the most popular being Microsoft's Remote Desktop Protocol [13] or Citrix's XenDesktop [5]. The architecture of XenDesktop is shown in Figure 14.3. A virtual desktop means in essence that applications run on a machine other than the user's. The desktop is presented to the user as an image of the one running remotely, and local keyboard and mouse are used to interact with it, in typical fashion. The virtual desktop server, located to a remote platform, is responsible for on-demand application provisioning, managing user settings and the running the Operating System's functions. The matching virtual desktop's image is rendered at the user's platform through a virtual desktop protocol. Since the user's platform has no permanent storage, the remote data centre works as its virtual storage.

Such a remote access presents multiple benefits, specifically in an enterprise setting:

1. *Reduced costs*: it is possible for several users to share licensed software on a server machine. This also means that licensed software does not need to be installed on all UEs.
2. *Simple UE*: the resource constraints on the UE are reduced, as it does not need to have a large compute/storage capacity. It is also possible to extend the lifespan of the UE.
3. *Simpler management*: There is a lesser need for upgrades on UEs, as they run less software.
4. *Improved security*: UEs can be more easily confined and present a lesser vector for attacks, as well as being less subject to being infected or compromised.

Such benefits have not been as clear for the general public, who needs to be able to hold onto its data. In this case, however, we have seen the emergence of the related *virtual application*, typically using an office suite remotely through a web browser. One such example is Google Docs [6]. Most of the benefits listed above remain, in terms of reduced overhead, operations and maintenance costs.

Low communication costs and high bandwidth have also contributed to the popularity of this model. It has become attractive because it is now possible and affordable to work remotely through the Internet, or to deploy high bandwidth within corporate networks. In the meantime, the cost of owning and operating individual physical personal computers has remained fairly stable.

14.3.3 Concerns

There are also clear elements of concern associated with this model, which may be more significant for the general public and the virtual application model. These are:

- *Data location*: users want to keep their data on their computer, just the same way corporations may want to restrict access to data by not allowing its storage on a local computer. Jurisdiction over data can be an issue and especially so for copyrighted materials.
- *Data compatibility*: independently operated remote applications may use different formats for data manipulation and this could lead to difficulties in data exchange.

- *Ownership*: since software is no longer "owned" the general user may find herself in situations of losing access to applications, or being locked into proprietary software.
- *Functionality*: virtualized applications may not provide the full set of features of an equivalent local application.
- *I/O*: interactive applications may have restrictions on the use of local I/O. Traditional virtual technology has been focused on streaming bitmaps while capturing keyboard and mouse only. Over time, such restrictions have slowly been overcome, but not in any standard way.

14.3.4 IMS-Specific Issues

Since we are interested in a general public model for IMS, the virtualized application model is more appealing to us. Again, for this model, the web browser has emerged as the generic support platform. It however requires extended support for I/O and this issue will be discussed more extensively below.

Another dimension of interest of virtualizing the IMS client is the opportunity to develop a *mash-up* model [18], that is, to give the user the opportunity to create her own set of services from a set of offered components, which would include telephony, conferencing, messaging, Video on Demand (VoD), etc. Such a feature would overcome some of the challenges of deploying new services directly on the UE and managing compatibility and interoperability issues in the UE. In the following section we study in more detail how the virtual IMS client can be realized.

14.4 IMS Client Virtualization

We now present an architecture and a prototype for a virtual terminal which fully integrates with the IMS architecture. The proposed architecture for IMS client virtualization is shown in Figure 14.4. In this architecture, most of the signalling load is transferred to a server co-located with a P-CSCF, which we shall call a *surrogate*.[1] The surrogate acts as a virtual server for the user to access, organize, provision and monitor her services.

[1] The term surrogate is also used in RFC 3040 [14] to address a different types of network node.

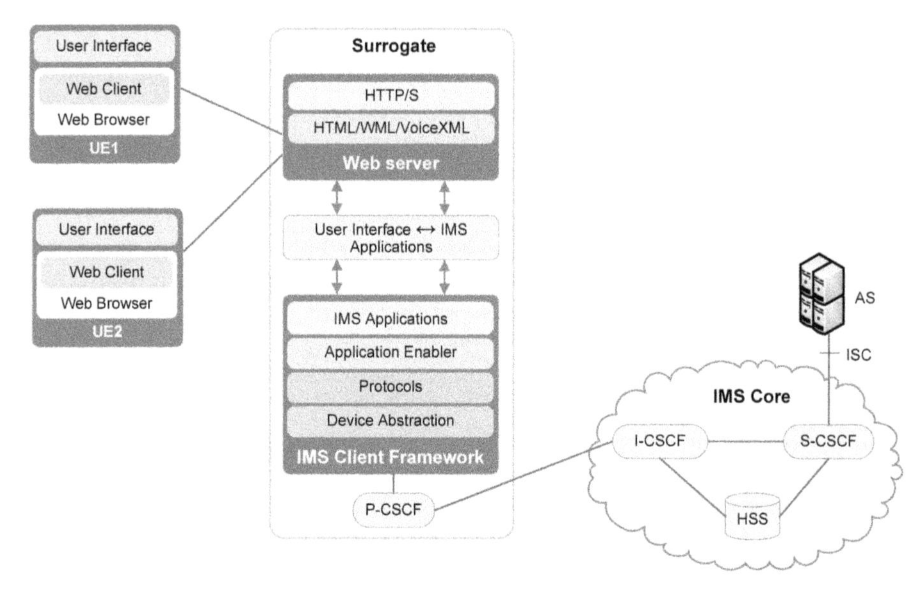

Figure 14.4 IMS client virtualization architecture.

14.4.1 The Surrogate Architecture

This surrogate is transparent for the IMS operations in the network and this deployment has no impact on the IMS core architecture. On the other hand, it creates new opportunities for extensions to the IMS model with increased benefits to the users, beyond those related to virtualization, such as session mobility. By transferring signalling and session management complexity to the surrogate, a simple UE with functionalities restricted to operating the GUI and audio/video media playback could be used. The surrogate presents to the core an IMS client, and acts as a server-side of the virtualized client for the user. Web-based GUIs are commonly used now for virtual client side and most UE platforms come equipped with some Web browsers. A surrogate implements a Web server, which receives the users' input through the GUI running on the Web client inside the UE. An interaction layer is needed between the Web server and the IMS client for establishing communication between them. This layer transfers the GUI's input to the IMS client applications and IMS session status (e.g. IMS registration success) to the Web server. It is worth mentioning that the IMS client framework inside the surrogate has a different top layer protocol stack from the generic client framework of Figure 14.2. Since the user interface layer is already implemented in the UE (through a Web-based GUI), we only require to implement an IMS Application layer to expose different IMS applications to the Web server.

14.4.2 Surrogate Implementation

Although the details of the surrogate implementation is outside the scope of this chapter, we would like to mention that the surrogate should be implemented in the close proximity of the users. The network access provider (e.g. the Internet Service Providers, ISP) could leverage the use of their existing infrastructure by implementing and maintaining surrogates in their Autonomous Systems (AS). Such implementation of surrogates could be used in other service models (e.g. in peer-to-peer (P2P) networks) where the users could access services through virtual clients. Since the surrogate works transparently to the IMS core, the implementation of a surrogate has no impact on the IMS core. On the other hand, the rationale of desktop virtualization (which is presented in Figure 14.3) is used as a concept in implementing virtual client; it has no implication in the implementation of the surrogate.

14.4.3 Authentication

Each IMS subscription is associated with an IP Multimedia Private Identity (IMPI) and one or more IP Multimedia Public Identity (IMPU). In UMTS networks, the identity is established through an authentication process based on an application, the IP Multimedia SIM (ISIM), which runs on the smartcard present on any terminal. This smartcard securely stores the IMS subscriber's identity and credentials.

In our proposed architecture, however, users are free to use any terminal with IP connectivity. They can switch terminal and connect to the IMS core network from anywhere they want. However, if the user wishes to use any terminal device and to connect from anywhere, it is not possible to deploy IMS information on the user's terminal device. An ISIM application depends on a smartcard for hosting the application, which – by its very essence – could not be deployed inside the surrogate. Hence, an IMS soft client that instantiates a virtual ISIM application is required for our architecture. The soft client that the surrogate deploys needs to perform more additional tasks than a state-of-the-art IMS client (installed in a PC or laptop) does. For example, multiple users from heterogeneous devices and networks communicate with this IMS client through the Web server. The IMS client must instantiate virtual ISIMs as required and also maintain separate states for each user.

14.4.4 Media Handling

This architecture supports different models for media handling. In the "straight through" model, media goes directly to the user's platform. On the contrary, in the "intercepted" model the surrogate processes the flow of information before delivery to the user's platform. Each presents specific benefits and each service can choose the more suitable model, transparently to the user and the IMS core, since the required SIP processing is performed by the surrogate itself, acting as the "real" SIP client. A direct media delivery method is fast and easy to implement. However, the UE might not have the appropriate audio/video codecs to decode the received media. In such circumstances, the surrogate intercepts the media and transcodes the media to another format to be understandable by the user's platform.

14.4.5 The User Services Portal

An IMPI is always permanently allocated to a user's subscription (it is not a dynamic identity), and is valid for the duration of the user's subscription with the home network. This identity is used for the registration of the subscriber, for authentication and authorization, for administration of the subscription and for billing. A subscription is always attached to one or more service profiles, which are stored in the HSS [1]. The S-CSCF, in accordance with the service profile, routes a SIP request to the appropriate Application Server(s). A service profile also stores the policies required by the S-CSCF to decide which types of media are authorized inside a SIP session.

Similar to the IMS subscription and the service profile, the surrogate may implement a user's subscription database and this database could be integrated to a user portal. However, we must insist that the proposed user portal and the IMS service profile are used for different purposes. IMS service profile is used for routing SIP request, authorizing media, etc. Through the user portal, users can subscribe to and also possibly assemble their services *à la carte*, according to their preferences. On successful registration, a user will get access to his own portal and add, delete, edit and create (through assembly) new services. Note that any change in this portal must be reflected in the IMS service profile. How this change could be updated to the IMS service profile is beyond the scope of this chapter.

14.5 A Prototype Using Portlet

We present a prototype implementation of a surrogate integrating the key concepts presented above. The GUI at the user end is developed using *portlets*, a Java technology based web component, which is defined by the JSR-168 specification [9]. In the following, an overview of portlet technology is presented before explaining our prototype implementation.

14.5.1 Portals and Portlets

A *portal* is a single point of access – a form of virtual front door in the case of the Web – to aggregated information from different sources. It is also responsible for the presentation of this aggregated information. The primary goal of most portals is ease-of-use by providing a rich navigation structure for information of greatest interest to its user. The news portal is the most common such occurrence. In general terms a portal is just a gateway, and a Web portal can be seen as a gateway to the information and services of the Web. A Web portal may provide different value-added services:

- Personalization
- Single Sign-on
- Content aggregation from various sources
- Secure search facilities
- Localization of content
- Applications integration

Portlets are software components which are assembled into a portal. Each portlet provides the user interface of a specific application, such as news flash or weather reports, in the form of markup (HTML) code generated by the portal and transferred to the UE. It also supports the backend communication with that application.

The standard portlet-based portal architecture is shown in Figure 14.5. Much like a Web application server (which has a Web container to manage the running Web components, such as servlets, JSPs, filters, etc.), a portal server has a portlet container to manage the running portlets. A portlet container is an extension of the Java servlet container.

14.5.2 Implementation Architecture

The IMS client's interface is meant to be customizable and flexible so that the end user can add or remove IMS applications at runtime. A portal-based IMS

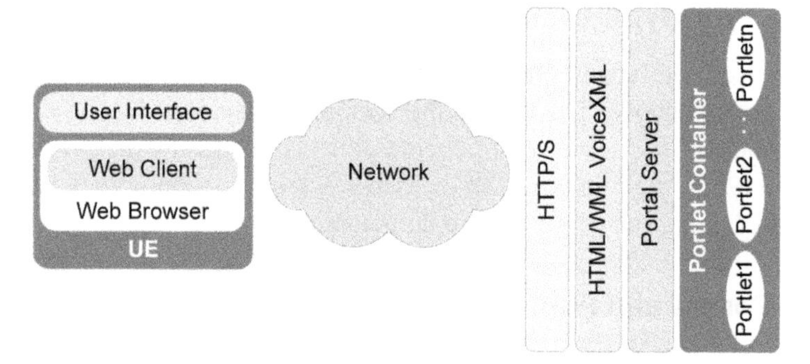

Figure 14.5 Portlet-based portal architecture.

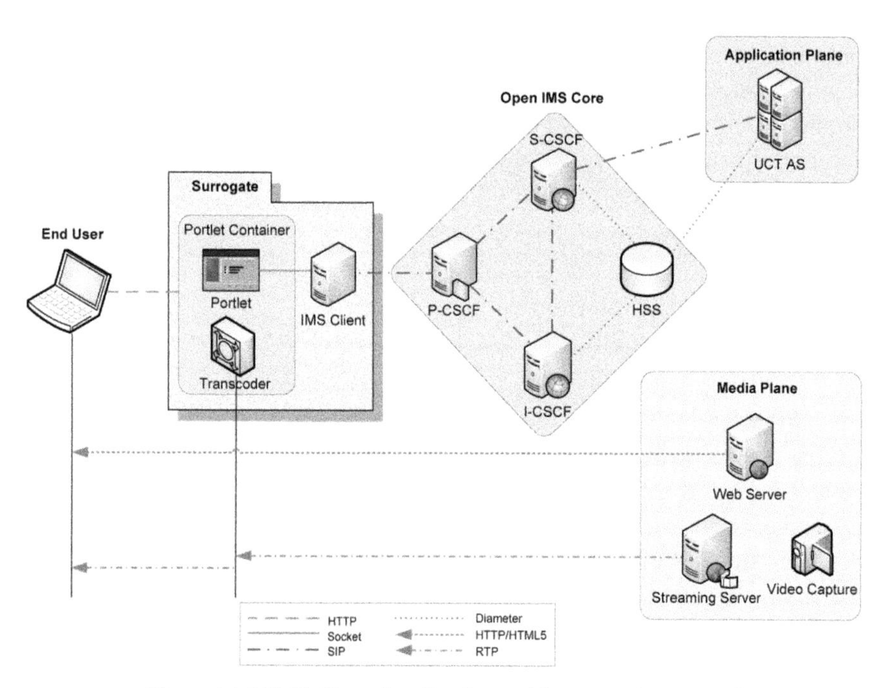

Figure 14.6 IMS client virtualization architecture using a portal.

client interface will offer a virtual IMS client at the UE, while the IMS client framework will be implemented inside the surrogate. Figure 14.6 shows our implementation architecture for IMS client virtualization using a Web portal, within the context of the surrogate.

The portlet container hosts the portlet that generates the GUI Web pages at the user's platform. We have implemented a simple IMS client using eXtended osip (eXosip) library in C programming. Our IMS client performs IMS registration to the IMS core, generates and forwards the SIP INVITE message in response to the user request. We have used OpenIMSCore [12], an open source IMS core implementation developed by the FOKUS group. The OpenIMSCore implements all three Call Session Control Functions (CSCFs) as well as a lightweight Home Subscriber Server (HSS), which altogether are the core elements of IMS or Next Generation Network (NGN) architecture.

In the application plane, we have deployed the UCT Advanced IPTV AS [24] as the application server. This programme has been developed by the Communications Research Group at the University of Cape Town as a standard implementation of an IMS based IPTV service.

In the media plane, two modes of media delivery – VoD and realtime video streaming – have been implemented. For VoD service, a Web server that hosts a number of video clips has been deployed in the media plane. For realtime video streaming, the Java Media Studio (JMStudio), a Java Media Framework (JMF) [8] based tool has been used to capture the video stream of a Webcam and to transfer that video stream in realtime. We have implemented both straight through and intercepted media delivery (mentioned in Section 14.4.4) in our experiment. The VoD stream has been directly delivered to the user's platform while the realtime stream has been intercepted by the surrogate. The ongoing specification of HTML 5 [7] supports "video" type tag and thus delivers video data through the Web browser (not all Web browsers support HTML 5 at this moment). For the intercepted media delivery, it is assumed that the UE is not equipped with the necessary video codec. Therefore, a transcoder programme has been implemented inside the surrogate. The transcoder receives the video stream broadcasted by the streaming server (i.e. the JMStudio tool), transcodes the stream (i.e. changes the video codec) to an understandable UE format, and transmits the resulting transcoded stream to the UE.

Finally, we need a mechanism to implement the middle layer that establishes communication between the portal and the IMS client. Since there is no standard protocol for this purpose, we have used the classical network socket APIs.

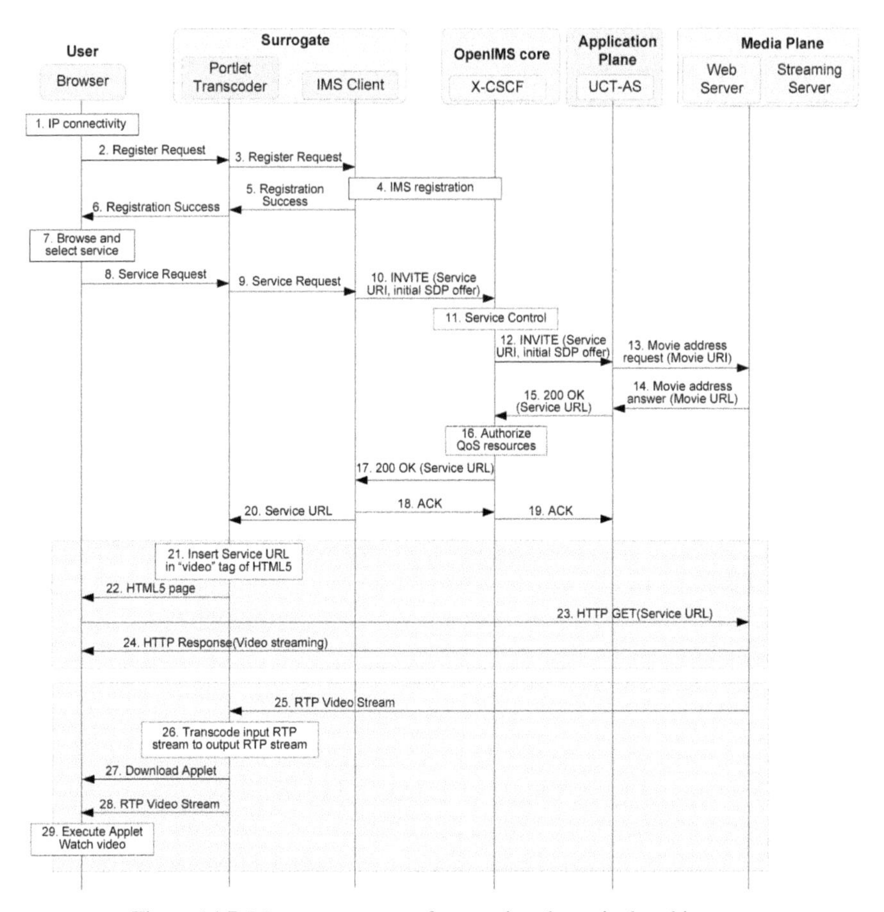

Figure 14.7 Message sequence for our virtual terminal architecture.

14.5.3 Message Sequence

The message sequence of our implementation is shown in Figure 14.7. We assume that all links are secured and therefore we do not deploy any user authentication except for IMS registration. In the following we briefly explain different steps of the message sequence.

14.5.3.1 IMS Registration (Steps 1–6)

First, the user accesses the Web portal-based user interface located at the remote server. Next, the user enters the necessary authentication and provisioning information (user identities, password, address of the P-CSCF, etc.)

for completing the IMS registration. This information is transmitted to the IMS client and triggers the IMS registration. The IMS client communicates with the IMS core and performs a regular IMS registration through IMS AKA (Authentication and Key Agreement) [3]. On successful registration, the IMS client sends a Registration Success message to the portlet. It is worthwhile to note that the UE or the portlet triggers the IMS registration and that neither ever maintains any IMS registration related information.

14.5.3.2 Service Request (Steps 7–20)

Next, the portal dynamically creates Web pages that lists available video clips. Some of the video clips would be directly delivered to the UE and some of them would be intercepted by the surrogate for transcoding. The user selects the video clip he wants to receive and a corresponding Service Request message is sent to the IMS client through the Web portal. The IMS client sends an INVITE message to the IMS core that contains the URI of the video (e.g. `sip:video1@iptv.open-ims.test`) and the initial Session Description Protocol (SDP) [16] offer for this call. In this URI, `video1` is the identity of the video clip and `iptv.open-ims.test` is the domain name of the UCT-AS.

The IMS core performs a Service Control operation, i.e., checks the registration status of the IMS client and verifies if the client is authorized to originate such an INVITE message. The HSS database must be properly updated so that the S-CSCF, on successful Service Control, forwards the INVITE message with `iptv.open-ims.test` to the UCT-AS.

The UCT-AS requests the media plane for the URL of the video clip. Using the identity of the video (i.e. `video1`), the media plane retrieves the URL and returns it back to the UCT-AS. The communication between the UCT-AS and the media plane is not shown in Figure 14.6 since it has been implemented through proprietary methods. Next, the requested service's URL is forwarded inside a 200 OK message by the UCT-AS to the IMS core.

On successful authorization, the IMS core forwards the 200 OK response to the IMS client, which forwards the service URL to the portlet. Note that although the initial SDP offer is sent by the IMS client to the AS, this SDP offer has no effect in our implementation (and hence no end-to-end resource reservation is required).

14.5.3.3 Direct Media Delivery (Steps 21–24)

For direct video delivery to the UE, the portlet dynamically creates an HTML 5 Web page by inserting the service URL inside a video tag. This

HTML 5 page is rendered to the UE. Thus, the user has the opportunity to start the video streaming with navigation (i.e. pause, stop, volume control, etc.) opportunities. Video streaming and navigation opportunities are built into the latest version of HTML 5. However, at the time of this writing, not all browsers support these facilities.

14.5.3.4 Intercepted Media Delivery (Steps 25–29)

For the intercepted model, the service URL is the UDP/RTP address to which the streaming server broadcasts live video stream (captured from a Webcam). This UDP/RTP address is used as the input for the transcoder programme, which transcodes the input video stream into another video codec and also changes the video resolution. In our experiment, a video in H263/RTP format with 320 × 240 resolution is transcoded to JPEG/RTP format with 160 × 120 resolution. Next, the output UDP/RTP stream is forwarded to the UE.

Since we assume a simple UE with minimum software support, we are not using Flash Media Player for playing back videos at the user's platform. Alternatively, an applet that could display RTP video stream has been created. The applet is downloaded at the UE in on demand fashion. The output stream of the transcoder programme is embedded inside the applet as the source of the RTP stream. Thus, the transcoded RTP stream could be viewed at the user's platform even in absence of Flash Media Player.

Given that HTML 5 supports video type tags, a UDP/RTP streaming address could be placed in the value of the video type tag. The HTML 5 aware browsers only support limited number of video codecs (e.g. H.264). On the other hand, our transcoder is based on JMF, which does not support any codec that is also understandable by the HTML 5-aware browsers. Hence, we could not take full advantage of HTML 5-based video streaming in our intercept model. The future versions of the browsers are expected to enhance the domain of the supported codecs. In that case, our intercepted model could be implemented without downloading the applet viewer at the user's platform.

14.6 Conclusions

We have introduced a new model to offer and deploy IMS services. Following an analysis of the current limits in IMS client platforms, we propose to move to a new form of client based on virtualization, which abstracts away issues of platform compatibility that tend to limit deployment of new services.

This model has many benefits, including shielding the user from IMS' complexity without compromising neither its architecture, nor its benefits.

Most importantly, it offers flexibility on the ways new services can be created and offered to users with minimal overhead, as well as natural support for mobility.

We have also shown through a proof-of-concept implementation that standard Web technology can be used to integrate the virtual server-side with an interface allowing users to choose and control their services. This model intrinsically supports the integration of media and data services, again in a way that is transparent for users and user equipment.

Finally, the current trend in Web browser evolution provides the means for an efficient and portable delivery of media, reducing the amount of platform-specific external support, even for multimedia functions. This in turn translates into straightforward multimedia support in the surrogate.

References

[1] 3rd Generation Partnership Project. Technical specification group core network and terminals; IP Multimedia (IM) Subsystem CX and DX interfaces; Signalling flows and message contents. 3GPP TS 29.228 V8.5.0, March 2009.

[2] 3rd Generation Partnership Project. Technical specification group services and system aspects; 3G security; Network domain security; IP network layer security. 3GPP TS 33.210 V8.1.0, October 2008.

[3] 3rd Generation Partnership Project. Technical specification group services and system aspects; Access security for IP-based services. 3GPP TS 33.203 V8.3.0, June 2008.

[4] 3rd Generation Partnership Project: Technical specification group services and system aspects; IP Multimedia Subsystem (IMS), stage 2. 3GPP TS 23.228 V8.5.0, June 2008.

[5] Citrix System. Available from http://www.citrix.com/.

[6] Google Docs. Available from http://docs.google.com.

[7] HTML5: A vocabulary and associated APIs for HTML and XHTML. W3C Working Draft. Available from http://www.w3.org/TR/html5/, March 2010.

[8] Java Media Framework (JMF). Available from http://java.sun.com/javase/technologies/desktop/media/jmf/index.jsp.

[9] JSR 168: Portlet specification. Available from http://www.jcp.org/en/jsr/detail?id=168.

[10] JSR 281: IMS services API. Available from http://jcp.org/en/jsr/detail?id=281.

[11] myMONSTER – Telco communicator suite. Available from http://www.monster-the-client.org/.

[12] The open source IMS core project. Available from http://www.openimscore.org/.

[13] Remote desktop protocol. Available from http://msdn.microsoft.com/en-us/library/aa383015(VS.85).aspx.

[14] I. Cooper, I. Melve, and G. Tomlinson. Internet web replication and caching taxonomy. RFC 3040, 2001.

[15] Nick Feamster, Lixin Gao, and Jennifer Rexford. How to lease the internet in your spare time. *ACM SIGCOMM Computer Communication Review*, 37(1):61–64, 2007.

[16] Mark Handley, Van Jacobson, and Colin Perkins. SDP: Session Description Protocol. RFC 4566, July 2006.

[17] Piotr Kessler. Ericsson IMS client platform. *Ericsson Review*, 2:50–59, 2007.

[18] Xuanzhe Liu, Yi Hui, Wei Sun, and Haiqi Liang. Towards service composition based on mashup. In *Proceedings of the IEEE Congress on Services*, pages 332–339, 2007.

[19] Susanta Nanda and Tzi-cker Chiueh. A survey on virtualization technologies. Technical report. Available from http://www.ecsl.cs.sunysb.edu/tr/TR179.pdf, 2005

[20] Eurescom Project P1656. Definition of an open and extendible IMS client framework. Available from http://www.eurescom.de/public/projects/P1600-series/P1656/.

[21] Tristan Richardson, Quentin Stafford-Fraser, Kenneth R. Wood, and Andy Hopper. Virtual network computing. *IEEE Internet Computing*, 2(1):33–38, 1998.

[22] Jonathan Rosenberg, et al. SIP: Session Initiation Protocol. RFC 3261, June 2002.

[23] Gregor Schaffrath, et al. Network virtualization architecture: Proposal and initial prototype. In *Proceedings of the 1st ACM workshop on Virtualized Infrastructure Systems and Architectures*, pages 63–72, 2009.

[24] Richard Spiers, Robert Marston, Richard Good, and Neco Ventura. The UCT IMS IPTV initiative. In *Proceedings of the Third International Conference on Next Generation Mobile Applications, Services and Technologies*, pages 503–508, 2009.

[25] Do van Thanh, et al. Personalised dynamic IMS client using widgets. A Joint White Paper by Telenor, Linus, Ubisafe and Oslo University College.

[26] Yaping Zhu, Rui Zhang-Shen, Sampath Rangarajany, and Jennifer Rexford. Cabernet: Connectivity architecture for better network services. In *Proceedings of the International Conference on Emerging Networking Experiments and Technologies*, 2008.

PART 4

EVENT DISTRIBUTION

15

A Scalable Publish/Subscribe Broker Network Using Active Load Balancing

Hui Zhang[1], Abhishek Sharma[2], Yunfei Zhang[3], Samrat Ganguly[1], Sudeept Bhatnagar[1] and Rauf Izmailov[1]

[1]*NEC Laboratories America, 4 Independence Way, Suite 200, Princeton, NJ 08540, USA; e-mail: {huizhang, samrat, bhatnagar, rauf}@nec-labs.com*
[2]*Embedded Networks Laboratory, Computer Science Department, University of Southern California, 3710 S. McClintock Avenue, Ronald Tutor Hall (RTH) 418, Los Angeles, CA 90089, USA; e-mail: absharma@usc.edu*
[3]*China Mobile Research Institute, Beijing 100045, China; e-mail: zhangyunfei@chinamobile.com*

Abstract

Publish/subscribe enables a loose coupling communication paradigm in service-oriented architecture, and a broker network handles message filtering (matching) and forwarding between publishers and subscribers. In this chapter, we take load balancing in publish/subscribe services as the target problem, and design a scalable broker network atop a Distributed Hash Table substrate: Chord. We first show that in a Chord network, a simple and optimal load balancing scheme can be implemented atop the aggregation tree rooted on a data sink when the data traffic has a *uniform* distribution over all nodes. Next we present a *load shuffling* scheme to achieve the uniform traffic distribution for any potential aggregation tree in the network. Combining these two together with the Chord substrate, we build a scalable broker network called *Shuffle*, which adaptively balances the publish/subscribe workload for optimal performance. Through extensive simulation, we validate the architecture design and show the scalable and efficient performance of Shuffle upon dynamic publish/subscribe workload.

Anand R. Prasad et al. (Eds.), Future Internet Services and Service Architectures, 317–337.

Keywords: service-oriented architecture, publish/subscribe service, load balancing, Distributed Hash Table, algorithms.

15.1 Introduction

Publish/subscribe (pub/sub) [14, 15] enables a loose coupling communication paradigm for information dissemination: information providers (publishers) and consumers (subscribers) are decoupled through a broker (overlay) network which stores the registered interests (subscriptions) of subscribers and does content-based routing on the messages (events) from publishers to subscribers based on the latter's interests.

Services suitable for pub/sub-style interaction are many, such as stock quotes [14], RSS feeds [26], online auctions [31], networked game [4], located based services [11], enterprise activity monitoring and consumer event notification systems [5, 8, 12], and mobile alerting systems [17, 18], and we expect more are coming. They exhibit significant diversity in terms of requirements for subscription expressiveness, and different pub/sub models have been proposed in the literature, including channel-based pub/sub, topic-based pub/sub, attribute-based pub/sub, keyword-based pub/sub, similarity-based pub/sub, and pattern-based pub/sub (for a survey, see refer to [34]). To allow a pub/sub broker network universally useful (e.g., as an enterprise service bus for diverse enterprise applications [28], or a generic event notification network for Web access [34]), the broker network has to be designed to accommodate a large spectrum of pub/sub models.

To support scalable operations under diverse pub/sub models, we take the *service hosting model* for the broker network design. We believe there is no one-size-fits-all solution but different optimized in-memory matching algorithms can be implemented for different pub/sub models. In the service-hosting broker network, initially a matching algorithm will be hosted by one server which handles the matching of the related events and subscriptions. When the workload (service popularity) increases and that single server cannot meet the demand, enough servers will be added to run the same algorithm and the workload is split among those servers in terms of event and/or subscription partition. Workload management in such a hosting platform is the main focus of this chapter.

We propose Shuffle, a scalable broker network with active workload management. In Shuffle, load balancing issues are addressed on all types of workloads including message parsing, matching, and delivery. Shuffle actively aggregates/distributes events and subscriptions among broker servers to avoid performance bottlenecks in terms of message parsing, and offers a simple and fast mechanism to achieve optimal load balancing on message matching and delivering with little overhead. While the active workload management incurs message forwarding overhead, the overhead is distributed among the servers and even declines when the load balancing mechanism is applied upon high event arrival rates. To the best of our knowledge, Shuffle is the first broker network designed to address all four workload issues within a service hosting framework.

This chapter is based on a conference paper the authors presented at ICC'08 [37]. The chapter is organized as follows: Section 15.2 presents the background information on content-based pub/sub model and the Chord protocol. Section 15.3 describes the related work. The architecture design of the Shuffle system is described in Section 15.4. Section 15.5 presents the simulation results. and Section 15.6 concludes the chapter with future work.

15.2 Background Information

15.2.1 Content-Based Pub/Sub Model

In a content-based pub/sub model, a multi-dimensional data space is defined on d attributes. An event e can be represented as a set of $\langle a_i, v_i \rangle$ data tuples where v_i is the value that this event specifies for the attribute a_i. A subscription can be represented as a filter f that is a conjunction of k ($k \leq d$) predicates, and each predicate specifies a constraint on a different attribute, such as "$a_i = X$", or "$X \leq a_i \leq Y$".

In the rest of this chapter, we focus on content-based pub/sub models for presentation concreteness. The extension of Shuffle to supporting topic-based pub/sub models is straightforward. For non-attribute based pub/sub models such as XML schema based, Shuffle may not be applicable.

15.2.2 Chord

Next, we briefly describe the Chord [32] DHT system for background information.

Like all other DHT systems, Chord supports scalable storage and retrieval of arbitrary $\langle key, data \rangle$ pairs. To be able to do this, Chord assigns each overlay node in the network an m-bit identifier (called the node ID). This identifier can be chosen by hashing the node's address using a hash function such as SHA-1. Similarly, each key is also assigned an m-bit identifier (following [32], we use key and identifier interchangeably). Chord uses *consistent hashing* to assign keys to nodes. Each key is assigned to that node in the overlay whose node ID is equal to the key identifier, or follows it in the *key space* (the circle of numbers from 0 to $2^m - 1$). That node is called the *successor* of the key.

The Chord protocol enables fast, yet scalable, mapping of a key to its assigned node. It maintains at each node a *finger* table having at most m entries. The i-th entry in the table for a node whose ID is n contains the pointer to the first node, s, that succeeds n by at least 2^{i-1} on the ring, where $1 \leq i \leq m$. Node s is called the i-th finger of node n.

Suppose node n wishes to lookup the node assigned to a key k (i.e., the successor node x of k). To do this, node n searches its finger table for that node j whose ID immediately precedes k, and passes the lookup request to j. j then recursively[1] repeats the same operation; at each step, the lookup request progressively nears the successor of k, and the *search space*, which is the remaining key space between the current request holder and the target key k, shrinks quickly. At the end of this sequence, x's predecessor returns x's identity (its IP address) to n, completing the lookup. Because of the way a Chord's finger table is constructed, the first hop of the lookup from n covers (at least) half the identifier space (clockwise) between n and k, and each successive hop covers an exponentially decreasing part. It is easy to see that the average number of hops for a lookup is $O(\log N)$ in an N-node network.

Our brief description of Chord has omitted several details of the Chord protocol, including procedures for constructing finger tables when a node joins, and maintaining finger tables in the face of node dynamics (the Chord *stabilization* procedure). The interested reader is referred to [32] for these details.

15.3 Related Work

In the recent past, a large body of work has emerged focusing on the problem of large scale selective dissemination of information to users. Several solutions were proposed based on using the multicast model. Using conventional multicast model is not scalable as the number of multicast trees can grow up to 2^n to capture all possible subscriber groups. The channelization problem formulated in [1] provides a solution to map sources and destinations to a limited set of multicast trees to minimize the unwanted message delivery.

[1] Chord lookups can also traverse the circle iteratively. In our system the lookup type is always set as "recursive".

Another category of work [7,23,27,34] creates a limited number of multicast trees by proper clustering of user subscription profiles. In the above solutions, filtering is done at the source, at the receiving point, or both. In contrast, some authors [21, 30] proposed the use of filters in the intermediate nodes in a given multicast tree for selective data dissemination. Shah et al. [30] provide a solution to filter placement and leak minimization problem. In multicast based approaches, the forwarding path of a message is restricted to predefined multicast tree topology. Although these approaches can be applied well in topic/subject based or messages with single attribute, they are not suitable to support general predicates over multiple attributes.

Instead of restricting to a multicast model, a general model is to create a routing network composed of content-based routers as proposed in Siena [9, 10] and Gryphon [2]. A content-based router creates a forwarding table based on subscription profiles and performs both data filtering and forwarding based on predicate matching. As with any data distribution network, the speed of matching the subscription predicates at each content-based router determines the sustainable throughput. The goal of content-based routing is to provide processing latency meeting the wire-speed. In certain solutions such as Siena [2, 9, 10], the message dissemination path is coupled with the subscription movement path and therefore lacks the routing flexibilities.

Existing work on the design of a scalable broker network targets two objectives: (a) how to minimize the subscription states maintained at each node in order to perform fast in-memory subscription matching [10] and (b) how to search the subscription states efficiently in order to reduce the number of messages [7]. Recently, there has been a significant interest in searching multi-dimensional subscription profiles using DHT [3, 16, 20, 24, 25, 33] They are different from our work in that their design goal is to maximize the in-networking filtering effectiveness through overlay routing in a single pub/sub service, while our design goal is to utilize overlay routing for optimal workload management in a service hosting model to support diverse pub/sub services.

SDIMS [35] is a scalable distributed information management system for large scale networked applications. While it also utilizes a DHT substrate to build a tree for network information aggregation, two key points distinguishes SDIMS from Shuffle: (1) the aggregation tree in SDIMS is a balanced binary tree, which is logically defined based on hierarchical aggregation abstraction [35]; while the aggregation tree in Shuffle is an unbalanced tree, which is physically formulated following the overlay routing protocol; and (2) load balancing is assumed to be addressed with random hashing in a DHT utilized

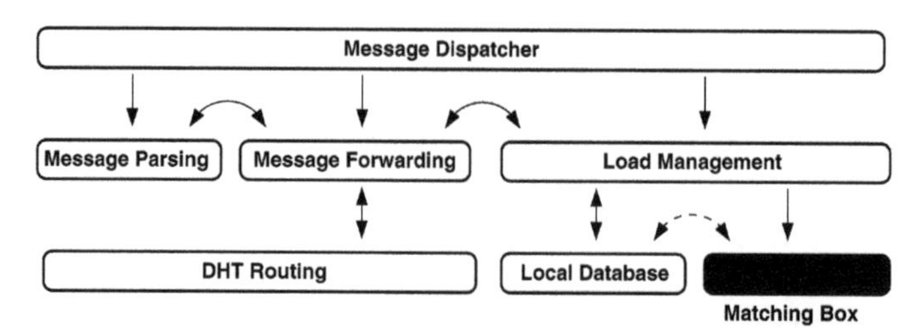

Figure 15.1 The Shuffle node architecture.

by SDIMS as the applications in it usually have large numbers of data attributes and small numbers of information publishers and consumers compared to pub/sub services, while the focus of Shuffle is load balancing on dynamic and heavy pub/sub workload.

A significant amount of research [6,10,13,15,29,36] has been carried out to find better in-memory algorithms for general predicate matching at a single node. As this is not the focus in this chapter, we skip the related discussion.

15.4 Shuffle Design

15.4.1 Architecture Overview

Shuffle supports a generic middleware framework independent of application-specific pub/sub matching algorithms. It is designed atop a set of *dedicated* servers (either dispersed over wide area or residing within a LAN) as broker nodes. The network usually starts with a small size (e.g., in tens) and may grow to thousands of nodes with the increasing popularity of the services.

Each Shuffle node has the implementation of some in-memory processing algorithm for message matching, and can work as an independent publish/subscribe server. In addition, all nodes organize into an overlay network and collaborate with each other in workload aggregation/distribution. The aggregation is defined on the attributes that events and subscriptions are associated with. While Shuffle supports in-network message filtering at the attribute name level in its overlay routing, the main functionality of the overlay network is to aggregate related messages for centralized processing if possible, and distribute the workload from overloaded nodes to other nodes in an *efficient*, and *responsive* way.

Figure 15.1 shows the software components in a Shuffle node. We describe their functionalities through an event distribution process:

1. When an event message arrives at some server in the broker network, the *message dispatcher* component decides on an action dependent on its type. If it is an original message arriving from outside the broker network, the dispatcher will send it to a randomly chosen server through the shuffling process (see Section 15.4.3 for details). Otherwise, the event message will be passed on to the message parsing component.
2. If the event message has been shuffled but not been parsed yet on its content, the *message parsing* component will parse it and generate the corresponding Shuffle message including its virtual ID.
3. The parsed event message will be passed on to the *message forwarding* component which implements all routing-related functionalities on top of the *DHT routing* component. The event message will be forwarded to the root node of the aggregation tree defined on its predicate attributes.
4. When arriving at the root node, the event message will be passed on to the *matching box* which implements the application-specific in-memory message matching algorithm. For a matched event against subscriptions stored in the *local database*, it will be delivered to the corresponding subscribers from the root node.
5. When the root node becomes overloaded due to either high event matching workload or delivery workload, the *load management* component will be activated to provide the load balancing functionalities including both *loading-forwarding* and *replicating-holding* (see Section 15.4.4 for details).

15.4.2 Overlay Construction

Shuffle uses the Chord protocol to map both nodes and pub/sub messages onto an unified one-dimensional key space. Each node is assigned an ID and a portion (called its range) of the key space, and maintains a Chord finger table for overlay routing. Both event and subscription messages are hashed onto the key space based on one of the attributes contained in the messages, and therefore are aggregated and filtered to a coarse degree (attribute level). For each attribute in the pub/sub services, the routing paths from all nodes to the node responsible for that attribute (called the *root node* in the rest of the chapter) naturally forms an aggregation tree, on which Shuffle manages the workload associated with that attribute.

In a Shuffle network with $n = 2^k$ nodes and equal partition of the overlay space among the nodes, there is a specific load distribution in an aggregation tree when all nodes generate the same amount of messages to the root

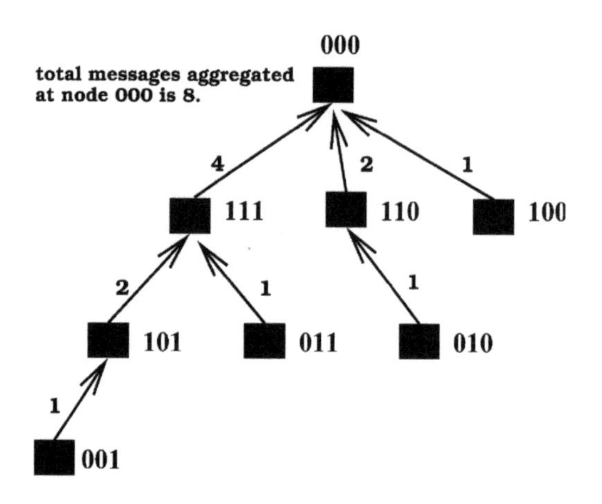

Figure 15.2 One aggregation tree example in an 8-node Shuffle network.

node. As shown in Figure 15.2, let us assume that every node generates one message which is then routed to the root node and some non-leaf node S has aggregated a total of $T(S)$ unit messages in the overlay forwarding. We can show (see [37] for the details) that S has $\log(T(S))$ immediate children in the tree, and the messages forwarded by the children follows the *half-cascading* distribution: one child forwards $1/2$ of the total messages $T(S)$, one child forwards $1/4$ of $T(S)$, and so on, until one child, which is a leaf node, contributes one message. As a numeric example, Figure 15.2 shows the aggregation tree rooted at node 000 in an 8-node Shuffle network with 3-bit key space. It is easy to verify that all non-leaf nodes have the *half-cascading* traffic distribution on their incoming links.

When the key space is not equally partitioned among the nodes, the aggregation load will not follow the exact half-cascading distribution. It is well known that the original node join scheme in Chord can cause $O(\log n)$ stretch in key space partition. In a Shuffle system, we propose a node joining/leaving scheme for equal key space partition:

Optimal Splitting (OS)

- *Joining*: a new node joins the network by finding the Shuffle node owning the *longest* key range and taking half of the key range from that node.

When there are multiple candidate nodes with the longest key ranges, the tie is broken with a random choice.

- *Leaving*: the leaving node x finds the Shuffle node owning the *shortest* key range and asks it to leave its current position and rejoin at x's position. When there are multiple candidate nodes with the shortest key ranges, x's immediate successor node is given the priority if it is one such candidate; otherwise, the tie is broken with a random choice.

It is easy to show that the OS scheme can partition the key space equally among the nodes, but it requires global information on current space allocation situation. We show in [37] that a natural extension of the original Chord joining/leaving scheme can achieve even key space partition close to that from the OS scheme.

15.4.3 Message Shuffling

As a proxy for a pub/sub overlay service, the first assignment of a Shuffle node x at receiving an original subscription message m is to re-distribute it in the system. x will pick a random key for m (e.g., by hashing some subscription ID contained in the message) and send it to the node y responsible for that key in the overlay space. y is responsible for parsing m, constructing a new message $\tilde{m}$ tailored for fast overlay routing and the in-memory message matching algorithm associated with the service, and sending $\tilde{m}$ to the destination node z, which y decides by arbitrarily picking an attribute A specified in m and hashing A onto the overlay space.[2] We call z the root of the aggregation tree associated with the attribute A, and y generates one message for the aggregation tree by sending $\tilde{m}$ to z through the overlay routing. For an event message m, each attribute specified in it will be used to decide a destination node to send one copy of $\tilde{m}$.

The above randomization procedure (called *message shuffling*) achieves two goals:

- The randomization makes the distribution of the input traffic for any potential aggregation tree uniform on the key space. Combing *message shuffling* and Chord with OS scheme, Shuffle can construct half-cascading aggregation trees.

[2] For unsubscribe consideration, the random keys for the paired subscribe/unsubscribe messages should be the same, and the choice of the attribute A has to be made consistently even if y is later replaced by another node for the key range containing the random key.

- The cost of message parsing on subscriptions is distributed evenly throughout the system so that Shuffle eliminates the potential performance bottleneck due to message parsing workload.

The next section shows how the half-cascading distribution enables simple and optimal load balancing in a Shuffle system.

15.4.4 Optimal Load Balancing in a Single Aggregation Tree

With the half-cascading distribution, a simple *pushing* mechanism can be used to achieve optimal load balancing with little overhead. In a Shuffle aggregation tree, initially only the root node processes the aggregated messages and all other nodes just forward messages. We say the root node is in *active* state and the rest of the nodes are in *hibernating* state. When the processing workload is excessive, the root node re-balances the workload by activating the child forwarding half of the workload and pushing that part of the workload back to that child for processing. This amounts to splitting the original aggregation tree into two with each node acting as one root. The pushing operation may happen recursively until each activated node can accommodate its assigned processing workload. With the *half-cascading* distribution, an active node can reduce half of its workload after each pushing operation. Therefore, the maximal number of pushing operations each node incurs is $\log n$, where n is the network size. If the workload after $\log n$ operations is still too high for a node, we can claim all other nodes are also overloaded. This is because the current workload this node has is due to the messages itself generates and all nodes generate the same amount of messages (workload).[3]

There are two types of messages in Shuffle: subscriptions and events. Accordingly, the load balancing on an aggregation tree is classified into two kinds.

15.4.4.1 Subscription Balancing

In Shuffle, a subscription message n will be replicated at each hop on the routing path from the message sender y to the root node z. In addition, every node in an aggregation tree will count the forwarding fraction of the subscriptions that each of its children contributes in message forwarding. When z is

[3] Due to message shuffling, the messages generated by each node have the same distribution in the multi-dimensional attribute space. Therefore, the workload for processing the set of messages generated by any nodes is probabilistically the same given the same set size.

overloaded with too many subscriptions from an aggregation tree A, it applies the *loading-forwarding* scheme for subscription balancing:

1. It ranks all the children in this tree based on their forwarding fraction, and marks them as "hibernating" initially.
2. Among all "hibernating" children, z picks the node l with the largest fraction. z will unload all subscriptions forwarded from l by sending the corresponding "unsubscribe" messages into z's local black box, and asks l to load the cached copies from its local database into its local box.
3. z marks l as "active", and forwards all event messages in the aggregation tree of the same attribute A to l in the future.
4. z repeats Step 2 if it needs to unload more subscriptions locally.

The network overhead of the *loading-forwarding* scheme is only one control message. In Shuffle, the *loading-forwarding* scheme is applied recursively on each active node for load balancing on subscriptions.

15.4.4.2 Event Balancing

In a Shuffle aggregation tree, each node also counts the forwarding fraction of the events that each of its children contributes in message forwarding. When z is overloaded with too many events from an aggregation tree A, it applies the *replicating-holding* scheme for load balancing:

1. z ranks all the children in this tree based on their forwarding fraction, and marks them as "hibernating" initially.
2. Among all "hibernating" children, z picks the node l with the largest fraction. z will replicate all subscription messages from the same aggregation tree at node l, and ask l to hold event forwarding and instead process all the events it receive at its local box in the future.
3. z marks l as "active", and goes to Step 2 if it needs to shed more arriving events.

Clearly, in Step 2 z only needs to transfer to l the subscription messages that were not forwarded by l in the subscription aggregation tree. As the fraction α of event forwarding from l to z in an aggregation tree is the same as the fraction β of subscription forwarding from l to z, reducing the event workload by a factor α on z requires transferring only $(1 - \alpha)$ of the subscription messages in z. This saving is obvious in the first few replication operations, which are most significant in workload reduction.

In Shuffle, the *replicating-holding* scheme is applied recursively on each active node for load balancing on events.

15.4.4.3 Load Balancing Scheduling Scheme

We propose a simple scheme for scheduling the order of load balancing operations on events and subscriptions at an overloaded node, n:

- Node n determines its target event processing rate r. Using the *replication-holding* scheme, node n stops as many of its children as required from forwarding events to itself so that its event arrival rate is at most r. If node n stops its child n_1 from forwarding events, then the aggregation tree rooted at n is split into two disjoint trees rooted at n and n_1, respectively.
- Based on r, node n determines the upper bound on the number of subscriptions it should manage, say, th. It then uses the *loading-forwarding* scheme to off-load subscriptions to its children so that its subscription load is at most th.

15.4.5 Load Balancing over Multiple Aggregation Trees

If the overloaded node n is part of multiple aggregation trees, it needs to decide along which aggregation tree should it off-load work(events/subscriptions) to its children.

Consider a pub/sub broker network with N nodes. Suppose node n is part of k aggregation trees. Let X_t^n be the total number of subscriptions that n is currently managing. Since n is part of k aggregation trees,

$$X_t^n = X_1^n + X_2^n + \cdots + X_k^n$$

where X_i^n is the number of subscriptions for attribute i currently being managed by n. Let X_{total} and X_i be the total number of subscriptions and the total number of subscriptions for attribute i managed by the pub/sub system, respectively, i.e.

$$X_{total} = X_1 + X_2 + \cdots + X_k$$

A threshold $th = r \times (X_{total}/N)$ can be written as

$$th = th_1 + th_2 + \cdots + th_k$$

where $th_i = r \times (X_1/N)$. Hence, the following always holds true in the case of multiple aggregation trees:

- If $X_t^n > th$ then $\exists$ at least one attribute $i \ni X_i^n > th_i$.

The above fact motivates the following heuristic, henceforth referred to as *Multiple Tree heuristic* for choosing an aggregation tree:

1. If the total number of subscriptions at node n, X_t^n exceeds the threshold th, it picks an attribute $i \ni X_i^n > th_i$ and off-load subscriptions along the aggregation tree for attribute i.
   When there are multiple attributes $i \ni X_i^n > th_i$, node n can either choose an attribute randomly or pick an attribute j for which $(X_j^n - th_j)$ is the highest.
2. Repeat (1) until $X_t^n \leq th$.

An overloaded node that needs to use the *replication-holding* scheme can use an analogous heuristic for picking an aggregation tree.

15.4.6 Discussions

15.4.6.1 Pub/Sub Service Hosting Model

The diverse requirements of multiple pub/sub applications hosted by a single broker network necessitates a design that can scale on multiple parameters – high event rate, large number of subscriptions, complex subscription profiles. We present an *active load shuffling* scheme to transfer arbitrary input traffic distribution to the desired uniform traffic distribution for any potential aggregation tree in the network. As a side effect, it also eliminates potential performance bottleneck due to message parsing by balancing the related workload throughout the network. For event overloading (high event arrival rate) and subscription overloading (excessive subscriptions), we design two half-cascading load balancing schemes respectively.

While we propose one aggregation trees for each separate attribute, having a single aggregation tree for all attributes is another interesting idea where our technology still applies. In that case, we can assign all events and subscriptions with the same ID (e.g., key 0), and the workload will be aggregated in one single root node and balanced to its child nodes if necessary. This brings the benefit on energy efficiency side, as the broker network can turn off unnecessary leaf nodes on the single tree under light load, while having minimal impact on the event matching process.

15.4.6.2 Shuffling Cost

Notice that the sole goal of message shuffling is to transform the unpredictable distribution of the input traffic coming from publishers/subscribers into uniform. Therefore, the extra hops due to overlay routing should be avoided if possible. When node dynamics are relatively low in the system and each node can maintain the list of all active nodes in the system, the routing cost for message shuffling can be reduced significantly by sending the original event/subscription messages directly to their destination nodes for processing.

15.5 Evaluation

15.5.1 Methodology

In this section, we examine the features of our architecture in details. For this purpose, we implemented the Shuffle simulator with the underlying DHT being a basic version of Chord, by extending the p2psim simulator [22].

For comparison, we consider two other possible load balancing schemes:

- *Random-Half*: In this scheme, an overloaded node keeps probing other nodes randomly until it finds an underloaded node. It then splits its load with that node by replicating only those subscriptions there which are not already present (the subscriptions could have originated there or have been replicated there by another node). The original node repeats the operation until its load is reduced below a target level. If the destination node also gets overloaded, it too searches for other underloaded nodes to shed its load. This scheme is similar to the DHT load balancing scheme proposed in [19].
- *Random-Min*: Random-Min is the same as Random-Half except when an overloaded node splits its load with an underloaded node, it just delegates a bare minimum load equal to the target value to the chosen node by replicating its subscription set (less the subscriptions originating from the chosen node) there and forwarding a commensurate fraction of event traffic there. Thus, the only node ever overloaded is the root node which keeps searching for a new node to shed its load until its load is less than the target value.

For the Random-Half and Random-Min schemes, we assume that the node trying to find a random underloaded node incurs a single control message overhead even if it were to use multi-hop DHT-based routing to reach that node. We do this to be as lenient as possible in measuring the cost of the competing systems.

We note that both these schemes are not scalable because a node may have to keep information about a large number of other nodes in a heavily loaded situation whereas Shuffle does not need to store information about any other nodes except its $\log(N)$ fingers (for a network with N nodes). Also, in the simulations these schemes take advantage of the half-cascading distribution that is guaranteed only in the Shuffle design. These schemes are meant for benchmark purposes only.

In the simulations, nodes join the system sequentially. The network sizes chosen for the simulations are from 48 to 1024 nodes. These choices are made for two reasons: (a) they represent both the size (power of 2) where half-cascading load distribution will be guaranteed, and other sizes where half-cascading distribution will be approximated; and (b) they represent the reasonable system size range of a broker network. We evaluate the Shuffle performance with the metrics on its control message overhead, message forwarding overhead, number of nodes affected to reach a load balancing target. The overhead is normalized as per event/subscription value so that the simulation results do not show the absolute number of events and subscriptions.

15.5.2 Load Balancing

15.5.2.1 Single Tree Load Balancing

For this set of experiments, we consider the load being incurred due to a high event rate so that the root node uses the *replicating-holding* scheme. Each node generates the same number of events to the root node. We set the target load level as multiples of the total events generated by each node. A lower target load level implies that a large number of nodes need to active to be able to manage the workload. We present results for control messages generated and event forwarding overhead. In Figures 15.3 and 15.4, the curve labeled "Cascaded" corresponds to the Shuffle system.

Control Message Overhead: Figure 15.3 shows the number of control messages generated to achieve the target workload level in a network with 768 nodes and an OS node joining scheme. The number of control messages increase with a decrease in the target load value. This is because a larger number of nodes need to be contacted to divide the load among them. However, the number of messages sent out by Random-Half and Random-Min schemes jumps significantly when the target load value is small. Once a large fraction of nodes in the network are *active* (i.e. managing workload offloaded from overloaded nodes), it becomes difficult for an overloaded node to find an *inactive* (or underloaded) node by random sampling and hence both Random-Half and Random-Min end up generating significant control traffic. In contrast, Shuffle generates significantly less control messages because it uses the aggregation tree to locate the underloaded children of an overloaded node.

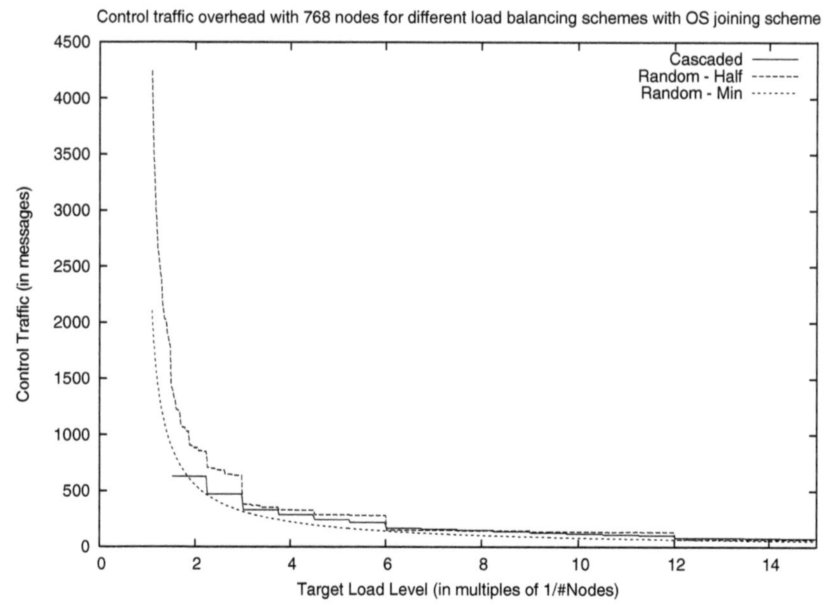

Figure 15.3 Control message overhead: 768-node network size.

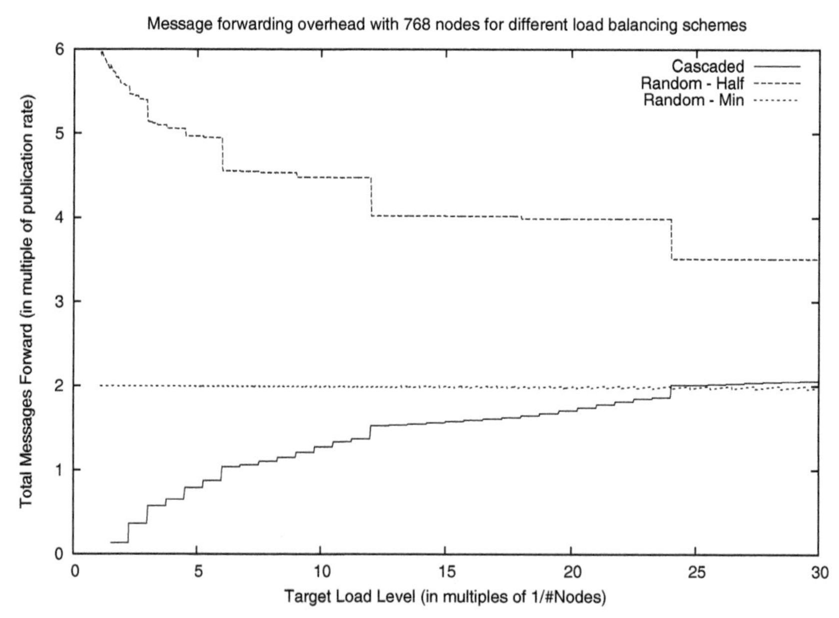

Figure 15.4 Message forwarding overhead: 768-node network size.

Forwarding Traffic Overhead: Figure 15.4 shows the event message forwarding overhead (in multiples of the total event rate) of the three load balancing schemes for 768 nodes. In Shuffle, even when the system is lightly loaded (target load level $\geq$ 25), events are forwarded to the root of the aggregation tree and hence, the Shuffle system incurs an initial message forwarding overhead. However, as the system becomes overloaded (low target load levels), the *replication-holding* scheme stops the children of the overloaded node from forwarding events. Thus, as the system gets overloaded due to higher even arrival rate, the Shuffle system reduces the message forwarding overhead by requiring more nodes to handle the events locally. In the extreme case (target load level close to 1), when each node manages only the events it generates, there is no message forwarding.

With the Random-Half scheme, as the event workload on the system increases, more nodes become overloaded and the forwarding overhead becomes higher due to a higher fraction events being off-loaded from the overloaded nodes to the underloaded nodes.

In the Random-Min scheme, since the root directly forwards the appropriate fraction of events to the nodes which share its load, the total event traffic is never more than twice the total event rate (factor of one corresponding to the nodes forwarding messages to the root and another factor of one for the root appropriately forwarding to the other load-sharing nodes).

15.5.2.2 Multiple Tree Load Balancing

We compare the combination of *loading-forwarding* scheme and the *Multiple Tree heuristic* (Section 15.4.5) of the Shuffle system (represented by the curve labeled "Shuffle" in Figure 15.5) with the Random-Half and Random-Min schemes. Starting with a set of overloaded nodes, the aim of all three load balancing schemes is to redistribute the workload so that the number of subscriptions managed by any node in the network is at most the target load level. The metric of interest is *nodes affected.* A node is affected during subscriptions load balancing if some overloaded node off-loads subscriptions to it. The event forwarding traffic in the case of *loading-forwarding* is proportional to the number of nodes affected during workload redistribution. Hence, to minimize the event forwarding traffic, it is essential to minimize the number of nodes affected.

The network size is 1024 nodes with 512 aggregation trees. The node joining scheme is OS. Each node generates the same number of subscriptions. A subscription is mapped to an aggregation tree using a zipf distribution. We set the target load level as multiples of the total subscriptions generated by

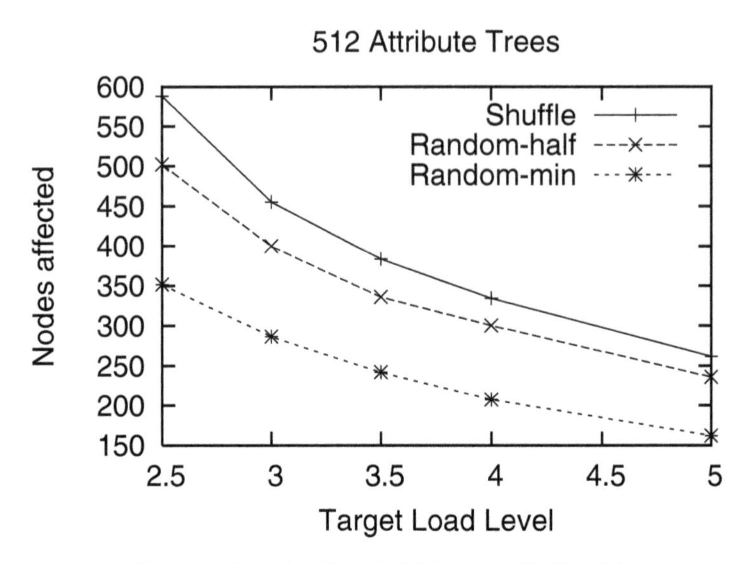

Figure 15.5 Node affected: 1024 nodes, OS, Zipf Dist.

each node. A lower target load level implies that a large number of nodes need to active to be able to manage the work load.

Nodes Affected: As shown in Fig 15.5, the Random-Min scheme is very efficient in terms of the number of nodes affected as it always transfers subscriptions to an underloaded node (it can be easily verified that it affects the minimum number of nodes in the single aggregation tree case). With multiple aggregate trees, the Shuffle system affects up to 20% (70%) more nodes compared to the Random-Half (Random-Min) scheme for target load level $r = 2.5$. For $r \geq 4$, its performance is comparable to the Random-Half scheme. The reason for a higher number of nodes being affected in the Shuffle system is that the *Multiple Tree heuristic* does *loading-forwarding* separately for each aggregation tree containing overloaded nodes. Hence, the load shed after one load balancing iteration over a tree is limited by the workload on that tree, which could be much smaller than the total workload on the overloaded node (when the overloaded node is part of several other trees). Compare this with the situation in case of Random-Min and Random-Half where a node sheds the workload freely without differentiating between the trees. While per-tree load balancing affects a higher number of nodes, the trade-off is the simplified overlay management in Shuffle with the existence of only one overlay – the original Chord topology. Random-Half and Random-Min create an overlay topology (the trees resulting from load balancing) for each overloaded node that is different from the underlying Chord topology, thus increasing the network management complexity significantly (this increased complexity is not captured by the nodes affected metric).

Note that in the Shuffle system, since a child node already caches the subscriptions forwarded to its parent, there is no subscription movement overhead. The Random-Half and Random-Min schemes can incur significant subscription movement overheads (especially in large network).

15.6 Conclusions

We present *Shuffle*, a scalable broker network designed in a service hosting model to support diverse pub/sub models. Shuffle offers an integral solution to efficiently manage all types of the workload in the network. We show how Shuffle uses Chord, a DHT substrate, to achieve load balancing in the system. Our simulation results clearly establish the effectiveness of Shuffle.

References

[1] M. Adler, Z. Ge, J. Kurose, D. Towsley, and S. Zabele. Channelization problem in large scale data dissemination. In *Proceedings of ICNP*, 2001.

[2] M. Aguilera, R. Strom, D. Sturman, M. Astley, and T. Chandra. Matching events in a content-based subscription system. In *Proceedings of Symposium on Principles of Distributed Computing*, 1999.

[3] R. Baldoni, C. Marchetti, A. Virgillito, and R. Vitemberg. Content-based publish-subscribe over structured overlay networks. In *Proceedings of ICDCS*, 2005.

[4] A. Bharambe, S. Rao, and S. Seshan. Mercury: A scalable publish-subscribe system for internet games. In *Proceedings of the 1st Workshop on Network and System Support for Games*, 2002.

[5] L. Cabrera, M. Jones, and M. Theimer. Herald: Achieving a global event notification service. In *Proceedings of HotOS-VIII*, 2001.

[6] A. Campailla, S. Chaki, E. Clarke, S. Jha, and H. Veith. Efficient filtering in publish-subscribe systems using binary decision. In *Proceedings of International Conference on Software Engineering*, 2001.

[7] F. Cao and J. Singh. Efficient event routing in content-based publish-subscribe service networks. In *Proceedings of Infocom*, April 2004.

[8] A. Carzaniga, D. Rosenblum, and A. Wolf. Design of a scalable event notification service: Interface and architecture. Technical Report CU-CS-863-98, Department of Computer Science, University of Colorado at Boulder, 1998.

[9] A. Carzaniga, M. Rutherford, and A. Wolf. A routing scheme for content-based networking. In *Proceedings of IEEE INFOCOM*, March 2004.

[10] A. Carzaniga and A. Wolf. Forwarding in a content-based network. In *Proceedings of ACM SIGCOMM*, August 2003.

[11] X. Chen, Y. Chen, and F. Rao. An efficient spatial publish/subscribe system for intelligent location-based services. In *Proceedings of the 2nd International Workshop on Distributed Event-Based Systems*, 2003.
[12] G. Cugola, E. Nitto, and A. Fugetta. The JEDI event-based infrastructure and its application to the development of the OPSS WFMS. In *IEEE Transactions on Software Engineering*, 2001.
[13] P. Eugster, P. Felber, R. Guerraoui, and S. Handurukande. Event systems: How to have your cake and eat it too. In *Proceedings of DEBS*, 2002.
[14] P. Eugster, P. Felber, R. Guerraoui, and A. Kermarrec. The many faces of publish/subscribe. Technical Report DSC ID:2001, 2001.
[15] F. Fabret, H.A. Jacobsen, F. Llirbat, J. Pereira, K.A. Ross, and D. Shasha. Filtering algorithms and implementation for very fast publish/subscribe systems. In *Proceedings of ACM SIGMOD*, 2001.
[16] Abhishek Gupta, Ozgur D. Sahin, Divyakant Agrawal, and Amr El Abbadi. Meghdoot: Content-based publish/subscribe over p2p networks. In *Proceedings of International Conference on Distributed Systems Platforms and Open Distributed Processing*, 2004.
[17] http://mobile.yahoo.com/wireless/alert.
[18] Y. Huang and H. Garcia-Molina. Publish/subscribe in a mobile environment. In *Proceedings of the 2nd ACM International Workshop on Data Engineering for Wireless and Mobile Access*, 2001.
[19] D. Karger and M. Ruhl. Simple efficient load balancing algorithms for peer-to-peer systems. In *Proceedings of ACM SPAA*, 2004.
[20] V. Muthusamy and H.-A. Jacobsen. Small-scale peer-to-peer publish/subscribe. In *Proceedings of P2P Workshop at MobiQuitous*, 2004.
[21] M. Oliveira, J. Crowcroft, and C. Diot. Router level filtering on receiver interest delivery. In *Proceedings of 2nd International Workshop on Networked Group Communication*, 2000.
[22] p2psim: A simulator for peer-to-peer protocols. http://pdos.csail.mit.edu/p2psim/.
[23] O. Papaemmanouil and U. Cetintemel. Semcast: Semantic multicast for content-based data dissemination. In *Proceedings of ICDE*, April 2005.
[24] Peter R. Pietzuch. Hermes: A scalable event-based middleware. In Technical Report UCAM-CL-TR-590, University of Cambridge, 2004.
[25] Peter R. Pietzuch and Jean Bacon. Peer-to-peer overlay broker networks in an event-based middleware. In *Proceedings of the 2nd International Workshop on Distributed Event-Based Systems*, 2003.
[26] V. Ramasubramanian, R. Peterson, and E.G. Sirer. Corona: A high performance publish-subscribe system for the world wide web. In *Proceedings of NSDI06*, 2006.
[27] A. Riabov, Z. Liu, J. Wolf, P. Yu, and L. Zhang. Clustering algorithms for content-based publication-subscription systems. In *Proceedings of ICDCS*, 2002.
[28] M.-T. Schmidt, B. Hutchison, P. Lambros, and R. Phippen. The enterprise service bus: Making service-oriented architecture real. *IBM Systems Journal*, 44(4), 2005.
[29] B. Segall and D. Arnold. Elvin has left the building: A publish/subscribe notification service with quenching. In *Proceedings of AUUG*, 1997.
[30] R. Shah, R. Jain, and F. Anjum. Efficient dissemination of personalized information using content-based multicast. In *Proceedings of Infocom*, 2002.

[31] D.X. Song and J.K. Millen. Secure auctions in a publish/subscribe system. Available at http://www.csl.sri.com/users/millen/, 2000.

[32] I. Stoica, R. Morris, D. Karger, F. Kaashoek, and H. Balakrishnan. Chord: A peer-to-peer lookup service for internet applications. In *Proceedings of ACM SIGCOMM*, 2001.

[33] D. Tam, R. Azimi, and H.-A. Jacobsen. Building content-based publish/subscribe systems with distributed hash tables. In *Proceedings 1st International Workshop on Databases, Information Systems, and P2P (DBISP2P)*, 2003.

[34] Y. Wang, L. Qiu, D. Achlioptas, G. Das, P. Larson, and H. Wang. Subscription partitioning and routing in content-based publish/subscribe networks. In *Proceedings of International Symposium on Distributed Computing*, 2002.

[35] P. Yalagandula and M. Dahlin. A scalable distributed information management system. In *Proceedings of the ACM SIGCOMM Conference*, 2004.

[36] T. Yan and H. Garcia-Molina. Index structures for selective dissemination of information under the Boolean model. In *ACM Transactions on Database Systems*, 1994.

[37] Hui Zhang, S. Ganguly, S. Bhatnagar, R. Izmailov, and A. Sharma. Optimal load balancing in publish/subscribe broker networks using active workload management. In *Proceedings of IEEE International Conference on Communications (ICC'08)*, pages 5892–5896, May 2008.

16

Enabling Publish/Subscribe Services in Sensor Networks

Duc A. Tran[1] and Linh H. Truong[2]

[1]*Department of Computer Science, University of Massachusetts, Boston, MA 02125, USA; e-mail: duc@cs.umb.edu*
[2]*IBM Zurich Research Laboratory, Säumerstrasse 4, 8803 Rüschlikon, Switzerland; e-mail: hlt@zurich.ibm.com*

Abstract

This past decade has seen sensor technologies increasingly deployed in many applications. Sensor networks are used to monitor the surrounding environment for important events such as climate changes, chemical leaks, early warnings of a natural disaster, or violations in a no-trespassing zone. The publish/subscribe paradigm is one that represents a large class of such applications. This chapter presents a survey of important techniques that enable publish/subscribe services in sensor networks, with special attention given to the issues of routing design, query aggregation, event matching, and middleware development.

Keywords: publish/subscribe, sensor networks, data dissemination.

16.1 Introduction

A sensor network for monitoring purposes basically involves two types of nodes: the "query" nodes and the "sensor" nodes. The query nodes are those that send queries into the network to inquire about sensor data of interest. A query node can be a sink node that needs to collect the data from the

Anand R. Prasad et al. (Eds.), Future Internet Services and Service Architectures, 339–363.

network or an actuator node whose operation is triggered based on events detected in the network. The sensor nodes are those that capture event data and need to send them, or notify of their existence, to interested query nodes. In many applications, a query needs to be submitted to the network in advance waiting for notification of the events that match the query. Because a sensor node does not know who may be interested in its data, and, vice versa, a query node does not know where in the network its events of interest may occur, a challenging problem is to design an effective mechanism for query subscription and the event notification so that an event can notify its inquirers quickly and efficiently. In this publish/subscribe problem, the sensor nodes and query nodes are respectively called the *publisher* nodes and *subscriber* nodes.

Many Internet-based publish/subscribe systems have been designed, e.g., PADRES [24], REBECA [25], SIENA [12], and XNET [14]. Some of these concepts may be useful, but enabling publish/subscribe services in sensor networks is fundamentally different due to unique constraints on communication, storage, and computation capacities. On the other hand, despite a lot of research and development efforts made for sensor networks that provide search mechanisms, the majority of these is focused on retrieving sensor data that have *already* been stored by the network. The publish/subscribe model poses a different challenge as a query of this model is to inquire about *future* events. Thus, the query must be stored proactively in the network to wait for those events. When such an event occurs, it needs to be published to the network to search for the matching queries. With the "proactive" storing of queries, events can be delivered "timely" to the subscribers, while in the "retrieving" case, subscribers only receive "past" events. This is of importance to applications that require a real-time monitoring of the events.

To enable publish/subscribe services in sensor networks, a desirable design should consider the following important issues:

- Routing design: To subscribe a query, without any knowledge about where a matching event could occur, the simplest way is to broadcast it to the entire network. This way, an event can find its matching queries immediately at the local node. Because the traffic due to broadcasting can be overwhelming for a large-scale network, other attempts have been made to reduce this communication with an efficient routing design. A key approach is to replicate a query to a set of select nodes and to publish an event to another corresponding set of select nodes such that, if the event matches the query, there exists a rendezvous node, certainly or at least with a high probability. This guarantee needs to take into account the communication cost due to transmissions of queries and events.

- Query aggregation: Because a sensor node's storage capacity is usually limited, it is desirable that a node stores as small a number of queries as possible. The broadcast approach aforementioned incurs a high storage cost because each node has to store every query. It is thus important to reduce the number of replicas for any given query. By doing so, it also helps reduce the amount of traffic in the network due to query forwardings. Query aggregation serves this purpose. For example, queries can be merged to produce fewer queries. Or, if a new query arrives at a node finding itself covered by a locally stored query, the new query may not necessarily be forwarded further. This is so because any event matching the new query will match the existing query as well, and thus this event will be returned anyway as a result of the existing query's earlier subscription.
- Event matching: When a node receives a publication of an event, the node may need to evaluate its locally stored queries to find those matching the event. Because a sensor node has limited processing capability, this procedure if not properly designed may create a computational burden too heavy for the node. Several techniques have been proposed to organize the queries into some indexing architecture that is convenient for event matching. Alternatively, one may employ a multi-phase checking algorithm where the earlier phases are to determine quickly whether a query should be ignored and the later phases are to evaluate the queries that pass the earlier phases.

While most techniques emphasize the above algorithmic issues, e.g., [16, 22, 27, 31, 34, 40, 48, 49], publish/subscribe middleware techniques aimed at high-level service abstraction have also been proposed [28, 33, 44, 52]. They provide a convenient middleware layer serving as an API for the application developer, hiding all the underlying network complexities and the implementation details of the publish/subscribe mechanisms.

A survey of representative publish/subscribe techniques for sensor networks is presented in this chapter, with attention given to the issues of routing design, query aggregation, event matching, and middleware development.

16.2 Preliminaries

16.2.1 Event and Query Representation

An event in a publish/subscribe system is usually specified as a set of d attribute-value pairs $\{(attr_1, v_1), (attr_2, v_2), \ldots, (attr_d, v_d)\}$ where d is the number of attributes, $\{attr_1, attr_2, \ldots, attr_d\}$, associated with the event. For example, if the sensor network is used to monitor temperature, humidity, wind speed, and air pressure of some area, d is four – representing these four sensor data attributes.

The constraints in a query, in general, can be specified in a predicate of the disjunctive normal form – a disjunction of one or more condition clauses, each clause being a conjunction of elementary predicates. Each elementary predicate, denoted by $(attr_i \ \gamma \ p_i)$, is a condition on some attribute $attr_i$ with γ being the filtering operator. A filtering operator can be a comparison operator (one of $\{=, <, >\}$) or a string operator such as "*prefix of*", "*suffix of*", and "*substring of*" if the attribute is of a *string* type. However, for simplicity of implementation, most schemes assume that a query is a single conjunctive clause of elementary predicates that can only use the comparison operators. This form of query can be called the *rectangular form* because if an event is modeled as a point in a d-dimensional coordinate system, each dimension representing an attribute, a query can be considered a d-dimensional box with the vertices defined based on the attribute constraint values provided in the query clause. Sometimes it can be a tedious process to specify all the lower and upper bounds for all the attributes of a query. In such a case, it is more convenient to specify a query in terms of an event sample and ask to be notified of all the events similar to this sample. For example, consider a camera-sensor remote surveillance network deployed over many airports to detect criminal suspects. If a particular suspect is searched for, his or her picture is submitted as a subscription to the network in hopes of finding the locations where similar images are captured. A query of this kind can be represented by a sphere, in which the sample is the center of the sphere and similarity is constrained by the sphere's radius. This query is said to have the *spherical form*.

16.2.2 Subject-Based vs. Content-Based

There are two main types of publish/subscribe designs [20]: subject-based or content-based. In the subject-based design, events are categorized into a small number of known *subjects*. There must be an event attribute called "*subject*", or something alike, that represents the type of the event and a query must include a predicate (*'subject'* $= s$) to search only events belonging to some known subject s. The occurrence of any event of subject s will trigger a notification to the query subscriber. The subscription and notification protocols are mainly driven by subject match rather than actual-content match.

The content-based design offers a finer filtering inside the network and a richer way to express queries. A subscriber wants to receive only the events that match its query content, not all the events that belong to a certain subject (which could be too many). A node upon receipt of a query or event message needs to extract the content and makes a forwarding decision based on this content. We can think of the subject-based model as a special case of the content-based model and because of this simplification, a subject-based system is less challenging than a content-based system to design.

16.2.3 Sensor Network Assumptions

There are different types of sensor networks that have been considered by existing publish/subscribe techniques. The techniques in [39,40] assume that a sensor node knows its location (e.g., by a built-in GPS-like device or by running a localization protocol [47]). A mapping from the query/event space to the location space can be designed, based on which a query and a matching event are sent to the corresponding nodes who can find each other within a short distance. For example, if a query is replicated at all the nodes along the vertical line crossing the subscriber node's location, and if an event is published to all the nodes along the horizontal line crossing the publisher node's location, a rendezvous node exists for every pair of queries and events. This is so because every vertical line must meet every horizontal line.

Several techniques do not require any knowledge of location information (e.g., [16, 27, 31, 34, 45, 49]). Besides the broadcast approach, another approach used by such techniques is based on some form of randomized dissemination to subscribe a query or publish an event. The main intuition is that if a query is replicated to a set of random nodes, an event published to its corresponding set of also random nodes, and if these two sets are large enough, then there exists a rendezvous node with a high probability. For example, one can use a random walk or a gossip-based protocol [16, 45] to visit these large sets of nodes. Some other techniques [27, 49] use a naming scheme to assign names to the network nodes and routing is driven by node names instead of nodes being chosen in random.

Most techniques assume that nodes are stationary. There are some techniques addressing publish/subscribe in general mobile networks that could apply to a mobile sensor network [4,15,18,23,26,32,53]. In this chapter only techniques designed for stationary networks are discussed.

16.3 Routing Design

Routing protocols for query subscription and event notification form the most important component of any distributed publish/subscribe system. While traditional routing protocols for sensor networks are designed for synchronous communication and are address-driven, routing for publish/subscribe purposes is asynchronous because subscribers and publishers are not aware of each other and the timing of the subscription and publication. There are various publish/subscribe routing approaches which are discussed below.

16.3.1 Centralized-Based

A simple approach is to employ a central brokerage station that receives the subscription of every query and publication of every event. This approach keeps the system design lightweight on each node. Routing of queries and events becomes trivial because the central broker is the only destination. Queries may be aggregated at the intermediate nodes on the way to the central broker. Event matching is conducted at the central broker to find all the queries matching a newly published event. The centralized-based approach, which has been implemented in the MQTT and MQTT-S middleware at IBM [33], can work with queries and events of any form. However, the central broker can easily be a severe communication bottleneck if queries and events are produced at high rates.

16.3.2 Broadcast-Based

Directed Diffusion [34] is one of the earliest publish/subscribe techniques for sensor networks. Given a new query, the first step in this technique is for the subscriber node to broadcast the query to all the nodes in the network. Each node upon receipt of this query creates a "gradient" entry in the routing table to point toward the neighboring node from which the query is received. Using a gradient path, a matching event can be sent toward the subscriber. Since there may be more than one such gradient path, the choice of which path to use is made based on some performance factors to improve energy efficiency.

Huang and Garcia-Molina [31] propose building a broadcast tree, called PST, spanning all the nodes in which events will be disseminated from a publisher (root node) to all their subscribers. Each node in PST that has a subscription needs to let its parent node know of this subscription. Thus a node can compute a combination of all the subscriptions downstream. When an event is published by the root, the event will follow the branches of PST that lead to all the matching subscribers. If there are multiple publishers, multiple trees are built each for a publisher, or alternatively, a shared root is selected that receives all the events from the publishers and a single tree is built based on this root to send the events to all the nodes. PST does not incur the cost to replicate queries at every node, but it is efficient only when there are a smaller number of publisher nodes.

In a different effort, Schönherr et al. [41] propose a hierarchical structure to organize the network into multiple clusters, in which each cluster forms a publish/subscribe sub-network and has a representative node to appear in the next layer of the hierarchy; the representative nodes form a layer of clusters which again are used to build the next layer in the hierarchy. The representative nodes are responsible for forwarding a query or event from one sub-network to another. Although this clustering idea can potentially reduce the subscription and notification traffic in the network compared to the pure broadcast-based approach, its complexity lies in the maintenance of the hierarchy under network dynamics.

16.3.3 Location-Based

Flooding the network can be expensive, especially for systems where there are a large number of queries and events generated by many potential subscriber and publisher nodes. Consequently, many distributed techniques have been proposed. If the location is known for every node, the location information can be useful. Ratnasamy et al. [39] proposed the method of Geographic Hash Tables (GHT) – to hash the data space to locations in the location space. Each data value is represented by an one-dimensional identifier called a “key” k, which corresponds to a geographic coordinate $h(k)$ based on the hash function. Thus, if both events and queries each can be identified by a single key, we can use GHT. For example, a subject-based system is a natural candidate for this technique because the subject value can be used as the key value for GHT. A query with key k will be stored at the node closest to the location $h(k)$. When an event with the same key k emerges, it will be routed to the node $h(k)$ and thus can find its matching queries. For routing from a sensor location to another sensor location, we can use a geographic routing protocol designed for sensor networks such as [35]. An extension of the GHT technique in combination with landmark-based routing to improve routing efficiency is proposed in [22].

Another location-based technique called Double-Ruling is proposed in [40], in which each node P is mapped to a point $f(P)$ on the surface of a 3D sphere and a key k is hashed to a point $h(k)$ also on this spherical surface. A query with key k subscribed by a node P will be sent to the node corresponding to the point $h(k)$. The routing follows the great circle connecting the points $h(k)$ and $f(P)$ on the 3D surface. An event is published to the corresponding node in a similar fashion and thus every event with key k will be sent to node $h(k)$ and can find all the matching queries there. A property of Double-Ruling is that it is distance-sensitive: the length of the notification path from a publisher to a matching subscriber is guaranteed to be within a small constant factor of the direct path connecting them. The GHT method does not offer this guarantee.

The above distributed methods (GHT and Double-Ruling) should not work efficiently with a content-based publish/subscribe system because it is difficult to represent a complex query by a single key. For events and queries of any dimensionality d, one can use the technique proposed in [48]. This technique assumes the spherical form for the queries. Given a query, a random-projection method is used to hash this query into a 2D rectangle in the location space; the query will be replicated at all the nodes inside this rectangle. Using the same random projection, an event is hashed a single location; the event will be sent to the node closest to this location. The random project method guarantees that this event will find all the matching queries. To reduce the number of query replicas, the subscription covering relationship is taken into account to avoid further replications of those queries that are subsumed by previously-subscribed queries.

16.3.4 Gossip-Based

Without location information, a common idea is to use some form of gossip to disseminate queries and events. A simple design is to use random walks [3,9]. Each query is replicated on all the nodes visited by a random walk starting from the subscriber node, and each event also follows a random walk from the publisher node to find the queries. If these random walks are long enough, it is highly likely that an event will find all the matching queries. To shorten the notification delay, multiple random walks can be used to propagate an event (or to replicate a query) [9].

Another approach is proposed in [16], where both queries and events are "selectively" broadcast to the network. A query is broadcast to an extent defined by the subscription horizon ϕ which limits the number of times the query is rebroadcast. In the broadcast of an event, only a fraction τ of the neighbor links at each current node is used to forward the event. By choosing appropriate values for ϕ and τ, we can control the overhead and effectiveness of the system.

We can also design a publish/subscribe mechanism by adopting BubbleStorm – a gossip idea proposed in [45]: each query is replicated in a random-walk based tree of $q = O(\sqrt{n})$ nodes rooted at the subscriber node, and each event is sent along a similarly-built tree of $c^2 n/q$ nodes rooted at the publisher node where n is the number of nodes and $c \geq 1$ is a certainty constant. It is shown that the probability to have a rendezvous node is $r = 1 - \exp(-c^2)$ (e.g., $c = 2$ means a hit probability of $r = 0.98$).

16.3.5 Naming-Based

Despite its simplicity, the gossip-based approach incurs significant communication, storage, and computation costs because each query or event needs to be disseminated to many nodes to have a good chance to meet its matches at rendezvous nodes. Also, the guarantee that a rendezvous node exists for every pair of queries and events, including those that do not match each other, is unnecessarily strong and thus leads to unnecessary traffic. Therefore, alternative techniques that requires no location information but is not based on gossiping are proposed [27, 49]. The common idea is to design a naming scheme that assigns *names* to the nodes in a way convenient for routing purposes. The technique in [27] clusters the network into a multi-level hierarchy and assigns to each node a name based on its position in this hierarchy. The resultant hierarchical naming scheme is used for routing during event publication and query subscription. Each node requires only $O(\log n)$ bits per node to store auxiliary routing information. Any event can find its subscribers in a distance-sensitive way. Further, by visiting only $O(k)$ nodes, the subscriber can collect all occurrences of a particular type of data within a similarity radius k (for spherical queries).

The above technique works for the subject-based model only. The Pub-2-Sub$^+$ technique [49] is also based on a naming scheme but designed for

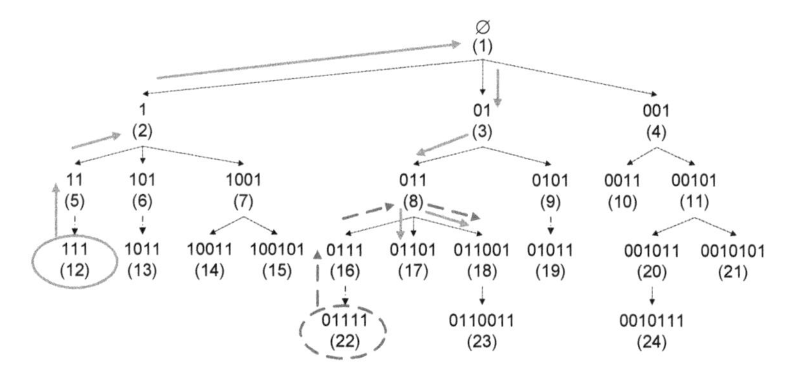

Figure 16.1 The Pub-2-Sub$^+$ scheme: Nodes are each assigned a name and the names form a prefix tree. Solid-bold path represents the subscription path of query ['0110001', '0110101'] initiated by node 12; dashed-bold path represents the notification path of event '0110010' published by node 22.

content-based services. Pub-2-Sub$^+$ maintains a set of m spanning trees each rooted at a node in the network. The root nodes are dedicated reliable nodes placed at random network positions. Each tree corresponds to a naming tree assigning a binary-string name to each node; hence, a node has m names. The names on a tree form a prefix tree. Based on the naming scheme, each node is assigned a "zone" of binary strings to own. The zone of a node is the set of all binary strings starting with this node's name but not with any child node's name. A query is subscribed to a random tree and an event is published to all the trees. Pub-2-Sub$^+$ formats an event as a binary string (e.g., '0110010') and a query an interval of binary strings (e.g., ['0110001', '0110101']). On the randomly chosen tree, a query is routed to, and stored at, all the nodes whose zone overlaps with the query's interval. On each tree, an event is published to the node whose name is the longest prefix of the event string. Figure 16.1 provides an example where $m = 1$. The query ['0110001', '0110101'] is stored at nodes 3, 8, 17, and 18. The event '0110010' is published to node 18; it will find all the matching queries there. In general, the notification path is bounded by two times the tree height which should be $O(\log n)$ in most cases. Also, because there are multiple paths for the event notification, the disconnection of a path due to some failure does not stop an event from finding its way to the matching queries.

16.4 Query Aggregation

Queries in a publish/subscribe system need to be stored in advance. Due to the resource constraints of a typical sensor node, it is desirable to limit the replication of each query in the network. Although the routing design dictates how to disseminate a query, its integration with a query aggregation mechanism may lead to more efficient query forwarding. Queries arriving at a node can be merged and/or pruned to produce fewer queries representing equivalent constraints. Query aggregation based on covering relationship is useful to deciding whether a query needs to be forwarded further. This section describes some common aggregation strategies.

16.4.1 Subscription Covering

Used in various Internet-based publish/subscribe systems [12, 14, 24, 25], one strategy is that a node, upon receipt of a new subscription query, does not forward it if it is already covered by an existing locally stored query. Figure 16.2 illustrates this strategy. Typically, when a new query s is routed to a node P, the node will store a copy of this query and forward it to the next node P_{next} according to the subscription's routing protocol. When the next node P_{next} receives a publication of an event x that matches s, it will forward it back to P, which in turn forwards x to the node that previously sent s to P. Now, let us assume that there is an existing query s_i stored at P such that s_i covers s (i.e., $s_i \supset s$). In this case, P may opt not to forward s to P_{next} because P knows that if there is an event x matching s, it must be returned to P as a result of P's forwarding s_i earlier. By not forwarding s, we avoid the costs of disseminating s further in the network and collecting duplicate events that satisfy both s and s_i.

Detecting coverings in a large set of queries can, however, be a computationally expensive procedure for any given sensor node. On the other hand, it is not mandatory that covered queries must not be forwarded. Certain techniques [38, 43, 46] have been proposed to detect *approximately* subscription coverings. Although such a technique may still forward a query even if the query is covered by an existing one, it is guaranteed that no matching event will be missed. These techniques are still much more efficient than having to detect all coverings.

The technique in [43] assumes that queries adopt the rectangular form. It is observed that if we map a rectangular query $s = [l_1, r_1] \times [l_2, r_2] \times \cdots \times [l_d, r_d]$ into a $2d$-dimensional point $p(s) = (-l_1, r_1, -l_2, r_2, \ldots, -l_d, r_d)$, the subscription covering problem in d dimensions becomes the point dominance

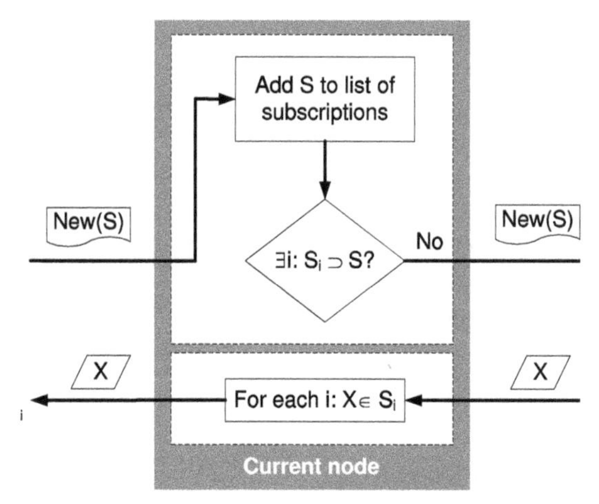

Figure 16.2 Subscription covering: Do not forward a new subscription s if it is already covered by an existing subscription s_i.

problem in $2d$ dimensions. A point p is said to *dominate* a point p' if every coordinate of p is no less than that of p'. Instead of solving this problem directly, Shen and Tirthapura [43] proposed to solve an approximate version of the problem:

> Find a data structure for a set Σ of n points in the m-dimensional box $[0, MAX_1] \times [0, MAX_2] \times \cdots \times [0, MAX_m]$ so that the following question can be answered efficiently: given a constant $\varepsilon \in (0,1)$, and a query point $p = (p_1, p_2, \ldots, p_m)$, search a subset of the box $[p_1, MAX_1] \times [p_2, MAX_2] \times \cdots \times [p_m, MAX_2]$, whose volume is at least $(1-\varepsilon)$ of the volume of the box, and report any point of Σ if it is in the subset.

A data structure was proposed that sorts the input points based on its positions on the Z space-filling curve [2]. For each query point, only a subset of segments of the Z curve is searched for the existence of any point of Σ that dominates the query point p.

A different approach has been proposed in [46] which can work with both spherical and rectangular queries. Unlike Shen and Tirthapura [43] which increased efficiency by searching only a subset of the set of queries, Tran and Nguyen [46] searched all queries but for each visited query the covering condition is quickly, however, tolerated by a probability of error. To illustrate the idea, suppose that queries are spherical. Firstly, a constant $k < d$ is predetermined. Secondly, k random orthonormal vectors, $\{u_1, u_2, \ldots, u_k\}$, are generated. Then, each query centered at s with radius r is mapped to the following k-dimensional rectangle: $u(s,r) = u_1(s,r) \times u_2(s,r) \times \cdots \times u_k(s,r)$ where each side $u_i(s,r)$ of this rectangle is the projection of the subscription on unit vector u_i, which is the following interval: ($\circ$ is the inner product) $u_i(s,r) = [u_i \circ s - r,\ u_i \circ s + r]$. To check whether a query s covers a query s', we check whether the projection rectangle $u(s)$ covers the projection rectangle $u(s')$. If it is found that $u(s)$ covers $u(s')$ (or not), it is concluded that s covers s' (or not). Although this conclusion is not always correct, it is shown that if queries are uniformly distributed in both center points and radii, the probability of error is roughly bounded by $(2/\pi)^k$. This probability approaches zero quickly as k increases.

The case of rectangular queries is handled similarly. Each original d-dimensional rectangular query s with 2^d vertices $v_1, v_2, \ldots, v_{2^d}$ is mapped to a k-dimensional rectangle where each side is the following interval: $u_i(s) = [\min_{j \le 2^d}(v_j \circ u_i), \max_{j \le 2^d}(v_j \circ u_i)]$. Using a similar random projection, with a low probability of error, the subscription covering problem in d dimensions can be mapped into that in $k \ll d$ dimensions. The value of k can be tuned to achieve any given success rate.

Ouksel et al. [38] addressed a more generic covering problem; that is, to find whether a new query s is covered by *the set* of existing queries $s_1 \cup s_2 \cup \ldots \cup s_n$, rather than by a single existing query. The basic idea is as follows. First, k points $\{x_1, x_2, \ldots, x_k\}$ that satisfy s are selected in random. Second, we check whether each of these points satisfies any of the existing queries. It is concluded that s is covered if all the selected points satisfy the set of existing queries, and otherwise not covered. Thus, the probability of erroneously concluding that a query is covered is upper-bounded by $(1 - p_w)^k$, where p_w is the probability that a random point x_i satisfying s also satisfies the set of existing queries. This error probability is quickly improved as k increases.

16.4.2 Subscription Merging/Pruning

Another technique to reduce the number of query replicas in the network is via *subscription merging* [17,25,36,50]. Given a set of queries S, we need to merge them to create a new set of queries S' such that (1) $|S| < |S'|$ and (2) the events satisfying S are the same as that satisfying S'. Instead of forwarding the queries in S, we forward the queries in S', thus reducing the amount of subscriptions in the network.

Although subscription merging is theoretically helpful, we cannot always find a perfect merging for a given set of queries. Even if such a merging exists, its algorithm can be computationally expensive. Indeed, Crespo et al. [17] showed that the merging problem is NP-complete. Consequently, imperfect merging algorithms are suggested [17,36,50]. These algorithms aim to merge the queries in S into a new set of queries S' such that (1) $|S| < |S'|$ and (2) the events satisfying S are a *subset* of that satisfying S'.

One such an algorithm [50] clusters S into groups of similar queries and then finds a merging for each group. In another algorithm [36], the representation of a query is based on the concept of ordered Binary Decision Datagram (BDD) [10]. A BDD is a rooted, directed acyclic graph designed for easy manipulations of Boolean functions. A query is expressed as a BDD in which each node is a predicate of the query and a solid (dashed) link from a node means that the corresponding predicate is satisfied (unsatisfied). For example, Figure 16.3(a) shows a BDD for the query $\{$('*temperature*' $>$ 100) AND ('*humidity*' $<$ 50)$\}$ OR $\{$('*temperature*' $>$ 100) AND ('*humidity*' $\geq$ 50) AND ('*wind speed*' $>$ 50) AND ('*wind speed*' $<$ 100)$\}$. Given an event, it traverses the BDD and if terminal node 1 is reached, it is concluded that the event satisfies the query. To address a large number of queries, thus a large number of BDDs, a modified version of BDD called MBD is introduced. An MBD is a forest of BDDs, each representing a query, with the property that if a predicate occurs in multiple queries only one common node is shared among these queries to represent the predicate; this predicate is evaluated only once. An MBD thus represents the merging of multiple queries. Within an MBD, two predicate nodes can also be merged to create a single node to represent the same condition, resulting in a simpler MBD. An event is evaluated on an MBD rather than on each individual query.

Imperfect merging has its tradeoff. At any node P where an imperfect merging of S into S' is performed, we have the problem that many false events, that satisfy S' but not S, could be returned to node P. This is unnecessary traffic which can be costly for sensor networks. In addition, the computational and space costs to perform subscription merging (perfect or imperfect) may exceed that a sensor node can afford. Thus, subscription merging is recommended only when its scope is small.

Subscription pruning [7] is another strategy to reduce the query storage load in the network. Each query is represented as a tree [6] in which each inner node represents a Boolean operator (e.g., AND, OR) and each leaf being a single-attribute predicate (e.g., ('*temperature*' $>$ 100), or ('*humidity*' $<$ 20). For example, Figure 16.3(b) shows the Boolean tree representing

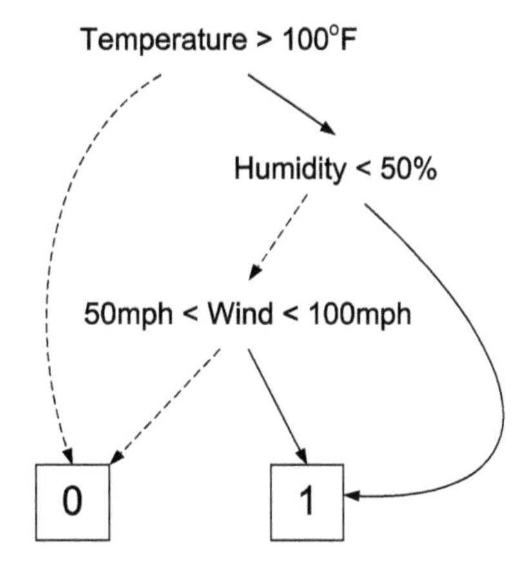

(a) A query as a binary decision diagram

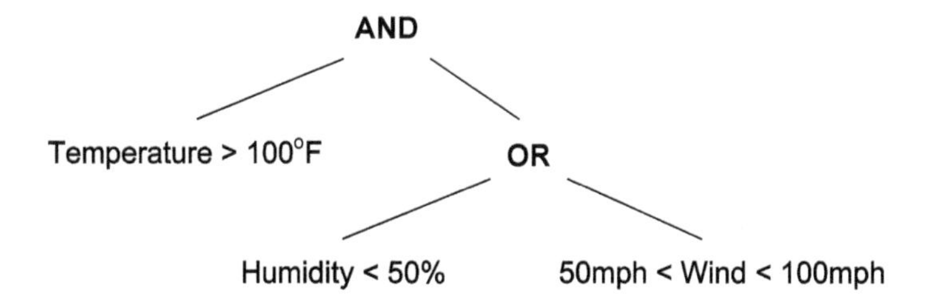

(b) A query as a Boolean tree

Figure 16.3 Examples of query representations.

the same query in Figure 16.3(a). The tree is then simplified by pruning off some predicates or by replacing them with simpler ones. Since the queries become simpler, data structures and algorithms for processing them require less memory and computation. On the other hand, similar to imperfect subscription merging, subscription pruning also results in notification of false events.

16.5 Event Matching

When a publication of some event reaches a node where the locally stored queries need to be evaluated against this event, simply testing every query and predicate may be computationally expensive for a sensor node, especially when there are numerous, complex queries and high volumes of events. Consequently, it has been proposed that the queries are organized into some data structure that enables faster than linear-time event matching.

The Matching Tree in [1] is such a data structure, where the matching time is sub-linear and the space complexity is linear. This tree allows incremental query updates (insert/delete) and is most suitable where events are published at a fast rate. Each tree node is a test on some of the attributes and each link eminating from a node is labeled with a result of the corresponding test. A link with label "*" means a "do not care" link. Each query corresponds to a leaf node and the path from the root to this leaf node consists of all the tests whose conjunction is equivalent to the query. For example, Figure 16.4 shows a simple matching tree storing three queries: sub_1: $(attr_1 = v_1)$ AND $(attr_2 = v_2)$ AND $(attr_3 = v_3)$, sub_2: $(attr_1 = v_1)$ AND $(attr_3 = v'_3)$, and sub_3: $(attr_1 = v'_1)$ AND $(attr_2 = v_2)$ AND $(attr_3 = v_3)$. In order to find the queries matching an event, the corresponding test is performed at each node starting at the root, and a link to the next node is followed if its label matches the result of the test. This step is repeated at the next node. The leaves that are finally visited correspond to the queries that match the event. In the case where a query consists of equality tests on the attributes, the expected time to match a random event is $O(n^{1-\lambda})$ where n is the number of queries and λ is a parameter dependent on the number and type of attributes (in some cases, $\lambda = 1/2$). The constants hidden behind the big-O notation are quite reasonable.

The matching tree structure follows the approach that the search for queries matching a given event starts from the attribute constraints derived from the full set of queries and moves through them consulting the attributes appearing in the event. The binary decision datagram (BDD) structure discussed in Section 16.4 also follows this approach. Alternatively, we can start with the attributes appearing in the event and move through them consulting the query constraints. An early method adopting this approach is the SIFT system [51], which is the basis for subsequently designed structures [13, 21]. SIFT is limited to strings only and the equality operator over strings. Le Subscribe [21] allows the integer type and its associated operators. [13] adds the prefix, suffix, and substring operators for strings and, especially, allows a query to be expressed as a disjunction of conjunctive clauses, not just a single conjunctive clause. The matching algorithm in this work is based on a counting algorithm also used in [21, 51] and when tested on a 950Mhz computer for a 10-attribute event and 20 queries consisting of 25 conjunctions could find the matching queries in 3 milliseconds. It took 48 bytes of storage for each elementary predicate. This algorithm is therefore simple and efficient enough to be implemented on a sensor node.

Another strategy for building an efficient data structure is based on the query covering relationship. It is observed that if an event x matches query s_1 (denoted by $x \in s_1$) and if we already know that $s_1 \subset s_2$ for some query s_2, then it must be true that $x \in s_2$ and we need not check whether x matches s_2. Based on this observation, we should build a data structure that captures the covering relationship among subscriptions. This is similar to the problem of *Orthogonal Range Search* in Computational Geometry, for which several structures exist such as the *kd*-tree [5] and the layered range tree [19]. Using the *kd*-tree, the complexities would be $O(n^{1-1/d})$ for time and $O(n)$ for storage, while that using the layered range tree would be $O(\log^d n)$ for time and $O(n \log^{d-1} n)$ for storage. If one of these two structures must be used, due to the high storage cost of the layered range tree, the kd-tree should be a better choice for sensor networks.

The random projection approach in [48], which was discussed earlier in Section 16.4.1, can be used to expedite the matching procedure. Using a random projection from d dimensions to k dimensions (d is the number of event attributes, k is some small constant), a query is mapped to a k-dimensional rectangle. To check whether an event x matches a query s (center u, radius r), a quick check can be to verify whether the projection x' of x in the k-dimensional space is inside the k-dimensional rectangular projection s' of s. If $x' \notin s'$, we can immediately ignore s; otherwise, we perform the check whether $x \in s$ as usual. The first check is performed in a k-dimensional space, thus much quicker than the second check which is in a d-dimensional space. A nice property of this method is that the number of queries that can be ignored after the first check increases quickly if a larger value for k is chosen.

16.6 Middleware Development

It is important to have a middleware component that can be integrated on top of a sensor network to allow for easy deployment of publish/subscribe applications. An application developer should know just how to call the publish/subscribe functions, not having to worry about the complexity of the

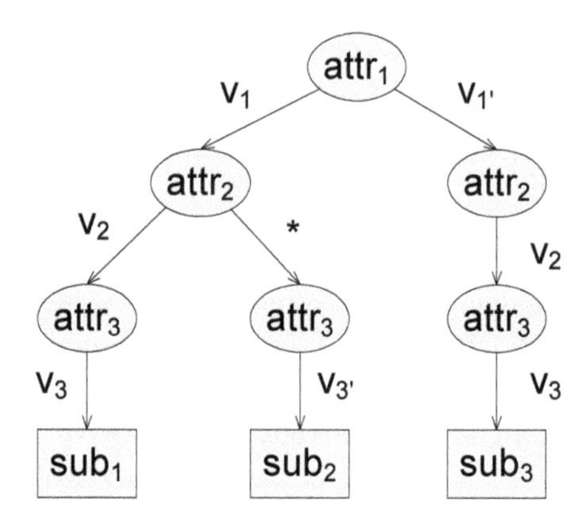

Figure 16.4 Example of a matching tree.

underlying network and the implementation details of the publish/subscribe mechanisms. There have been various approaches toward providing middleware services in sensor networks, and according to the categorization in [30], these approaches can be placed in one of the following groups: (1) database-inspired approaches (TinyDB [37], COUGAR [8], SINA [42]); (2) tuple space approaches (TinyLIME [11]); (3) event-based approaches (Mires [44], MQTT-S [33], TinyCOPS [28]; and (4) service discovery based approaches (MiLAN [29]). Publish/subscribe middleware belongs to the group of event-based approaches.

Mires [44] is such a publish/subscribe middleware design, which has been implemented using nesC on top of TinyOS 1.x. Mires supports *subject-based* publish/subscribe applications and provides an architecture that allows sensor nodes to advertise the subject of sensor data they can provide, user applications to select subjects of interest from the advertised services, and sensor nodes to publish their data to the corresponding subscribers. The subscription and notification protocols integrated in this middleware are similar to that of Directed Fusion. Mires also offers data aggregation services for some common aggregation functions, such as min, max, and average, to help reduce the event traffic in the network.

Another middleware design is the MQTT-S architecture by IBM [33]. MQTT-S is an extension of MQTT originally developed for telemetry applications using constrained devices.[1] This architecture is applicable to

[1] See http://mqtt.org.

subject-based publish/subscribe applications that run in a network integrating multiple wireless sensor networks (WSN). Its aim is to hide end-point details: an application running on either the backbone network or inside a WSN does not know whether the data is coming from a device in a WSN or the backbone network. A query can be submitted anywhere in the global network subscribing to events that may belong to one or more participating WSNs. Broker nodes are placed in the backbone network to provide a publish/subscribe service to the nodes in the backbone network; the broker nodes run the original MQTT middleware. MQTT-S is used to provide a publish/subscribe service within a single WSN, with two main entities: MQTT-S Client running on a sensor node and MQTT-S Gateway running on a gateway node connecting its WSN to the global network. An MQTT-S Client contains both a publisher and a subscriber, thus allowing sensor nodes to not only publish their data, but also receive, e.g., control information sent by nodes residing in the global network. The main function of the gateway is to translate between MQTT-S and MQTT protocols. This architecture is illustrated in Figure 16.5. MQTT-S provides a method to search for a local gateway and allows for multiple gateways used per WSN, thus increasing the reliability of the publish/subscribe system under deployment. MQTT-S has been implemented in a testbed which comprises two wireless sensor networks of different types, a ZigBee-based and a TinyOS-based one. Both implementations are lightweight with about 12 kB of codes. The testbed devices each have only 64 kB of program memory available. The MQTT-S Gateway is written in Java and uses the Java Communications API to communicate with the gateway devices over the serial port. The Gateway connects to a broker using the MQTT protocol. Nodes residing on the TinyOS-based network can communicate with nodes on the ZigBee network via the broker.

Mires and MQTT-S focus on architectural and networking issues and provide support for simple subject-based subscriptions and publications. The middleware development in [52] is aimed for a rich expressiveness of query and event description. A set of filtering operators is proposed to support not only standard comparison and string operators, but also allow for spatiotemporal constraints. For example, a query can be subscribed to receive the average temperature at a specific location during a future period of time. A problem with this middleware development is due to the complexity involved in the implementation of the subscription and notification protocols. Indeed, a complex timestamping scheme is needed to support the temporal operators and, further, the spatial operators requires location information which is not always available for every sensor network [30].

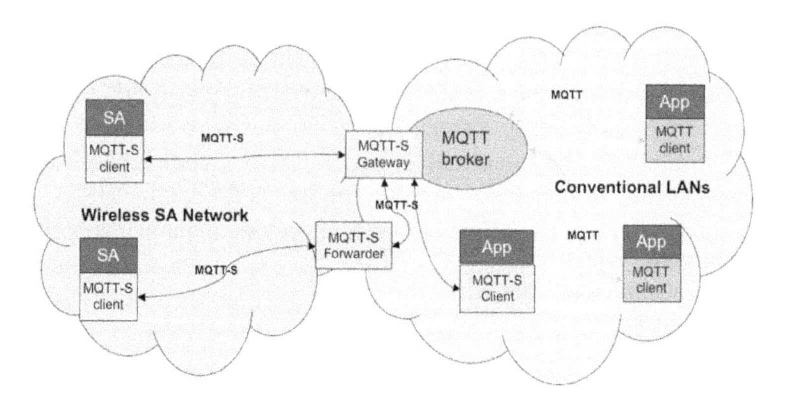

Figure 16.5 The MQTT-S middleware architecture.

The above-mentioned middleware designs tightly integrate filtering, routing and forwarding mechanisms resulting in more optimized, but less flexible solutions. TinyCOPS [28] is aimed to be a unified middleware architecture enabling *content-based* publish/subscribe applications, not just subject-based, that gives the application developer a wide range of orthogonal choices about the communication protocol components to use for subscription and notification, the supported data attributes, and a set of service extension components. This allows the adaptation of the publish/subscribe service to the specific needs of the application to deploy. TinyCOPS also introduces the concept of *metadata* included in the description of each query to influence the communication and sensing process. For example, the metadata can specify the expiration time for a given query or request a sampling rate events to be sent to the subscriber. TinyCOPS has been implemented to run on top of TinyOS 2.0, in which two communication protocols are integrated (broadcast-based or gossip-based).

16.7 Concluding Remarks

The publish/subscribe paradigm represents a large class of applications in sensor networks as sensors are designed mainly to detect and notify upon events of interests. This important paradigm allows a user to subscribe in advance a query specifying the early warnings of a wildfire, so that any event matching these warnings when detected by a sensor can be published to the network quickly to notify the subscriber. In industrial applications, we can provide a more secure working environment by deploying a sensor network that warns workers upon the detection of dangerous events.

Many publish/subscribe techniques for sensor networks have been inspired by that for traditional Internet-based networks. For example, some gossip-based routing schemes, query aggregation schemes, and event matching algorithms that have been designed for the Internet can also be used in sensor networks, as we discussed earlier in the chapter. There, however, remains much room for future research. For a large-scale sensor network where broadcast-based and gossip-based routing approaches may not be the best fit, the routing design should be driven by the content of the message being routed so as to limit the scope of propagation. In networks where location information is available, it should be a main factor in the routing design. In other networks without location information, the naming-based approach seems a promising direction, the challenge, however, being how to maintain the naming structure efficiently under network dynamics. Since sensor networks may be of different types (small vs. large, location-aware vs. location-unaware, static vs. dynamic) and the application to deploy may also have its own characteristic (low vs. high query rate, low vs. high event rate, subject-based vs. content-based, etc.), it is difficult to choose a publish/subscribe design that works well in every case. Thus, rather than trying to find a universally "perfect" design, it would be better to categorize the networks and applications into similarity-based groups and design the "best" technique for each group.

It is then natural for the next step to develop a publish/subscribe middleware package that provides a set of common services to most publish/subscribe network/applications no matter their categories, and another set of services each customized toward a specific category. This middleware should provide convenient tools for the middleware designer to add new service components to the existing architecture, such as a new language for query and event description and a new implementation for routing, data aggregation, or an event matching algorithm. It should also give the application developer freedom and a convenient API to choose the publish/subscribe service configuration that is best for the context of the deployment. Middleware development for publish/subscribe applications in sensor networks remains ad hoc and isolated. It should be a high-priority item in the future research towards publish/subscribe services in sensor networks.

References

[1] M.K. Aguilera, R.E. Strom, D.C. Sturman, M. Astley, and T.D. Chandra. Matching events in a content-based subscription system. In *PODC'99: Proceedings of the Eighteenth Annual ACM Symposium on Principles of Distributed Computing*, pages 53–61. ACM Press, 1999.

[2] T. Asano, D. Ranjan, T. Roos, E. Welzl, and P. Widmayer. Space-filling curves and their use in the design of geometric data structures. *Theoretical Computer Science*, 181(1):3–15, 1997.

[3] C. Avin and B. Krishnamachari. The power of choice in random walks: An empirical study. *Comput. Networks*, 52(1):44–60, 2008.

[4] M. Avvenuti, A. Vecchio, and G. Turi. A cross-layer approach for publish/subscribe in mobile ad hoc networks. In *MATA*, Lecture Notes in Computer Science, Vol. 3744, pages 203–214. Springer, 2005.

[5] J.L. Bentley. Multidimensional binary search trees used for associative searching. *Commun. ACM*, 18(9):509–517, 1975.

[6] S. Bittner and A. Hinze. On the benefits of non-canonical filtering in publish/subscribe systems. In *ICDCSW'05: Proceedings of the Fourth International Workshop on Distributed Event-Based Systems (DEBS)*, pages 451–457. IEEE Computer Society, Washington, DC, 2005.

[7] S. Bittner and A. Hinze. Pruning subscriptions in distributed publish/subscribe systems. In *ACSC'06: Proceedings of the 29th Australasian Computer Science Conference*, pages 197–206. Australian Computer Society, Darlinghurst, Australia, 2006.

[8] P. Bonnet, J.E. Gehrke, and P. Seshadri. Towards sensor database systems. In *Proceedings of 2nd International Conference on Mobile Data Management (MDM)*, pages 3–14, 2001.

[9] D. Braginsky and D. Estrin. Rumor routing algorthim for sensor networks. In *WSNA'02: Proceedings of the 1st ACM International Workshop on Wireless Sensor Networks and Applications*, pages 22–31. ACM, 2002.

[10] R.E. Bryant. Graph-based algorithms for Boolean function manipulation. *IEEE Trans. Comput.*, 35(8):677–691, 1986.

[11] M.G.C. Curino, M. Giorgetta, A. Giusti, A.L. Murphy, and G.P. Picco. Tinylime: Bridging mobile and sensor networks through middleware. In *Proceedings of 3rd IEEE International Conference on Pervasive Computing and Communications (PerCom)*, pages 61–72, March 2005.

[12] A. Carzaniga, D.S. Rosenblum, and A.L. Wolf. Design and evaluation of a wide-area event notification service. *ACM Transactions on Computer Systems*, 19(3):332–383, 2001.

[13] A. Carzaniga and A.L. Wolf. Forwarding in a content-based network. In *ACM SIGCOMM*, pages 163–174, 2003.

[14] R. Chand and P. Felber. Xnet: A reliable content-based publish/subscribe system. In *SRDS'04: Proceedings of the 23rd IEEE International Symposium on Reliable Distributed Systems*, pages 264–273. IEEE Computer Society, Washington, DC, 2004.

[15] P. Costa, C. Mascolo, M. Musolesi, and G.P. Picco. Socially-aware routing for publish-subscribe middleware in delay-tolerant mobile ad hoc networks. *IEEE Journal on Selected Areas in Communications*, 26(5), 2008.

[16] P. Costa, G.P. Picco, and S. Rossetto. Publish-subscribe on sensor networks: A semi-probabilistic approach. In *Proceedings of the 2nd IEEE International Conference on Mobile Ad-hoc and Sensor Systems (MASS05)*, Washington DC, USA, November 2005.
[17] A. Crespo, O. Buyukkokten, and H. Garcia-Molina. Query merging: Improving query subscription processing in a multicast environment. *IEEE Transactions on Knowledge and Data Engineering*, 15(1):174–191, 2003.
[18] G. Cugola and H.-A. Jacobsen. Using publish/subscribe middleware for mobile systems. *SIGMOBILE Mob. Comput. Commun. Rev.*, 6(4):25–33, 2002.
[19] M. de Berg, M. van Kreveld, M. Overmars, and O. Schwarzkopf. *Computational Geometry: Algorithms and Applications*. Springer-Verlag, New York, 1997.
[20] P.T. Eugster, R.G.P.A. Felber, and A.-M. Kernmarrec. The many faces of publish/subscribe. *ACM Computing Surveys*, 35(2):114–131, 2003.
[21] F. Fabret, H.A. Jacobsen, F. Llirbat, J. Pereira, K.A. Ross, and D. Shasha. Filtering algorithms and implementation for very fast publish/subscribe systems. In *SIGMOD'01: Proceedings of the 2001 ACM SIGMOD International Conference on Management of Data*, pages 115–126. ACM Press, New York, 2001.
[22] Q. Fang, J. Gao, and L.J. Guibas. Landmark-based information storage and retrieval in sensor networks. In *Proceedings of IEEE INFOCOM*, 2006.
[23] U. Farooq, E.W. Parsons, and S. Majumdar. Performance of publish/subscribe middleware in mobile wireless networks. *SIGSOFT Softw. Eng. Notes*, 29(1):278–289, 2004.
[24] E. Fidler, H.-A. Jacobsen, G. Li, and S. Mankovski. The padres distributed publish/subscribe system. In *Proceedings of FIW*, pages 12–30, 2005.
[25] L. Fiege, M. Cilia, G. Muhl, and A. Buchmann. Publish-subscribe grows up: Support for management, visibility control, and heterogeneity. *IEEE Internet Computing*, 10(1):48–55, 2006.
[26] D. Frey and G.-C. Roman. Context-aware publish subscribe in mobile ad hoc networks. In *Proceedings of International Conference on Coordination Models and Languages (Coordination 2007)*, pages 37–55, 2007.
[27] S. Funke, L.J. Guibas, A. Nguyen, and Y. Wang. Distance-sensitive information brokerage in sensor networks. In *Proceedings of DCOSS*, pages 234–251, 2006.
[28] J.-H. Hauer, V. Handziski, A. Kaopke, A. Willig, and A. Wolisz. A component framework for content-based publish/subscribe in sensor networks. In *Proceedings of European Workshop on Wireless Sensor Networks (EWSN)*, Bologna, Italy, January 2008.
[29] W.B. Heinzelman, A.L. Murphy, H.S. Carvalho, and M.A. Perillo. Middleware to support sensor network applications. *IEEE Network*, 18(1):6–14, August 2004.
[30] K. Henricksen and R. Robinson. A survey of middleware for sensor networks: State of the art and future directions. In *Proceedings of ACM Workshop on Middleware for Sensor Networks*, Melbourne, Australia, 2006.
[31] Y. Huang and H. Garcia-Molina. Publish/subscribe tree construction in wireless ad-hoc networks. In *MDM'03: Proceedings of the 4th International Conference on Mobile Data Management*, pages 122–140, Springer-Verlag, London, 2003.
[32] Y. Huang and H. Garcia-Molina. Publish/subscribe in a mobile environment. *Wirel. Netw.*, 10(6):643–652, 2004.

[33] U. Hunkeler, H.-L. Truong, and A. Stanford-Clark. MQTT-S: A publish/subscribe protocol for wireless sensor networks. In *Workshop on Information Assurance for Middleware Communications (IAMCOM'08)*, Bangalore, India, January 2008.

[34] C. Intanagonwiwat, R. Govindan, and D. Estrin. Directed diffusion: A scalable and robust communication paradigm for sensor networks. In *MobiCom'00: Proceedings of the 6th Annual International Cconference on Mobile Computing and Networking*, pages 56–67. ACM Press, 2000.

[35] B. Karp and H.T. Kung. GPSR: Greedy perimeter stateless routing for wireless networks. In *MobiCom'00: Proceedings of the 6th Annual International Conference on Mobile Computing and Networking*, pages 243–254. ACM Press, New York, 2000.

[36] G. Li, S. Hou, and H.-A. Jacobsen. A unified approach to routing, covering and merging in publish/subscribe systems based on modified binary decision diagrams. In *ICDCS'05: Proceedings of the 25th IEEE International Conference on Distributed Computing Systems*, pages 447–457. IEEE Computer Society, Washington, DC, 2005.

[37] S. Madden, M.J. Franklin, J.M. Hellerstein, and W. Hong. TinyDB: An acquisitional query processing system for sensor networks. *ACM Transactions on Database Systems*, 30(1):122–173, 2005.

[38] A.M. Ouksel, O. Jurca, I. Podnar, and K. Aberer. Efficient probabilistic subsumption checking for content-based publish/subscribe systems. In *Proceedings of ACM/IFIP/USENIX 7th International Middleware Conference*, pages 121–140, 2006.

[39] S. Ratnasamy, B. Karp, S. Shenker, D. Estrin, R. Govindan, L. Yin, and F. Yu. Data-centric storage in sensornets with ght, a geographic hash table. *Mob. Netw. Appl.*, 8(4):427–442, 2003.

[40] R. Sarkar, X. Zhu, and J. Gao. Double rulings for information brokerage in sensor networks. In *MobiCom'06: Proceedings of the 12th Annual International Conference on Mobile Computing and Networking*, pages 286–297, ACM, New York, 2006.

[41] J.H. Schönherr, H. Parzyjegla, and G. Mühl. Clustered publish/subscribe in wireless actuator and sensor networks. In *MPAC'08: Proceedings of the 6th International Workshop on Middleware for Pervasive and Ad-hoc Computing*, pages 60–65, ACM, New York, 2008.

[42] C.-C. Shen, C. Srisathapornphat, and C. Jaikaeo. Sensor information networking architecture and applications. *IEEE Personal Communications*, 8(4):52–59, August 2001.

[43] Z. Shen and S. Tirthapura. Approximate covering detection among content-based subscriptions using space filling curves. In *Proceedings of IEEE International Conference on Distributed Computing Systems*, Toronto, Canada, June 2007.

[44] E. Souto, G. Guimaraes, G. Vasconcelos, M. Vieira, N. Rosa, C. Ferraz, and J. Kelner. Mires: A publish/subscribe middleware for sensor networks. *Personal Ubiquitous Comput.*, 10(1):37–44, 2005.

[45] W.W. Terpstra, J. Kangasharju, C. Leng, and A.P. Buchmann. Bubblestorm: Resilient, probabilistic, and exhaustive peer-to-peer search. In *SIGCOMM'07: Proceedings of the 2007 Conference on Applications, Technologies, Architectures, and Protocols for Computer Communications*, pages 49–60. ACM, New York, 2007.

[46] D.A. Tran and T. Nguyen. Subscription covering detection in publish/subscribe systems based on random projections. In *Proceedings of IEEE International Conference on Collaborative Computing (Collaboratecom 2007)*. IEEE Press, White Plains, NY, October 2007.

[47] D.A. Tran and T. Nguyen. Localization in wireless sensor networks based on support vector machines. *IEEE Transactions on Parallel and Distributed Systems*, 19(7):981–994, July 2008.

[48] D.A. Tran and C. Pham. Cost-effective multidimensional publish/subscribe services in sensor networks. In *IEEE Workshop on Localized Communications Algorithms and Networks – IEEE Conference on Mobile Ad hoc and Sensor Systems (MASS 2008)*. IEEE Press, Atlanta, GA, September 2008.

[49] D.A. Tran and C. Pham. A content-guided publish/subscribe mechanism for sensor networks without location information. *Journal on Computer Communications (COMCOM)*, 33(13):1515–1523, August 2010.

[50] Y.-M. Wang, L. Qiu, C. Verbowski, D. Achlioptas, G. Das, and P. Larson. Summary-based routing for content-based event distribution networks. *SIGCOMM Comput. Commun. Rev.*, 34(5):59–74, 2004.

[51] T.W. Yan and H. Garcia-Molina. The SIFT information dissemination system. *ACM Trans. Database Syst.*, 24(4):529–565, 1999.

[52] E. Yoneki and J. Bacon. Unified semantics for event correlation over time and space in hybrid network environments. In *Proceedings of IFIP International Conference on Cooperative Information Systems (CoopIS)*, pages 366–384, 2005.

[53] Q. Yuan and J. Wu. Drip: A dynamic voronoi regions-based publish/subscribe protocol in mobile networks. In *Proceedings of the 27th IEEE Conference on Computer Communications (INFOCOM 2008)*, pages 2110–2118, 2008.

PART 5

VANETs

17

Service and System Architectures for Vehicular Networks

Stephan Eichler

Institute of Communication Networks, Technische Universität München, D-80290 München, Germany; e-mail: s.eichler@mytum.de

Abstract

In this chapter an architecture concept for Vehicular Network services is presented. A brief introduction to the concepts of Vehicular Networks (VNs) is given first, in order to understand the presented architecture requirements. Based on these system requirements the architecture concept which is described has been designed. The chapter explains both aspects of a VN service environment, the infrastructure part and the mobile ad hoc part, and gives examples of the service provisioning and service usage. Moreover, the security aspects of a Vehicular Network service architecture are explained.

Keywords: vehicular network, VANET, architecture, backend.

17.1 Introduction

As interest in and capabilities of Internet services and mobile technologies grow, also their range of application widens. A very promising application field for mobile and location-based services is the automobile. Installing wireless communication technology into vehicles opens up new business opportunities for service providers, while increasing comfort and safety for vehicle passengers. Wireless communication between vehicles, also referred

Anand R. Prasad et al. (Eds.), Future Internet Services and Service Architectures, 367–390.

to as Vehicular Networks (VNs) are a prominent example for emerging next generation services and the corresponding architectures.

A VN relies on wireless communication to both other vehicles and a service infrastructure, e.g., operated by a service provider. The member nodes of a VN use their communication means to exchange information between each other in order to get up-to-date traffic information and local danger warnings. Moreover, the communication can be used to interact with a service infrastructure which can provide centralized services such as toll collection or infotainment. While communication to a service infrastructure can rely on a cell-based communication technology such as Universal Mobile Telecommunications Standard (UMTS), the vehicle-to-vehicle (V2V) communication needs a wireless communication technology such as Wireless Local Area Network (WLAN) or the designated IEEE802.11p/Wireless Access in Vehicular Environments (WAVE) [15]. Communication between vehicles needs to rely on an ad hoc network concept, which can automatically identify and update the current network topology. This networking concept is also known as Mobile Ad Hoc Network (MANET) or Vehicular Ad Hoc Network (VANET) if the nodes are vehicles.

The introduction of a VN in reality requires the definition of a complete service and system architecture concept, including aspects like communication, security, and management. Such an architecture needs to be complete while still being very adaptable for future trends and enhancements. Therefore, a modular concept including modules for security and communication, applicable in centralized as well as decentralized settings has to be followed. Only a modular concept can fit both the infrastructure-based services and the V2V applications. A good introduction on VN aspects has been presented in [7].

Whereas a distinct mechanism such as an ad hoc message dissemination protocol needs to meet only a few requirements, a system architecture for a VN needs to comply with many requirements from diverse contexts. Thereby security and communication aspects play an important yet not an exclusive role in determining the cornerstones for such an architecture [3, 8]. Also organizational aspects, management issues, and practical implementation-related matters influence the layout for such a system architecture significantly. This is why the definition of one standardized architecture concept is extremely difficult. However, defining a modular concept allows us to continuously adapt and improve such an architecture, keeping it compliant to the latest requirements. In the following section the requirements, areas of application, and main building blocks and their interactions of a modular VN service architecture are presented.

17.2 Vehicular Networks: Definitions and Introduction to Typical Services

Before explaining the types of services and the system architecture of a VN the most important terms and definitions are introduced and differentiated. The following definitions are important building blocks of a vehicular service architecture, as described in this chapter.

- *Vehicular Network*: In contrast to the term VANET the term VN is used to describe the full network scenario including the Mobile Entities (MEs), gateways, and all backend components. This corresponds to an extended "ad hoc access" network, which consists of mobile nodes and static gateways. Gateways are required for a VN, since the backend servers are relevant for the scenario.
- *VANET*: Even though the term VANET is somewhat misleading, since the name suggests a close relation to MANETs, while the characteristics of MANETs and VANETs are different (for example, no energy constraints exist in a vehicular context) [11], it is still used very often. In this chapter the term is used to refer to decentralized and uncoordinated wireless networks mainly used for inter-vehicle communication. The VANET consists of several independent MEs and can operate without infrastructure support in most cases.
- *Backend*: The term *backend* refers to all architecture components located in the fixed network part of the system architecture. Hence, all servers and databases, mainly needed as a prerequisite for the whole system to run, compose the backend. In most cases these components are connected via the Internet and are provided by a service provider.
- *Mobile Entity*: Usually, nodes in a VANET are vehicles. Nevertheless, also other devices equipped with the respective wireless communication interface – mobile phones or Personal Digital Assistants (PDAs) – could be nodes in such a scenario. Thus, the term ME is used as a synonym for all compatible devices in a VANET.
- *On-Board Unit*: The core component of an ME is the On-Board Unit (OBU). Basically an OBU is a special computer system installed in the vehicle. The OBU is the main hard- and software component running all platform- and V2V-related service applications. It requires a defined

system setup, especially for the telematics platform services part, to be compatible with the systems in the backend.
- *Roadside Unit*: Little sensor or computer unit placed at the side of the road, e.g. at intersections. A Roadside Unit (RSU) has the task to support and extend safety and efficiency applications for VNs.

In principle a VANET can be operated totally independent of infrastructure-based services. However, for the required security services in such an environment a supporting service infrastructure is not absolutely required, but it enables more security features. Moreover, the vehicle Original Equipment Manufacturers (OEMs) need a business model in order to provide the wireless communication technology. The typical V2V services will be free of charge, in order to attain a high number of participants. However, the vehicle-to-backend (V2B) and vehicle-to-infrastructure (V2I) services can and will be charged in most cases. Hence, this completes the business model for the OEMs. Thus, a VANET will always be supported by infrastructure services leading to a VN scenario.

Besides the business model there are also technical reasons for the extension of VANETs with a backend. The central services can increase the range of functionalities significantly and ease the usability for the customers by providing features like Single Sign-On (SSO) and a unified invoicing procedure.

17.2.1 Services in Vehicular Networks

Before explaining the main requirements and the architecture design for a VN service architecture, some typical services shall be described. Based on these services the requirements and the architecture setup will become more clear. The services in a VN can be categorized into two main groups.

Infrastructure-Based Services

All services provided by service providers inside the backend or services provided by roadside infrastructure belong to this category. These infrastructure-based services rely on V2I and V2B communication and will usually have the vehicle as data sink and service consumer.

Examples for backend services are support applications provided by the vehicle's OEM. This could for example be a wireless maintenance application. A second example is a wireless toll road invoicing. Vehicles approaching a toll road entrance can identify themselves and receive a digital access token. The toll fees can then be billed to the service user.

Roadside infrastructure can also be equipped with communication technology in order to provide services to vehicle drivers. Hence, traffic lights can signal the red and green light phase durations, electronic traffic signs can signal their status, and obstacles like construction sites or railroad crossings can send warning signals to surrounding vehicles. These services are generally provided by RSU which interact with the OBU of the connected MEs. For most cases RSUs provide interactive and non-interactive safety services.

Ad Hoc Services (VANET Services)

All services running in the VANET context, hence, without infrastructure and between MEs, belong to this category. These ad hoc V2V services rely on direct communication between the MEs. In this case the MEs act as data sources, data aggregators, data forwarders, and data sinks. The wireless ad hoc network is used for V2V cooperation. The more vehicles participate the better, since the data basis increases with every additional participant.

Typical examples for VANET services are all kinds of information distribution applications. The main focus is on traffic safety. Hence, vehicles collect information on traffic status, e.g. traffic jams or local danger situations and provide this information over the wireless network to surrounding vehicles. In addition, vehicle position tracking can be used to realize collision avoidance services or cooperative driving in general.

The ad hoc services do not rely on any infrastructure, however, the availability of additional infrastructure-based services can improve the quality and reliability of the services.

Overall V2V and V2I services complement each other. Only a VN providing both infrastructure-based and decentralized ad hoc services will provide a maximum number of features and a high service quality. However, both types of services need to be harmonized in a consolidated service architecture.

17.3 Requirements for Vehicular Network System Architectures

Since a VN is a complex system with numerous components, many different prerequisites and requirements exist in order to provide, operate, and maintain such a service environment. These requirements can be grouped in the following categories.

17.3.1 General Requirements for a Vehicular Network Service Architecture

A core challenge integrating mobile communication technology in vehicles is the fundamentally different lifecycle of the technologies. While common vehicles have a product lifetime between seven to ten years, todays electronics and mobile devices have a product lifetime between six to eighteen months. Therefore, the OBU inside the vehicles must continuously keep up with the technological progress for comparable technologies. In addition, it needs to integrate with the latest mobile devices. Thus, service applications and operating software has to be easily replaceable and upgradeable. The users need to be able to run the latest services even on older OBUs. Further, a user should to be able to sign up for services provided by different service providers. Otherwise the market will not reach the size necessary in order to be a viable business enterprise. The openness of the platform as well as the market itself is a crucial requirement for the success of the whole VN concept.

Information security mechanisms like authentication, confidentiality and integrity protection, e.g., used for service data, communication links, and invoicing information are absolutely necessary in order to install a service infrastructure for VNs. The security functionality is required for all parts of the system, the backend actors, the MEs, and the different communication technologies. The selected security concept has to be usable for the infrastructure-based services as well as the VANET data applications. A combined security concept is required in order to keep the complexity, overhead, and costs as low as possible, while reaching a maximum level of security. In addition to the security mechanisms certain services in a VN require privacy protection in order to reach a high customer acceptance [1,9].

17.3.2 Requirements for the Backend System and Infrastructure Components

A VN will be supported by a backend architecture which will most likely be heterogeneous. Hence, multiple service providers and operators will compete against each other and provide their services in the same service environment. This requires a common operating environment with standardized interfaces, communication protocols, and security mechanisms. Only this standardization will enable service providers to deploy their services to several different vehicle brands.

The heterogeneity of the platform operator and service provider structure has to be hidden from the customers. The customers need to have a single contract with one operator, getting one aggregated invoice for all services, no matter which provider delivered it. Moreover, the usage needs to be as easy as possible, hence, the security concept needs to provide a SSO mechanism. This will allow the user to enter the individual access credentials only once. All service usages after a successful log in can be done without further authentications. Hence, a user will be able to seamlessly use services from different providers.

To realize a backend service infrastructure with such a high usability, joint efforts by vehicle and OBU manufacturers, operators, and service providers is necessary.

17.3.3 Requirements for the Mobile Entities

An ME participating in a VN needs to be equipped with wireless communication technologies, capable for V2V and V2I interactions. The OBU inside the ME requires a specific communication stack inside the operating environment, in order to operate both types of services.

Concerning security the OBU needs to support an additional requirement beside the general ones stated above. Since the OBU is operated in the field outside the operator's sphere of influence, it has to be tamper-proof. Hence, a hardware security module needs to be integrated. This helps to ensure the integrity of the OBU itself. Only unmodified equipment will be able to operate and participate in the VN. Further, the security module is required to protect the user's private data stored on the device.

17.3.4 Wireless Communication Requirements

To operate a VN, first of all two types of wireless communication technologies are required: a connection to the backend infrastructure, usually a cellular communication technology, and an ad hoc communication technology. Especially the ad hoc communication technology needs to fulfill certain characteristics in order to provide the necessary flexibility and quality. It has to have a distributed channel allocation scheme, for example Carrier Sense Multiple Access with Collision Avoidance (CSMA/CA). It needs to operate in rural as well as urban environments. In order to be able to operate safety critical services it needs to have a high data rate with low latency and a robust data transmission. Further, it has to provide or at least be compatible

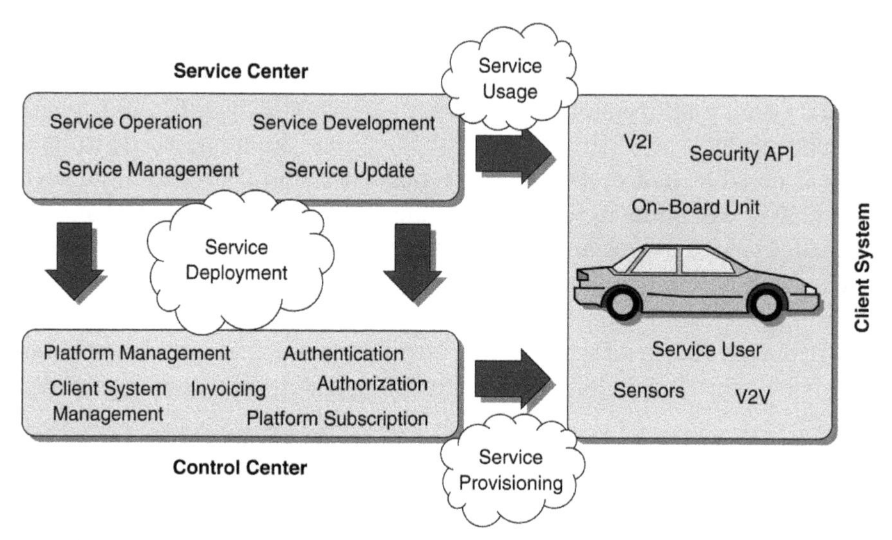

Figure 17.1 Vehicular network high level architecture and component interactions.

with a configurable Quality of Service (QoS) scheme to prioritize the ad hoc data traffic. The communication concept needs to be scalable, thus, it has to operate in both scarce and dense network topologies.

17.4 A System Architecture Concept for Vehicular Networks

The architecture concept for a VN has to take all of the requirements introduced in Section 17.3 into account. Two main parts have to be regarded, the backend system architecture and the architecture of the MEs. Both parts combined realize the full service architecture of a VN.

In Figure 17.1 an overview of the system architecture for a typical VN is shown. The figure concentrates on the infrastructure-based services, the required actors, and the main information flows. The backend consists of at least one Control Center (CC) and one Service Center (SC). The vehicles represent the Client Systems (CSs) and are the service consumers. The core platform functionalities like management, security, user subscription, and invoicing is provided by the CC components. All service related features like service development, management, and operation is provided by the SCs. The SCs need the CC to gain access to the service platform and to deploy the service on the VN service platform. The accc is responsible for the service

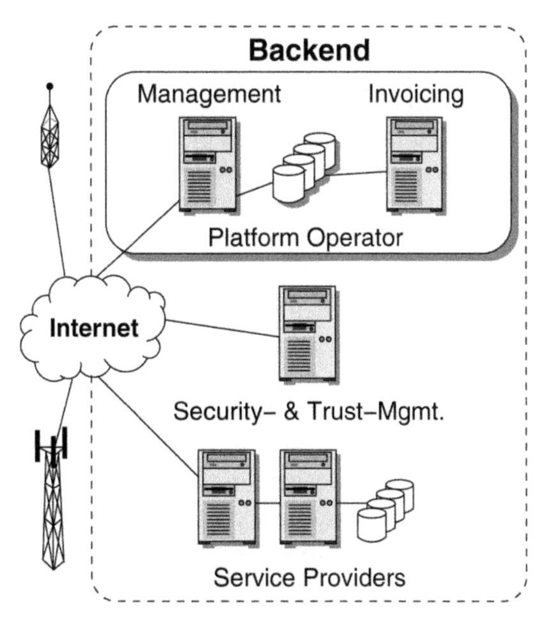

Figure 17.2 Components of the backend architecture in a vehicular network.

provisioning to the CSs. Once the service is provided, users can subscribe to them and use the service by directly interacting with the corresponding SC.

17.4.1 Backend System Architecture

A backend architecture can provide a great variety of features to VN nodes. These range from management to service provisioning functionalities. A VN is a very heterogeneous network with many different stakeholders, however, the operator of such a network uses the backend to control the network functionalities.

To better understand the different service interactions in a VN the backend architecture is introduced in greater detail. In Figure 17.2 the three main components of the backend are the platform operator with a management and an invoicing component, the security and trust management center, and the service provider(s). These three backend actors are required to offer a full service infrastructure, however, they can be operated by different stakeholders.

Features of the Backend Architecture

The most important feature located in the backend of future VNs is the main component of the system's trust environment, the Public Key Infrastructure (PKI). Closely connected to this feature is the user and subscription management. A server infrastructure in the backend network provides these features and issues certificates for new users in combination with a Registration Authority (RA) and a Certificate Authority (CA). Every VN using any form of trust environment needs a user management to provide the important feature of certificate revocation to maintain the system's security. Moreover, a platform operator will want to have a certain control and supervision on the activities and the members using the system.

The user management and subscription handling is also important from another point of view, telematics services offered by a service provider will be provisioned by server nodes located in the backend. To ease system management on the one hand and usability on the other hand, service providers will have a contractual relationship to the platform operator. Hence, they can use the user data held by the platform management. Thus, the respective data needs to be maintained in one position only. Further, the users do not need to hold individual contracts with the operator and the service providers. Moreover, a transparent service consumption involving several stakeholders can be realized due to this user subscription policy.

For the billing of services the sole responsibility for user data and its administration at the operator is also beneficial. Having an extended user and subscription management in the backend allows for the realization of a single invoicing procedure. Thus, no matter how many service providers are involved in the provisioning of applications for a single user, only one single invoice is issued. This increases the usability of the system for the users and makes the billing procedures much simpler.

Besides these more or less administrative features, the backend is required to realize any kind of server-based application or services relying on the Internet. The backend can, for example, be used to provide Internet connectivity as a service to the MEs. In addition, all platform-based services for telematics and infotainment require the backend architecture. They will make up a great deal of commercial applications needed to finance such a VN in the future.

Components of the Backend Architecture

The system setup of the backend architecture consists of several components. Three main components can be differentiated: The operator servers, the security and trust servers, and the servers of the service providers. A general setup of these components is shown in Figure 17.2. The components of the backend are all connected using the Internet or a similar network and their functionalities are distributed as follows:

- *Platform operator*: A VN using a backend with services requires a user and platform management. This is done by a platform operator. The most important task of the operator is to handle the user data and the related service subscriptions. It closely cooperates with the trust management to provide security credentials to its users. Moreover, the operator itself or a closely related stakeholder will do the invoicing of the platform. Hence, all service consumptions are logged and accounted by or at least in close cooperation with the platform operator. Further, the operator has a contractual relationship to one or several service providers. A service provider gets access to the users of the operator and in return offers its services to the users. The providers do not have to manage their own user database, however, they can solely rely on the user data provided by the operator. In a practical VN implementation not one operator necessarily needs to handle all potential users. Nevertheless, several operators can run in parallel. The users can interact as long as the trust environments are interconnected through cross-certification, establishing an extended circle-of-trust.
- *Trust management*: Closely connected to the platform operation is the trust environment and its management. In most settings a RA will be used as a mediator between the platform operator and the CA. It handles all certificate and revocation requests and bundles them to be handled by the CA. The CA will provide the certificates and the up-to-date Certificate Status Information (CSI) for the platform. Further, it will provide cross-certified certificates to realize trust relations to other CAs or trust environments. All members of the circle-of-trust, the platform operator, the service providers, and the users need to have a valid trust credential issued by the trust management to be able to interact with other platform actors.
- *Service providers*: The functionalities experienced by the users are provided by the service providers. Hence, they are a crucial part of a VN. Their server infrastructure will be located in the backend of the system. They are integrated in the platform on a contractual and technical basis. The providers offer their services through the platform of the operator, thus, they can not directly offer services to the users. However, as multiple platform operators can co-exist in a VN, a service provider can offer

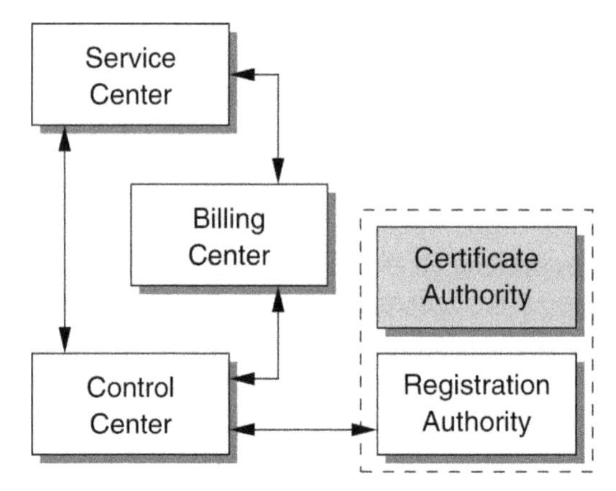

Figure 17.3 Interactions in the vehicular network backend.

its applications to multiple platform operators and their users. They only need to adapt the software to the requirements of the respective MEs and their OBUs.

Besides these three main components the backend needs to rely on a network infrastructure which will most likely be the Internet. In addition, the connectivity to the MEs needs to be realized with both a Inter-Vehicle Communication (IVC) technology as well as a General Packet Radio Service (GPRS)/UMTS network provider.

Interactions of the Backend Components

Since the backend components are not acting on their own behalf, the interactions to other components are crucial. Hence, the most important interactions are explained in the following.

The CC is the most important component in the interactions diagram (see Figure 17.3). Everything starts and ends with the CC. Many of its tasks can be handled without the help of other components. One exception is the management of certificates for the security mechanisms on the platform. To issue a certificate a CA is required. In most cases the operator of the CC will not operate a CA as well. However, if the CC operator does provide its own CA it will still be an extra component. Hence, any existing CA will be used via a RA.

Since the CC owns and manages the end-user data, each SC needs to interact with the CC. The operator of an SC needs a contract with one or more CC operators to offer the services to the end-users. Thus, besides the technical connection, also a contractual connection needs to be established between CC and SC.

The Billing Center (BC) needs to interact with both, the CC and the SC. If an end-user subscribes to a commercial service the SC interacts with the CC to check the validity of the end-user. If the validity is confirmed the SC can subscribe the end-user and charge the respective costs using the BC. The CC provides the relevant end-user data for the invoicing process to the BC.

17.4.2 In-Vehicle System Architecture

The most important components in a VN are the MEs, which make up the core of the network. In general, an ME in a VN is a vehicle equipped with the required communication and processing capability called OBU. Nevertheless, other types of MEs are conceivable, for example mobile phones or after market terminals, similar to the very popular navigation systems. Independent of the realization, an ME needs to provide a multitude of features and capabilities to fulfill the needs of the architecture concept.

Features of the Mobile Entity

The ME is the component closest to the user, hence, it needs to have all features the user wants to have. This includes both distributed V2V services and centralized portal applications. To be able to provide IVC and portal services the ME needs to be equipped with several different communication technologies. At least one type of infrastructure-based communication technology (for example GPRS/UMTS) and one IVC communication device needs to be available. The mobile client will support three types of communication relations: vehicle-to-vehicle (V2V) communication, vehicle-to-infrastructure (V2I) communication, and vehicle-to-backend (V2B) communication.

A crucial feature for the ME is the user interaction with a Human-Machine Interface (HMI). Especially an OBU inside a vehicle needs to provide visual as well as auditory user interaction. The OBU will practically be integrated in the on-board computer providing all types of information and entertainment features.

An ME integrated into a vehicle will have access to the on-board sensors of the vehicle. This is an important feature to collect status information especially for the IVC services. Due to the integration into the vehicle architecture, the OBU will be able to access all types of sensors, communication devices, and control units. Thus, it can aggregate information from different sources and use it for the event detection and as context information.

To be fully integrated into the architecture concept the ME needs to be able to handle security credentials. Further, the entity itself needs to be secured using soft and hardware security components. The security mechanisms need to be usable for broadcast, point-to-point (P2P), and end-to-end (E2E) communication, hence, all different communication relations that typically occur in a VN.

Components of the In-Vehicle Architecture Setup

The components of a typical ME architecture setup are shown in Figure 17.4. The central component in the architecture is the On-Board Unit (OBU), which is the execution platform for all services and related applications. The execution platform can, for example, be realized using Java combined with the Open Services Gateway Initiative (OSGi) framework [6, p. 74]. This combination allows remote servicing and service application provisioning as well as multiple networked service applications running in parallel.

The OBU is integrated into the vehicle architecture, thus, it can access and use the vehicle's sensors and control units. In addition, it can use the communication devices installed in the vehicle. Several support functions are provided by components such as Global Positioning System (GPS), which provides positioning and time information for the architecture. To extend the number of traffic information sources, broadcast services like Radio Data System (RDS), Traffic Message Channel (TMC), and the Transport Protocol Experts Group (TPEG) protocol [16] can be integrated into the architecture and provide their data to the OBU. Supplementary to these broadcast services the digital broadcast technologies (Digital Audio Broadcast (DAB) and Digital Video Broadcast (DVB)) can be used in the architecture to provide both entertainment and information services.

The integration of different communication and security mechanisms primarily influences the design of the OBU and its system stack. In Figure 17.5 the integration of security and communication mechanisms in the stack is shown. Depending on the service, application layer security mechanisms can be applied. Before the application data is sent, either V2V or platform service relevant communication protocols are used. In a final step the Secure Communication Layer is passed which is used to add the required security features.

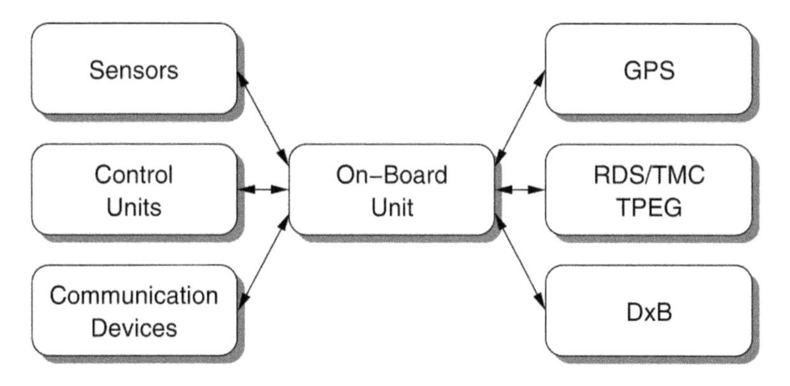

Figure 17.4 In-vehicle component interaction.

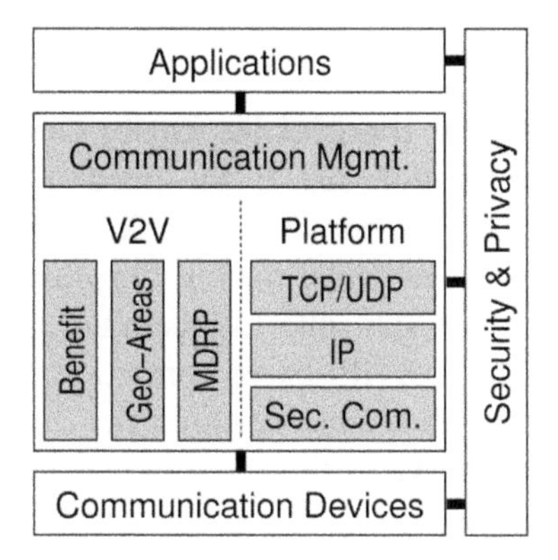

Figure 17.5 On-board unit communication stack.

Interactions between Mobile Entity Components

Like in the backend sub-scenario, the components in an ME use certain interactions. The core interactions are shown in Figure 17.4.

As described above, the OBU is the central component inside an ME. All information available to the ME is managed by the OBU. It uses the sensors to acquire new information about the current vehicle status. The collected data is provided to the service applications having sufficient access rights. The link to the control units is required for applications that are allowed to influence the driving behavior of the vehicle.

Since several applications run on the system in parallel, the OBU needs to control the access to the wireless interfaces. This is primarily the case for the cell-based communication devices. A managing component, which provides access to the available wireless communication technologies, similar to the communication gateway presented in [10], could be used to coordinate the service access and optimize the connectivity. Moreover, the OBU collects information from components like the GPS or the traffic information systems and provides the information to its applications. This reduces the load on the in-vehicle communication architecture and still provides up-to-date information to all service applications.

Besides managing the access to communication devices, the OBU uses a policy database to control the access to vehicle sensors, control units, and support functions like GPS. Service applications are not necessarily allowed to access all resources of the ME, therefore, a rulebase is used to define the applications' access rights for the in-vehicle components. The security implementation in the execution environment of the OBU is enforcing these access rules and allows or denies resource access to applications.

Security and Privacy for the Mobile Entity Communication

A very crucial implementation task is the integration of security and especially privacy into the communication stack. In Figure 17.5 the typical communication stack for a vehicular node is depicted. As introduced above, the communication stack is split into two parallel paths: The V2V and the conventional Internet Protocol (IP)-based communication path.

A security integration is mainly required for the application layer. Especially for the VANET services the security can be handled solely in the application layer, as long as no secure routing mechanism shall be used in the lower layers. But for a mere broadcast application a security integration in the application layer is sufficient. However, security for the platform services also has to be integrated in the network layers. One possibility to realize this is to use a specific security layer in the communication stack, as suggested by the Global System for Telematics (GST) project [4]. Characteristic for this security layer is the *secure communications engine*, which manages the security sessions of the node.

Whereas the security integration into the communication stack is straightforward, the privacy integration needs more attention and care. The privacy mechanisms require strong links to *all* communication stack blocks. This is crucial for the success of the privacy concept. To provide privacy effectively, the whole system architecture has to be evaluated and included in the privacy concept. In contrast to security mechanisms it is not sufficient for privacy to be included in one layer (for example, the application layer). To provide a high degree of anonymity the privacy concept has to ensure the unlinkability of "traces" a node leaves behind. Since *all* layers in the communication stack generate characteristic labels, such as IP-addresses, Medium Access Control (MAC)-addresses or session identifiers, the privacy component has to have control over every communication layer and the generation of labels. If any label changes *all* other labels have to change as well, otherwise the messages can be linked by the unchanged label(s). Hence, a single privacy component has to control all involved communication- and label-generating blocks in the system. Otherwise no reliable privacy concept can be provided.

Integration of Security – Security API and Hardware Security Module

The most important component for a secure vehicular network node is the security component. The security component has to fulfill all security and privacy requirements and provide respective mechanisms. It should be implemented with two parts, a security software Application Programming Interface (API) and a hardware security module referred to as Security Module. As the Security Module can not function alone it can be seen somewhat integrated into the API, which provides all necessary functionality to use the hardware component. Refer to Figure 17.6 for the full security API definition.

The API can be split into five blocks. The four software components (PKI, Secure Communication, Privacy, and Platform Security) use the security mechanisms and the secure storage functionality of the fifth component, the Security Module. The Security Module is the most crucial component of the system concerning security. It is a tamper-proof hardware security component similar to a Trusted Platform Module (TPM) [17] or a smart card. For the security level of the whole system it is crucial to implement the core security functions using a tamper-proof hardware component. This is the only way to sufficiently secure credentials, certificates, and key material on a platform being used in the field. The key materials will be generated and stored within the Security Module and cannot "leave" it at any time without destroying the component. This feature of the system is required to maintain a high degree of security and minimize the number of overall certificate revocations throughout the system. However, this requires a hardware component which is also capable of using the credentials and keys, generate for example secure random values, and encrypt/decrypt messages. The overall policy must be: No security credentials and keys that originate from a Security Module may leave the module in any way, all security operations have to run on the module itself.

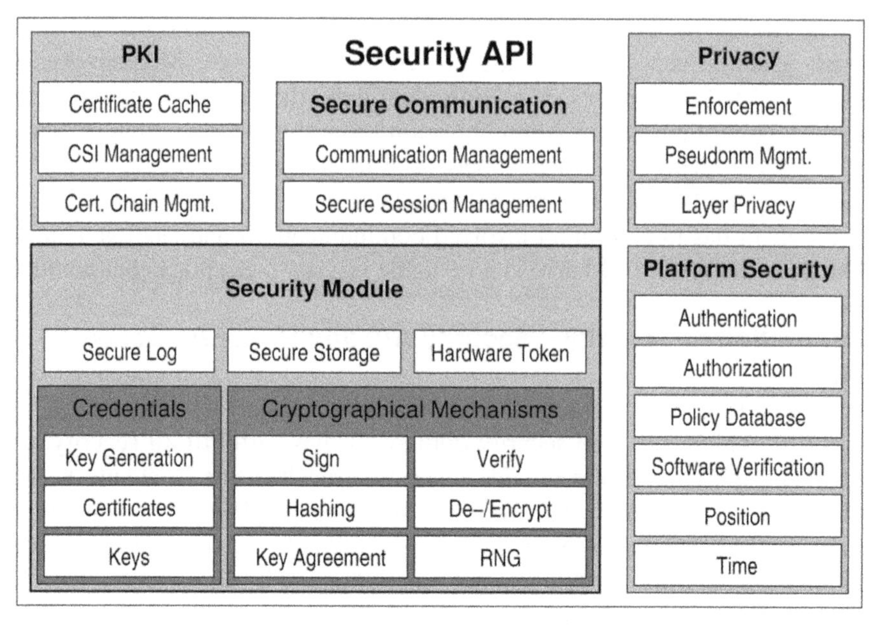

Figure 17.6 Security API and Security Module for a mobile entity.

The trust of the system and the related credentials for each node are handled in the PKI API-component. All security-related communication is handled by the Secure Communication component. The general security features (authentication, confidentiality, policy enforcement, secure software updates, etc.) are handled by the Platform Security component of the API. The management of privacy and the handling of pseudonyms as well as the layer management is handled by the Privacy component.

17.5 Application and Management of the Vehicular Network System Architecture

After having introduced the backend and ME architecture concepts in the previous sections, the application and management of the full VN architecture is discussed. Based on a few platform and IVC service examples the functionality of the system architecture concept is exemplified. This shall make the concept more clear and point out some of its benefits.

17.5.1 Application Examples for Platform Service Provisioning

To exemplify the use of platform services with the architecture two settings are discussed: Using Single Sign-On (SSO) for a transparent service usage and the general service provisioning when multiple service providers are involved. These two examples show the component interaction and the application of security mechanisms for platform services in VNs.

Realization of Single Sign-On for Transparent Service Usage

A very important feature for service platform architectures is the so-called SSO, where a user has to authenticate to the system architecture only once. After a successful authentication the user can transparently use all subscribed services without further interaction. The main goal is to reduce the required user interaction to a minimum and provide a trust environment which includes all relevant actors of the system.

To be able to use SSO in a VN several requirements need to be fulfilled. The main prerequisite is the setup of a circle-of-trust, both on a contractual and a technical level. A commercial agreement of all participants permits a seamless interaction between different entities belonging to the circle-of-trust. The contractual basis is mapped to the technical level using security mechanisms and protocols, for example the Security Assertion Markup Language (SAML). A second requirement is the federation of information, especially identities. A user will be identified with several different identities or credentials by many different entities in the system. However, to provide a seamless interaction after one authentication these identities need to be linked with each other, which is meant by *federation*. To realize a circle-of-trust for SSO all actors in the backend need to participate. The platform operator will be the aggregator, thus, the federating entity managing all user data. It takes in a central role in the SSO procedures since all authentication and verification interactions involve the aggregator.

The process of an SSO session establishment works as follows. The user authenticates with its personal identification data to the aggregator. The aggregator validates the authentication and issues an authentication credential to the respective user. The credential is stored in the OBU of the ME and used for each interaction within the circle-of-trust. Hence, for each interaction with an SC, for example a service usage, the credential needs to be presented to the peer. The SC receiving the credential requests a validation from the aggregator and only if a valid credential has been presented the service is provided to the user. In the course of the credential validation the aggregator will identify the user to the SC by presenting the user's identity known to the SC. Platform tasks, for example invoicing, can be handled easily using this identity management.

A more detailed description of SSO and the related federation concepts has been published in [18]. In addition, further information on SSO concepts, the related security protocol (SAML), and software realizations can be found in [5, 13, 14].

Service Provisioning and Usage with Multiple Service Providers

To be able to distribute service applications to MEs and configure their settings, a provisioning procedure is required. An important requirement is that multiple service providers can be used by the same ME. The service provisioning and usage follows several steps. These are explained in the following section, leaving out the security interactions in order to simplify the example. Nevertheless, the SSO example already detailed the typical security interactions for platform services.

The service-related interactions have already been pointed out in Figure 17.1; they are relevant for the following example. An SC needs to register its service at the platform operator, providing the application software, the requirements on the OBU, and the SC network addresses. The platform operator will add the new service to the service portfolio presented to the users. Therefore, all users having a compatible OBU will be able to subscribe to the service. All services offered to the users on the platform are listed in the service portfolio, no matter which SC is providing it.

After a service is listed in the portfolio it can be provisioned to the users. As soon as a user selects a service at the CS, the provisioning and subscription process is triggered. The CS requests the service application software and required components from the platform management (the CC). The CC registers the service for the respective user and provides the necessary software components to the CS. After finalizing the provisioning process the service is available to the user at the CS. All interactions required during the service usage will take place with the respective SC.

This task distribution allows the use of several different SCs, which can be used in parallel. During the provisioning process the CS receives the access data for the SC accountable for the respective service application. The SCs do not need any specific knowledge on the CSs since the CC is taking care of the software and system configurations of the subscribing CSs. Once the services are provisioned they can be used seamlessly at the same time irrespective of the SC.

17.5.2 Application Examples for Inter-Vehicle Communication

The communication settings for V2V communication applications are much more diverse than the settings for platform services. The reason for this is that IVC services can be both fully decentralized or use central server interaction from the backend architecture. Hence, the protocol capabilities need to be much more multifunctional.

Inter-Vehicle Data Diffusion using Security and Distribution Strategies

The main task for IVC is the distribution of data in a VANET. Many different distribution and communication strategies exist to increase the effectiveness and usability of these services [12]. In addition, security mechanisms and protocols are used to increase the reliability and reputation of the distributed data. All these mechanisms have been integrated into the VN system architecture concept and lead to a specific protocol sequence.

The applications generating and processing the data can rely on different security and communication mechanisms as shown in Figure 17.5. Especially many of the dissemination strategies coexist and the application as well as the default system settings influence which strategies are used. Therefore, depending on the network load and other context parameters the best matching dissemination modules are used. Moreover, the application will communicate its preferences to the Communication Management, which selects the respective modules. If the system settings require an application level security, most likely a content authentication and/or reputation, the respective security mechanisms are applied to the content before the data packet is passed to the communication modules.

In case an application layer security mechanism requires IVC, like it is the case for the Content Reputation System (CoRS), the Communication Management is involved. It provides the single-hop V2V packet communication and distributes the request to the neighboring nodes. All replies are directly handed back to the respective application.

The data packets for the dissemination are passed through the required communication modules before they are handed to the communication device. In addition, Communication Management adds communication security and privacy if this is required. Hence, it interacts with the security API to realize the security functionalities. The interaction with the security API is also important to control and manage the privacy settings of the communication modules. This is required to increase the unlinkability of packets to the sending node.

Content Distribution with Backend Server Involvement

Besides the V2V multihop communication, content distribution from the backend to the MEs is also an important application. The distribution of up-to-date CSI can be realized using this approach. In this case the backend servers initiate the distribution process.

In the stated example the trust environment servers generate the CSI and sign it, to provide authenticity and integrity protection. Next the data is distributed to the gateway servers, which can distribute the CSI to the MEs. The trust servers are unloaded of the traffic as soon as the gateway nodes hold the data. Depending on the distribution mechanism the gateway nodes either send out the data on request (for example using Mobile Data Request Protocol (MDRP) [2]) or they broadcast the data to the MEs within radio-range.

Once the data is distributed to the MEs they themselves are involved in the distribution process, which is similar to the IVC. However, the relevance of data, distributed by a backend server, is in many cases not bound to any context information other than its age. Therefore, the content distribution needs to be carried out quickly to be of any use. If an ME needs a certain piece of information provided by backend servers it can request it. However, the request will not reach the servers, since the closest gateway node will be able to reply and provide the requested information. This strategy helps to reduce the load on the main servers, similar to the content distribution used in the Internet.

Content distribution with backend server involvement is a simple store-and-forward type application. The MEs are only validating the security information, however, other data manipulation will not occur. Further, special communication protocols such as the MDRP protocol will be used to distribute this kind of information. This is mainly due to the fact that it is content addressable and in most cases sent on a regular basis. An alternative to this distribution method would be to use a platform-based service application, nevertheless, this excludes all MEs not participating in the platform service framework. Moreover, the platform-based distribution would put high network load on the communication networks and the server infrastructure, due to the centralized approach. Thus, the distributed V2V diffusion strategy is more scalable.

17.6 Conclusion and Summary

In this chapter an example for a service architecture in the vehicular context has been presented. The main actors in a Vehicular Network (VN) have been presented and their core functionalities have been presented. Two fundamentally different aspects make up a service architecture of a VN, infrastructure-based services and mobile ad hoc services provided and used by the MEs, the vehicles, themselves. Before introducing the architecture setup a selection of the most crucial requirements was presented. It has been pointed out that distinct requirements exist for the different parts of the architecture. However, also common requirements exist, especially concerning management and security aspects. These common requirements and the different capabilities of vehicle-to-infrastructure (V2I) and vehicle-to-vehicle (V2V)-based services together lead to the fact, that both aspects need to be realized in a vehicular service architecture in order to provide a usable and extensible platform.

Based on the desired service features and the system requirements the system architecture concept for a VN service platform has been presented. The concept includes the essential features for infrastructure-based and ad hoc services as well as the indispensable security mechanisms. The presented usage and service interaction examples point out some of the typical interactions between the platform actors. The presented service architecture concept shows that fully decentralized mobile services and infrastructure-based services can be deployed and offered in one combined service platform, taking into account all key requirements, especially extensibility and security. In combination with an open access to such a platform, for operators, service providers, and users, a VN service platform can be a successful business case and provide added value for all stakeholders.

References

[1] Sastry Duri, Marco Gruteser, Xuan Liu, Paul Moskowitz, Ronald Perez, Moninder Singh, and Jung-Mu Tang. Framework for security and privacy in automotive telematics. In *Proceedings of the 2nd International Workshop on Mobile Commerce*, pages 25–32. ACM Press, 2002.

[2] Stephan Eichler. MDRP: A content-aware data exchange protocol for mobile ad hoc networks. In *Proceedings of the IEEE International Symposium on Wireless Communication Systems (ISWCS)*, pages 742–746, October 2007.

[3] Stephan Eichler. A security architecture concept for vehicular network nodes. In *Proceedings of the 6th International Conference on Information, Communications and Signal Processing (ICICS)*, December 2007.

[4] Stephan Eichler, Jérôme Billion, Robert Maier, Hans-Jörg Vögel, and Rainer Kroh. On providing security for an open telematics platform. In *Proceedings of the 5th International Conference on ITS Telecommunications*, June 2005.

[5] Entróuvert. LASSO – Liberty alliance single sign-on. Website, http://lasso.entrouvert.org/, April 2008

[6] Fabio Ferro, Cristina Chesta, and Enrico D'Acquisto. High level architecture. Project Deliverable 3.1, European IP – Global system for telematics. Available at http://www.gstforum.org/en/downloads/deliverables/, October 2006.

[7] H. Füßler, S. Schnaufer, M. Transier, and W. Effelsberg. Vehicular ad-hoc networks: From vision to reality and back. In *Proceedings of the 4th IEEE/IFIP Conference on Wireless On demand Network Systems and Services (WONS)*, January 2007.

[8] M. Gerlach, A. Festag, T. Leinmüller, G. Goldacker, and C. Harsch. Security architecture for vehicular communication. In *Proceedings of the 4th International Workshop on Intelligent Transportation (WIT)*, March 2007.

[9] Jean Pierre Hubaux, Srdjan Capkun, and Jun Luo. The security and privacy of smart vehicles. *IEEE Security & Privacy*, 4(3):49–55, May 2004.

[10] Wolfgang Kellerer, Christian Bettstetter, Christian Schwingenschlögl, Peter Sties, Karl-Ernst-Steinberg, and Hans-Jörg Vögel. (Auto)mobile communication in a heterogeneous and converged world. *IEEE Personal Communications Magazine*, 8(6):41–47, December 2001.

[11] Wolfgang Kiess, Jedrzej Rybicki, and Martin Mauve. On the nature of inter-vehicle communication. In *Proceedings of the 4th Workshop on Mobile Ad-Hoc Networks (WMAN)*, pages 493–502, March 2007.

[12] Timo Kosch, Christian J. Adler, Stephan Eichler, Christoph Schroth, and Markus Strassberger. The scalability problem of vehicular ad hoc networks and how to solve it. *IEEE Wireless Communications*, 13(5):22–28, October 2006.

[13] Liberty Alliance Project. Liberty alliance project homepage. Website, http://www.projectliberty.org/, April 2008.

[14] OASIS. Oasis – Advancing open standards for the information society. Website, http://www.oasis-open.org/, April 2008.

[15] Task Group p. *IEEE P802.11p: Wireless Access in Vehicular Environments (WAVE)*. IEEE Computer Society, draft standard edition, 2006.

[16] Transport Protocol Experts Group. TPEG Forum. Website, http://www.tpeg.org/, April 2008.

[17] Trusted Computing Group. Trusted platform module (TPM) specifications. Website, https://www.trustedcomputinggroup.org/specs/TPM/, June 2008.

[18] Hans-Jörg Vögel, Benjamin Weyl, and Stephan Eichler. Federation solutions for inter- and intradomain security in next-generation mobile service platforms. *AEÜ – Journal of Electronics and Communications*, 60(1):13–19, January 2006.

18

Trends in Routing Protocols for Vehicle Ad Hoc Networks

Jasmine Chennikara-Varghese and Wai Chen

Telcordia Technologies, Inc., Piscataway, NJ 08854, USA;
e-mail: jchennik@telcordia.com, waichen@ieee.org

Abstract

There is a growing interest in vehicle ad hoc networks as an emerging communications option on roadways. However, the unique characteristics of such networks present challenges in achieving service delivery for a wide range of applications such as time-critical safety applications and content dissemination. To satisfy the performance expectations of target vehicle applications, routing protocols tailored to handle dynamic high speed mobility and frequent network partitioning are necessary. There are a number of innovative approaches proposed in recent years to handle information dissemination in vehicle ad hoc networks. This chapter highlights the trends in routing protocols applicable to vehicle safety and non-critical content delivery applications.

Keywords: VANET, routing, information dissemination.

18.1 Introduction

Vehicle Ad hoc NETworks (VANETs) have emerged as a potentially useful but challenging networking environment. Vehicle ad hoc networks consist of vehicles and roadside infrastructure units equipped with radios to support communications. To handle high mobility in the roadway environment, high

Anand R. Prasad et al. (Eds.), Future Internet Services and Service Architectures, 391–411.

bandwidth and short range communications is envisioned to support vehicle network applications. One leading wireless technology solution for vehicle networks is Wireless Access for Vehicular Environment (WAVE). WAVE uses Dedicated Short Range Communications (DSRC) radios based on IEEE 802.11p and designates seven 10-MHz channels (1 control plus 6 service channels) in the 5.9 GHz frequency band. Additionally, there is an increased interest in other wireless technologies such as smart antennas and millimeter radios to support a more controlled and directed radio communications in VANETs.

While VANETs are still in research and early implementation phases, it is already foreseeable that the decentralized organization of vehicles in VANETs will require different delivery options based on the application requirements. The two emerging types of applications are the delay-sensitive services such as safety applications and the delay-tolerant services such as non-critical content distribution. There has been significant push worldwide to develop advanced Vehicle Safety Communications (VSC) technologies to support dissemination of safety information such as road and traffic-related events and conditions. These include the ITS America VII (Vehicle Infrastructure Integration) effort on 5.9 GHz WAVE, the USDOT Intelligent Vehicle Initiative (IVI) program to accelerate research and development of advanced crash avoidance technologies, and initiatives such as the CAMP (Crash Avoidance Metrics Partnership) and the VSC Consortium (VSCC).

Vehicle ad hoc networks are envisioned to support ITS (Intelligent Transport System) services which detect and share traffic and environment conditions among neighboring vehicles. Safety applications include navigation, assisted lane changing and road information as well as warnings for accidents and road obstacles [3]. The key safety communication requirements are:

- Small latency: The one-hop message transfer latency among vehicles is generally required to be less than 100 milliseconds.
- Sustained throughput: The end-to-end multi-hop throughput should not deteriorate significantly with the number of hops.
- Reliability: In order to effectively utilize safety applications, delivery of information to vehicles of interest should be ensured.

Although the key driving force for VANETs communications is VSC, the availability of vehicle network access can be leveraged by other types of applications including localized service directories, content dissemination and infotainment as well as other location-aware services [8]. These applications have more flexible delay requirements than those required by safety services. Since the requirements among different VANET applications can vary greatly, a single routing protocol may not be capable of satisfying all applications needs under varying network densities.

In this chapter we identify the major trends in routing methods designed to distribute information in the roadway environment to support safety applications as well as more traditional content forwarding applications. Section 18.2 highlights the characteristics and architecture of vehicle ad hoc networks. Section 18.3 summarizes the routing protocols to support traditional unicast applications and Section 18.4 reviews routing approaches to support safety applications. In Section 18.5, we conclude with the summary.

18.2 VANET Architecture

VANET communications can be categorized into Vehicle-to-Vehicle (V-V) communications and Vehicle-to-Infrastructure (V-I) communications. In V-V communications, vehicles exchange information with each other without support from the infrastructure when they are within the same radio range, or when multiple-hop relay via other vehicles is possible. The network connectivity and routing is set up in an ad hoc manner. Chapter 17 provides additional details on VANET architectures.

Traditional ad hoc approaches are not applicable to vehicle-to-vehicle communications because those schemes were designed for a limited number of nodes moving at moderate speed with uniform node distribution and some pre-determined trajectory. Similarly, hot spot technologies, designed to support portability of nodes between access points, are not tailored to handle the mobility of high-speed vehicles. There are several VANET characteristics that distinguish it from typical ad hoc networks. These include:

- Dynamic topology: VANETs consist of non-uniformly distributed, uncoordinated vehicles in a roadway traffic environment. As a result, neighboring vehicles constantly change.
- Restricted mobility: Vehicle movement is limited to the roadway structures as shown in Figure 18.1.
- Frequent network partitioning: Vehicles tend to disconnect frequently and possibly for long periods of time due to sparse network connectivity or low market penetration of equipped vehicles.
- Localized high network density: During initial deployments, sparsely connected vehicle networks are expected. However, the network can

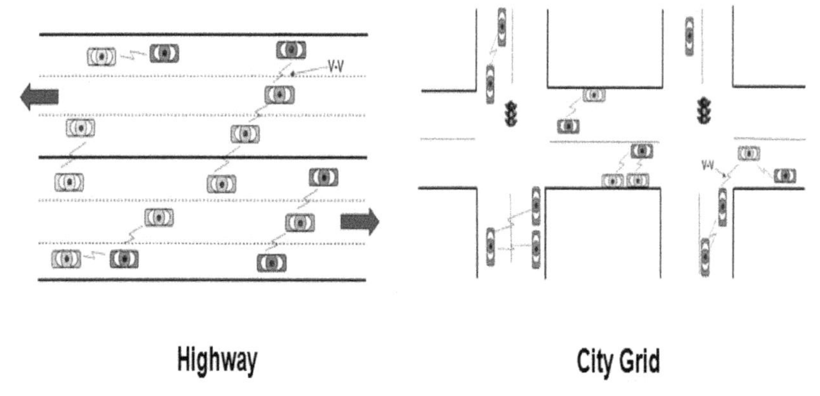

Figure 18.1 VANETs in highway and city grid roadway structures.

become densely connected, such as when a large number of equipped vehicles are on a slow-moving roadway.

Due to these network characteristics, VANET routing protocols should be designed with consideration of the following requirements:

- Respond quickly to network topology changes.
- Leverage roadway information.
- Maintain routing and forwarding in sparsely connected or disconnected networks.
- Mitigate collisions and interference in densely connected networks.

In early deployments, it is expected that vehicle-to-vehicle communications will be predominant, with initially only a small percentage of vehicles equipped to support vehicle communications. As vehicle networks mature and the infrastructure is deployed, eventually vehicles should flexibly handle both V-V and V-I communications options. In vehicle-to-infrastructure communications, vehicles communicate with each other with the support of roadside wireless access points or RoadSide Units (RSUs), located at fixed points along the roadway. V-I communications is not expected to be supported until later deployments because it requires time and investment to build the roadside infrastructure. The deployment of RSUs will most likely follow a phased approach similar to cellular base station deployment. It is envisioned that the final target RSU architecture will be one where the RSUs are networked to each other and to the backbone via wireless or wireline links [6]. Although developing solutions for integrated V-I and V-V communications is a continuing challenge, this chapter focuses on the routing protocols mainly applicable to V-V communications.

18.3 Routing Trends for Non-Safety Applications

In this section, we consider delivery methods for non-critical applications from a single source to a single destination vehicle, or node. Traditional content delivery applications require first locating a destination and then delivering information to the destination. To this end, we consider topology-based and position-based routing protocols for connected vehicle networks which have end-to-end connectivity between source and destination. We also consider methods to support Disruption/Delay Tolerant Networks (DTNs) which have the capability to deliver information to a destination across partitioned or disconnected networks. A high-level classification of routing approaches as well as the most applicable network connectivity scenario and the most applicable roadway structures (e.g. city grid, highway), if any, are summarized in Table 18.1.

18.3.1 Topology-Based Routing Protocols

While routing methods for ad hoc networks have existed for a number of years, a majority of the approaches focus on tactical Mobile Ad hoc NETworks (MANETs). In typical MANETs, nodes move in a coordinated manner and have some a priori relationship or knowledge about each other. MANET routing protocols are typically topology-based schemes which proactively and/or reactively determine routes. Reactive unicast schemes such as Ad hoc On-Demand Vector (AODV) [40] and Dynamic Source Routing (DSR) [13] create and maintain routes as needed to deliver packets to a destination node. In AODV, the route path is created and cached at intermediate nodes during route discovery. DSR discovers the end-to-end routing path and includes in the packet header the intermediate nodes to which the packets must be forwarded. Proactive unicast schemes such as Optimized Link State Routing OLSR [7] continuously maintain up-to-date routes for valid destinations and use periodic updates to capture network topology changes.

Since the traditional MANET view is node-centric, the path from source to destination is associated with specific nodes and this leads to broken routes when there are highly mobile vehicles. MANET routing protocols have been enhanced for VANETs by incorporating more roadway and vehicle information. For example, in Modified AODV [22], vehicle speed and Global

Table 18.1 Summary of VANET protocols for content dissemination.

Routing Protocol	*Routing Type*	*Routing Approach*	*Network/Roadway Characteristics*
AODV DSR	Unicast	Reactive topology-based routing	Connected network
OLSR	Unicast	Proactive topology-based routing	Connected network
Modified AODV PRAODV PRAODV-M LAR	Unicast	Topology and position-based hybrid routing	Connected network
Greedy Forwarding GPSR	Unicast	Geographic routing	Connected network/ highway
CBF	Unicast	Beaconless geographic routing	Connected network/ highway
DDOR	Unicast	Geographic routing without location service	Connected network/ highway
MORA MOPR	Unicast	Geographic routing with prediction	Connected network/ highway
GPCR GSR SAR	Unicast	Roadway-constrained geographic routing	Connected network/ city grid
A-STAR CAR GyTAR RBVT	Unicast	Roadway and traffic-constrained geographic routing	Connected network/ city grid
EAEP DAER Spray and Wait	DTN	Epidemic routing	Partitioned network
NIMF FIMF	DTN	Message ferry routing	Partitioned network
FFDRV	DTN	Velocity-based message ferry routing	Partitioned network/ highway
VADD	DTN	Delay-based geographic routing	Partitioned network/ city grid
MDDV GeOpps	DTN	Delay-based geographic routing	Partitioned network
MRP	DTN	Hybrid DTN routing	Connected & partitioned network
GeoDTN+Nav	DTN	Hybrid DTN routing	Connected & partitioned network/ city grid

Positioning System (GPS) parameters are incorporated in AODV routing decisions. PRAODV [36] enhances AODV by using node direction and speed to predict link lifetime and create a new link before the estimated lifetime of the current link is reached. As an extension, PRAODV-M [36] computes the maximum lifetimes of each route and selects a route based on link stability (longer lifetimes) rather than shortest path.

Location Aided Routing (LAR) [18] uses position information to enhance the route discovery phase of reactive ad hoc routing approaches. In order to establish a route, the sending node computes an Expect Zone containing the destination based on available position information. The node will also define a Request Zone, which is a set of nodes contained in the Expect Zone that should forward the route discovery packet.

18.3.2 Geographic Routing Protocols

A number of Position-Based Routing (PBR) or geographic routing protocols have been proposed in recent years for vehicle ad hoc networks. Such schemes do not require routing tables or store routes. Instead, in PBR schemes vehicles use local information such as current position, obtained via GPS, as well as information about neighboring vehicles to make forwarding decisions for each packet independently. Geographic forwarding schemes are considered more scalable and robust against topological changes compared to topology-based routing schemes.

Geographic or position-based routing generally consists of three basic components:

- A location service which maps vehicle ID (i.e. IP address) to a geographical position. Vehicles update their position information with the location service via beacons. The location service is then used by vehicles to resolve a destination vehicle to a position. Location service requests involve some flooding to obtain the information, adding complexity. There are several approaches for tracking vehicle location including proactive schemes such as Grid Location Service [26], and Hierarchical Location Service [16] as well as reactive approaches such as Reactive Location Service [15].
- One-hop neighbor position beacons: All vehicles send out beacons so one-hop neighbors, i.e. nodes within the transmission range, can be discovered. Beacons include information such as current position, velocity and direction.

- Geographic forwarding scheme: The source vehicle includes the destination ID and destination position information in each packet. The forwarding scheme uses this information, along with the geographical information of the sending vehicle and its one-hop neighbors, to select a suitable next-hop neighbor toward the destination.

There are a number of geographic forwarding strategies which can be applied. In the greedy packet forwarding scheme [32], the sender of a packet includes the approximate position of the destination vehicle and intermediate vehicles forward the incoming packet to a suitable one-hop neighbor that meets some criteria of geographical progress towards the destination. Greedy packet forwarding, however, may fail to find a path to make positive progress from the sender to the destination, even if one does exist, due to topology holes. In the Greedy Perimeter Stateless Routing (GPSR) [14] approach the packets are re-routed around the perimeter of the topology hole, taking a longer path to reach the destination.

Contention-Based Forwarding (CBF) [9] is a greedy forwarding scheme which requires a location service but does not utilize neighbor position beacons. In CBF, the forwarding node transmits packets as a single-hop broadcast to all neighbors and the neighboring vehicles contend to forward the packet. The neighbors initiate wait timers such that the timer for the node with the largest progress toward the destination will expire first, triggering packet forwarding. Upon hearing the packet transmission, other neighbors will suppress their transmission.

Destination Discovery-Oriented Routing (DDOR) [24] is an example of an approach which does not utilize location services but does use neighbor beacons. The scheme uses a unicast destination discovery process which directs destination discovery requests away from sender to intermediate nodes on the same road segment in a highway environment. Once the destination location is known, it is cached at intermediate nodes along the path of the reply message. This creates a virtual path from the source to the destination position. As the destination moves, the last hop on the virtual path detects position changes through neighbor beacons and will update the segment of the virtual path accordingly. For forwarding, DDOR selects the farthest one-hop neighbor vehicle that will remain in transmission range of sender for the duration of the packet transmission.

Prediction schemes such as Movement-based Routing Algorithm (MORA) [10] take into account physical location of vehicles as well as direction of vehicle movement in selecting a suitable next-hop neighbor.

Similar to PRAODV, the scheme is based on the assumption that neighboring vehicles moving in a certain direction would not maintain the communication link long so such neighbors are not viable candidates for next-hop selection. Movement Prediction Routing (MOPR) [34] adds an extra dimension by including a metric for link stability which would be utilized in selecting the next-hop. MOPR can work in conjunction with other routing schemes such as GPSR.

Most geographic routing protocols resolve the destination node to a target position using a location server. However, position resolution incurs overhead and delays. In addition, delayed position beacons can hinder accuracy in routing especially for very mobile nodes. Position-based routing protocols also suffer from delays caused by collision and contention of the underlying shared wireless medium which use contention-based medium access control (MAC) protocols such as IEEE 802.11. This is especially true when many nodes within the same radio coverage are sending out packets or advertising beacons, incurring large overhead in densely connected network scenarios.

Discovering a suitable forwarding neighbor using position can be problematic, due to sparsely connected VANETs, frequent network partitioning, high mobility, or physical structures. Typical position-based routing schemes using only destination and neighbor vehicle positions tend to perform better in open highways with few obstacles and with evenly distributed equipped vehicles. In a city grid environment, recovery methods such as GPSR perimeter mode can circumvent the topology hole but since such algorithms are stateless, it is inefficient when the topology hole is permanent (e.g., physical road constraints and buildings).

18.3.3 Roadway-Constrained Geographic Routing Protocols

There are a number of protocols which tailor geographic forwarding for the city grid roadway characteristics by choosing a next-hop node along a specified trajectory. These schemes promote forwarding decisions at intersection or junction points since more forwarding opportunities with vehicles in different directions are present at intersections. The following schemes are examples of roadway-constrained geographic routing:

- The Greedy Perimeter Coordinator Routing (GPCR) [30] protocol enhances GPSR to handle city grid road structures. A packet reaches the junction using restricted greedy packet forwarding along the roadway. At the junction, a next-hop neighbor is chosen which makes the most progress towards the destination.

- The Geographical Source Routing (GSR) [29] protocol uses position-based routing supported by a map of the city for more correlated routing to the physical topology. GSR discovers the destination position using location services and includes in the packet header the intermediate geographic points to be traversed in order reach the destination position. Intermediate vehicles greedily forward the packets to the intermediate geographic points along the path toward the destination.
- The Spatially-Aware Routing (SAR) [47] scheme uses a spatial environment model to proactively avoid permanent topology holes and includes a list of intermediate geographic locations in the packet header, similar to GSR.

There are drawbacks to using schemes which only leverage roadway structure. For example, GSR and SAR provide a list of intersections that define the path from source to destination but they may select empty roads. In GPCR, there is no guarantee that the forwarding node can find a suitable neighbor along the given road segment using the greedy approach. Some schemes enhance geographic source routing not just with roadway structure information but also consider factors such as vehicle type and vehicle traffic flow on the roadway which can vary based on roadway congestion, road construction, traffic lights, etc. Several approaches leverage additional vehicle traffic flow information to improve the basic GSR scheme and are more applicable to city grid structures:

- The A-STAR [44] approach modifies GSR by giving preference to streets served by transit buses, increasing the chances of selecting non-empty roads.
- The Connectivity Aware Routing (CAR) [37] approach uses an AODV-based path discovery to find routes. The path discovery packet includes not all nodes traversed but only some select nodes near intersections or road curves.
- The Greedy Traffic Aware Routing (GyTAR) [12] scheme uses a GSR approach but dynamically adds intermediate intersections points, choosing road segments based on vehicle traffic density and distance to destination.
- The Road-Based using Vehicular Traffic information (RBVT) [39] approach uses real-time traffic information to create the geographic source route consisting of road intersections with high probability of network connectivity by selecting intersections with higher vehicle densities. In

addition, for densely connected networks, the scheme uses a beaconless, receiver-based selection of the next-hop neighbor.

18.3.4 Delay/Disruption Tolerant Routing Protocols

As vehicle networks continue to evolve, it is foreseeable that in the near future VANET deployments will be limited and equipped vehicle penetration may be low, resulting in sparse or disconnected networks. Even with fully deployed VANETs, the vehicle network density can be relatively sparse during certain times of day and in less urban areas. It is expected that frequent network partitioning will occur in VANETs and routing protocols capable of handling long network disruptions and delays in network connectivity are necessary.

Delay-tolerant networks (DTNs) are dynamic networks where vehicles have intermittent link connectivity. In DTNs, the delays in reaching the destination are predominantly influenced by vehicle speed and vehicle trajectory as well as geographic distance between source and destination. Long disconnections and shifted delays in the DTN networks cause problems for traditional routing schemes such as AODV and DSR. Similarly, typical geographic routing schemes assume the underlying network is fully connected from source to destination to support normal behavior of greedy geographic forwarding with some strategies to deal with temporary lack of network connectivity, such as perimeter modes. However, most of these schemes will suffer in DTNs due to inability to further route to destination across a network partition or to query the destination's latest location (via location services) leading to low delivery ratios for such partitioned networks.

While schemes such as SAR and GPCR store packets only in recovery mode, DTN routing schemes select vehicles to deliver packets across the network partitions using approaches which leverage a "store, carry and forward" method to cache packets for multiple forwardings. Simple DTN solutions include epidemic routing where intermediate nodes forward to all or some neighbors based on some probability [49]. In the Spray and Wait method [45], the source transmits duplicate messages which are stored, carried and forwarded by neighbors. By adjusting the number of messages sent out by the originating node, the delivery rate changes. Although this approach can provide high network delivery ratio, the disadvantage is the sizing of the cache at each vehicle to maintain a large number of packets for forwarding.

Conditional epidemic routing protocols have been proposed to reduce the tradeoff in cache size. In Distance-Aware Epidemic Routing (DAER) [31], the number of hops and duplications of a packet are limited. Packets are forwarded to nodes in a greedy manner based on their distance to the final destination. In Edge-Aware Epidemic Routing (EAEP) [38], when a vehicle receives a packet, it waits for a period calculated based on its distance from the sender. During this wait period, it counts duplicate packets received from vehicles farther away from the sender. If some threshold of duplicates has not been met, the vehicle will then re-broadcast the packet. The wait time is biased so nodes farther away from the sender are more likely to re-broadcast packets.

In dealing with network partitions, vehicle direction and speed can be leveraged to bridge partitions. The idea of a message ferry vehicle or node whose mobility may be controlled to collect and relay messages or packets across network partitions is used in the following schemes:

- In Node-Initiated Message Ferrying (NIMF) [56], vehicles chosen to be the message ferries move according to a specific route and other vehicles take proactive movement to meet with the ferry. The message ferry will then forward the messages to these vehicles.
- In Ferry-Initiated Message Ferrying (FIMF) [56], to relay messages the message ferry proactively moves toward nodes. In this case, the message ferry is assumed to be faster than the other vehicles.
- Fastest-Ferry Routing in DTN-enabled VANET (FFRDV) [54] is a maximum velocity-based routing approach designed for highway road structures. In FFRDV, roads are segmented into logical blocks and at each logical block, the fastest-moving vehicle is detected and chosen to ferry the message across the block.
- In [17], the authors propose to use buses with scheduled routes to be message ferries since bus routes and movement are predictable.

Some DTN routing schemes select forwarding nodes which could achieve the shortest delay in reaching the destination. Vehicle-Assisted Data Delivery (VADD) [55] and its variations use data traffic flow history to predict traffic patterns and vehicle mobility in city environments. At each intersection, the selection of the forwarding branch road is based on the expected delay to reach the destination on any given road. Mobility-centric Data Dissemination for Vehicles (MDDV) [51] defines multiple paths from source to destination using geographic routing and trajectory-based routing. Along the path, MDDV considers road conditions and vehicle trajectory as well as number of lanes to select best next-hop vehicles to reduce propagation delay. In both

VADD and MDDV schemes when no suitable neighbor exists, packets are stored and carried for forwarding at a later time.

The Geographical Opportunistic Routing (GeOpps) [25] approach selects the next-hop node based on the suggested routes provided by the vehicle on-board navigation system. The one-hop neighbors provide their navigation-based suggested vehicle traveling routes, or trajectories to the forwarder. The forwarding node determines the Nearest Point (NP) to the destination location along each of the suggested trajectories. It then selects the neighbor whose NP is geographically closest to the destination. The scheme chooses vehicles which will provide shorter estimated arrival time to destination and re-iterates this process if it meets a more suitable vehicle to reach the destination. This approach requires exposure of traveling routes and assumes compatible map encoding for usability between different navigation systems.

Hybrid approaches attempt to handle routing in both connected and partitioned networks. For example, the Mobile Relay Protocol (MRP) [35] can be used with any typical VANET protocol. It uses the normal routing mode until the route to destination is unavailable. It then does a localized broadcast to immediate neighbors. These neighboring vehicles store the packets until the cache is full and then forward the packets to a single random neighbor.

Geographic DTN routing with Navigation prediction (GeoDTN+Nav) [5] is a hybrid approach for city environments which uses vehicle mobility and navigation systems to predict network partitions. The scheme also uses a unified framework to share navigation data between different navigation systems. When the network is connected, the vehicle uses GPCR-like forwarding. If a suitable neighbor is not found, the scheme goes into perimeter mode. GeoDTN+Nav uses heuristics, based on packet statistics such as number of hops traversed in perimeter mode, to detect network partitioning and switches to DTN mode if needed. In DTN mode, it chooses a neighbor with path or direction towards the destination, if available, without going into perimeter mode.

18.4 Routing Trends for Safety Applications

Due to the nature of most safety communications, data dissemination to all vehicles in a region of interest is necessary. A number of safety applications are only viable in VANETs if the network is also well-connected or densely connected in order to support near-real-time data dissemination. Thus the majority of protocols described for traditional unicast content delivery are not directly applicable to the more broadcast/multicast nature of safety traffic.

Table 18.2 Summary of VANET protocols for safety applications.

Routing Protocol	*Routing Type*	*Routing Approach*	*Network/Roadway Characteristics*
MAODV ODMRP MOLSR	Multicast	Topology-based routing	Connected network
PBM	Multicast	Geographic routing	Connected network/ highway
LBM	Geocast	Direct flooding	Connected network
GeoGRID	Geocast	Grid-based routing	Connected network
GeoTORA	Geocast	Graph-based routing	Connected network
HV-TRADE	Broadcast	Sender-based relay selection	Densely connected network
IVG DV-CAST RMDP	Broadcast	Timer-based relay selection	Densely connected network
LPG LORA-CBF PPC DDP	Broadcast	Group-based routing	Densely connected network
Streetcast UMB	Broadcast	Cross-layer MAC routing	Densely connected network/ city grid
VRR DBA-MAC	Broadcast	Cross-layer MAC routing	Densely connected network

Note that simply flooding time-critical information to vehicles in a desired region of interest cannot meet stringent safety communication requirements. The underlying reason is that the radio bandwidth is limited, and can be easily jammed due to redundancy, collisions and contention when multiple vehicles in the same radio coverage attempt to broadcast warning messages in an uncoordinated manner. In this section, we consider methods such as localized broadcast protocols, geocasting and group-based communications to deliver time-critical safety services in a specific geographic region. A high-level classification of routing protocols as well as the most applicable network connectivity scenarios and the most applicable roadway structures, if any, are summarized in Table 18.2 for safety applications.

18.4.1 Multicast and Geocast Routing Protocols

There are MANET multicast ad hoc schemes such as MAODV [41], ODMRP [53] and MOLSR [11] which are designed to reach a number of nodes more efficiently than ad hoc unicast routing. Such MANET routing protocols face several challenges when applied to safety applications. MANET protocols can incur high latencies in configuration time with uncoordinated vehicles. For example, in addition to routing convergence delays, the average periodic hello interval in the neighbor discovery procedure is about one second, which already exceeds the 100 milliseconds bound for safety applications. The collision and contention of the underlying MAC protocol such as IEEE 802.11 may also increase the routing convergence and neighbor discovery delays, especially when many nodes within the same radio coverage advertise their routing information simultaneously. As a result, these schemes work better in well-connected but not in densely connected networks such as congested roadways.

Extensions of position-based unicast protocols include the Position-Based Multicasting (PBM) [33] scheme which uses a generalized GPSR-type algorithm that generally works better in highway environments. However, for emergency safety information, such multicasting may not be feasible since vehicles must communicate quickly and reliably. Multicast route creation and location services will incur delays in discovering next-hop nodes.

Safety messages tend to be targeted for nodes in a specific region of interest (near accident) and have a limited temporal scope since road conditions or emergency information is only valid for some time. In such cases, protocols for geographically-defined information dissemination are useful. Geocast is a variation of conventional multicast, specifying destination as a geographic position rather than a specific node or multicast address. Unlike traditional multicast schemes, the geocast group is implicitly defined as the set of nodes within a specified area. That is, if a node is within the geocast region at a given time, it will automatically become a member of the corresponding geocast group at that time. There are various approaches to forward information to nodes in the geocast destination region:

- In direct flooding methods such as Location-Based Multicast (LBM) [19], flooding is limited to those nodes within a defined forwarding zone to reach the geocast area.
- GeoGRID [27] uses location information to define grids and elect a gateway in each grid which is responsible for forwarding packets to the geocast region.
- GeoTORA [20] maintains a graph for the geocast groups which is used to find the direction to the geocast region without flooding.

18.4.2 Broadcast Routing Protocols

Within a geocast or local region, broadcast routing approaches can be utilized to efficiently disseminate safety information in densely connected networks. In History-enhanced Vector-based TRacing DEtection (HV-TRADE) [46] based on position and movement, each vehicle classifies its neighbors into different forwarding groups and selects a few nodes in each group to re-broadcast packets. To avoid sender-based selection of forwarding node, some schemes use a defer time which allows each vehicle to individually determine if the received broadcast message should be dropped or re-broadcast after some calculated wait time. In timer-based schemes the defer time is adjusted to optimize performance and reduce broadcast storms:

- Distributed Vehicle broadCAST (DV-CAST) [48] determines the probability of a receiver re-broadcasting based on the distance between the transmitter and receiver, with higher probability given to nodes farther away.
- Received Message Dependent Protocol (RMDP) [42] changes the wait time of a vehicle before it re-broadcasts based on the number of received duplicates.
- Inter-Vehicle Geocast (IVG) [1] re-broadcasts via relays, one in each driving direction, with defer time generated randomly or computed according to the distance between vehicles.

Group-based routing approaches can also be utilized to achieve efficient and reliable safety communications. By introducing suitable boundaries in dense vehicle networks, one can group vehicles into more manageable units to coordinate the vehicles in message transmissions and relaying, and to control the range and direction of message propagation. Local Peer Groups (LPGs) [4] is one group-based approach which builds tight coordination of vehicles in the immediate vicinity (i.e., intra-LPG communication) to support near-instantaneous safety applications. A Group Header (GH) vehicle is designated to support grouping vehicles together into an LPG and provides identification of the group via periodic beacons. Group Node (GN) vehicles periodically broadcast messages to the GH in order to maintain membership within the LPG. The overheard LPG maintenance messages are used within the group to proactively create the unicast and multicast routes to all nodes within the LPG, which can then be used for safety message dissemination. Some other routing schemes also leverage grouping:

- LOcation Routing Algorithm with Cluster-Based Flooding (LORA-CBF) [43] reactively discovers routes as needed and forwards between and within the cluster.
- Directional Propagation Protocol (DPP) [28] leverages directionality of data between groups to localize network collisions. For each cluster, DDP tracks the vehicles located at the cluster front and cluster tail. Data is forwarded within and between clusters in specific directions by passing between the head and tail of cluster as needed.
- Position-based Prioritized Clustering (PPC) [50] uses node position and priority as well as broadcast information priority to create the cluster structure.

Broadcast routing protocols can leverage the MAC layer to support a cross-layer approach. VANETs are expected to use some variation of IEEE 802.11 for medium access control. For example, DSRC/WAVE is envisioned to use IEEE 802.11p. However, these access protocols are contention-based and were built to optimize unicast performance. Since collision and interference are due to the shared wireless medium, several broadcast routing protocols utilize the MAC layer to provide more reliable broadcast:

- In Urban Multi-hop Broadcasting (UMB) [21], the transmitter assigns the farthest node to re-broadcast using new MAC layer messages such as Request To Broadcast and Clear To Broadcast. At intersections, roadside units act as repeaters to ensure all road segments at junctions receive the broadcast.
- In Streetcast [52], city street maps are used in selecting relay nodes, which can include roadside infrastructure points, and multicast RTS (Request to Send) is used at the MAC layer to ensure reliable transmission.
- Dynamic Backbone-Assisted MAC (DBA-MAC) [2] periodically selects high priority nodes within an interconnected set of nodes to be backbone nodes which are able to re-broadcast messages. The backbone nodes are chosen to be nodes farther apart to reduce hop count for broadcasts.
- Vehicle Reactive Routing protocol (VRR) [23] takes advantage of multiple channels supported by DSRC. It utilizes one channel for signaling such as route discovery and maintenance and another channel is used for data dissemination. In this protocol, each receiving node re-broadcasts to all nodes in its transmission range which were not in range of the transmitter node. For more dense and slow-moving vehicle scenarios,

the scheme implements relays which are chosen based on direction, speed and distance from the transmitter.

18.5 Conclusions

The emergence of vehicle ad hoc networks poses new challenges for information dissemination due to the unique network behavior of vehicles on roadways. This chapter provided an overview of the trends in routing protocols applicable to vehicle safety communications and non-critical content delivery applications. While there are a number of viable approaches, it is unlikely that one individual protocol will emerge as the leading solution for all applications and roadway environments. Instead a more adaptive approach is promising, where the routing protocol is dynamically selected to meet application needs as well as vehicle network and roadway characteristics. Thus there is still a need for solutions designed for vehicle ad hoc networks, leveraging innovative techniques.

References

[1] A. Bachir and A. Benslimane. A multicast protocol in ad hoc networks inter-vehicle geocast. In *Proceedings of the IEEE Vehicular Technology Conference*, 2003.

[2] L. Bononi and M. Di Felice. A cross layered mac and clustering scheme for efficient broadcast in vanets. In *Proceedings of the IEEE Conference on Mobile Adhoc and Sensor Systems*, 2007.

[3] W. Chen and S. Cai. Ad hoc peer-to-peer network architecture for vehicle safety communications. *IEEE Communications Magazine*, 43(4):100–107, 2005.

[4] W. Chen et al. Local peer group organization and architecture for vehicle communications. In *Proceedings of the Vehicle-to-Vehicle Communcations Workshop*, 2005.

[5] P. Cheng et al. Geodtn+nav: A hybrid geographic and dtn routing with navigation assistance in urban vehicular networks. In *Proceedings of the International Symposium on Vehicular Computing Systems*, 2008.

[6] J. Chennikara-Varghese et al. Local peer groups and vehicle-to-infrastructure communications. In *Proceedings of the IEEE Workshop on Automotive Networking and Applications*, 2007.

[7] T. Clausen and P. Jacquet. Optimized link state routing protocol (OLSR). IETF RFC 3626, 2003.

[8] M. Dikaiakos et al. Location-aware services over vehicular ad-hoc networks using car-to-car communication. *IEEE Journal on Selected Areas in Communications*, 25(8), 2007.

[9] H. Fuessler et al. Contention-based forwarding for mobile ad hoc networks. *Ad Hoc Networks 1 Journal*, 1(4), 2003.

[10] F. Granelli et al. Mora: A movement-based routing algorithm for vehicle ad hoc networks. In *Proceedings of the IEEE Workshop on Automotive Networking and Applications*, 2006.

[11] P. Jacquet et al. Multicast optimized link state routing. IETF Internet Draft, draft-ieft-manet-olsr-molsr-01.txt, 2001.

[12] M. Jerbi et al. Gytar: Improved greedy traffic aware routing protocol for vehicular ad hoc networks in city environments. In *Proceedings of the ACM International Workshop on Vehicular Ad hoc Networks*, 2006.

[13] D. Johnson et al. The dynamic source routing protocol for mobile ad hoc networks (DSR). IETF Internet Draft, draft-ietf-manet-dsr-10.txt, 2004.

[14] B. Karp and H. T. Kung. Gpsr: Greedy perimeter stateless routing for wireless networks. In *Proceedings of the ACM Conference on Mobile Computing and Networking*, 2000.

[15] M. Kasemann et al. A reactive location service for mobile ad hoc networks. Technical Report TR-14-2002, Department of Computer Science, University of Mannheim, 2002.

[16] W. Kiess et al. Hierarchical location services for mobile ad hoc networks. *Mobile Computing and Communications Review*, 1(2), 2004.

[17] T. Kitani et al. Efficient vanet-based traffic information sharing using buses on regular routes. In *Proceedings of the IEEE Vehicular Technology Conference*, 2008.

[18] Y. Ko and N. Vaidya. Location-aided routing (LAR) in mobile ad hoc networks. In *ACM MobiCom*, 1998.

[19] Y. Ko and N. Vaidya. Geocasting in mobile ad hoc networks: Location-based multicast algorithms. In *Proceedings of the Workshop on Mobile Computing Systems and Applications*, 1999.

[20] Y. Ko and N. Vaidya. Geotora: A protocol for geocasting in mobile ad hoc networks,. In *Proceedings of the International Conference on Network Protocols*, 2000.

[21] G. Korkmaz et al. Urban multihop broadcast protocol for intervehicle communication systems. In *Proceedings of ACM VANET*, 2004.

[22] T. Kosch et al. Information dissemination in multihop inter-vehicle networks - adapting the ad-hoc on-demand distance vector routing protocol (aodv). In *Proceedings of the IEEE 5th International Conference on Intelligent Transportation Systems*, 2002.

[23] M. Koubek, S. Rea, and D. Pesch. Effective emergency messaging in wave-based vanets. In *Proceedings of the International Conference on Wireless Access in Vehicular Environments*, 2008.

[24] P. Kumar et al. Ddor: Destination discovery oriented routing in highways/freeways vanets. In *Proceedings of the IEEE Asia-Pacific Service Computing Conference*, 2008.

[25] I. Leontiadis and C. Mascolo. Geopps: Geographical opportunistic routing for vehicular networks. In *Proceedings of the IEEE International Symposium on a World of Wireless Mobile and Multimedia Networks*, 2007.

[26] J. Li et al. A scalable location service for geographic ad hoc routing. In *Proceedings of the ACM Conference on Mobile Computing and Networking*, 2000.

[27] W-H. Liao et al. Geogrid: A geocasting protocol for mobile ad hoc networks based on grid. *Journal Internet Technology*, 1(2), 2000.

[28] T. Little and A Agrawal. An information propagation scheme for vanets. In *Proceedings of the IEEE Conference on Intelligent Transport Systems*, 2005.

[29] C. Lochert et al. A routing strategy for vehicular ad hoc network in the city environments. In *Proceedings of the IEEE Intelligent Vehicles Symposium*, 2003.

[30] C. Lochert et al. Geogrpahic routing in city scenarios. *Mobile Computing and Communications Review*, 9(1), 2005.
[31] P. Luo et al. Performance evaluation of vehicular dtn routing under realistic mobility models. In *Proceedings of the IEEE Wireless Communications and Networking Conference*, 2008.
[32] M. Mauve et al. A survey of position-based routing in mobile ad-hoc networks. *IEEE Network Magazine*, 15(6), 2001.
[33] M. Mauve et al. Position-based multicast routing for mobile ad-hoc networks. Technical Report TR-03-004, Department of Computer Science, University of Mannheim, 2003.
[34] H. Menouar, M. Lenardi, and F. Filali. Movement prediction-based routing (MOPR) concept for position-based routing in vehicular networks. In *Proceedings of the IEEE Vehicular Technology Conference*, 2007.
[35] D. Nain et al. Integrated routing and storage for messaging applications in mobile ad hoc networks. *Mobile Networks and Applications*, 9(6), 2004.
[36] V. Namboodiri, M. Agarwal, and L. Gao. A study on the feasibility of mobile gateways for vehicular ad-hoc networks. In *Proceedings of the ACM International Workshop on Vehicular Ad Hoc Networks*, 2004.
[37] V. Naumov and T. Gross. Connectivity-aware routing (CAR) in vehicular ad hoc networks. In *Proceedings of the IEEE International Conference on Computer Communications*, 2007.
[38] M. Nekovee and B. Bogason. Reliable and efficient information dissemination in intermittently connected vehicular ad hoc networks. In *Proceedings of the IEEE Vehicular Technology Conference*, 2007.
[39] J. Nzouonta et al. Vanet routing in city roads using real-time vehicular traffic information. *IEEE Transactions on Vehicular Technology*, 58(7), 2009.
[40] C. Perkins et al. Ad hoc on-demand distance vector (AODV) routing. IETF RFC 3561, 2003.
[41] E. Royer and C. Perkins. Multicast operation of the ad-hoc on-demand distance vector routing protocol. In *Proceedings of Mobicom*, 1999.
[42] M. Saito et al. Evaluation of inter-vehicle ad-hoc communication protocol. In *Proceedings of the IEEE International Conference on Advanced Information Networking and Applications*, 2005.
[43] R.A. Santos, O. Alvarez, and A. Edwards. Performance evaluation of two location-based routing protocols in vehicular ad-hoc networks. In *Proceedings of the IEEE VTC*, 2005.
[44] B-C. Seet et al. A-star: A mobile ad hoc routing strategy for metropolis vehicular communications. In *Proceedings of Networking2004*, 2004.
[45] T. Spyropoulos, K. Psounis, and C. Raghavendra. Spray and wait: An efficient routing scheme for intermittently connected mobile networks. In *Proceedings of the ACM SIGCOMM Workshop on Delay Tolerant Networking*, 2005.
[46] M. Sun et al. GPS-based message broadcasting for inter-vehicle communication. In *Proceedings of the International Conference on Parallel Processing*, 2000.
[47] J. Tian et al. Spatially aware packet routing for mobile ad hoc inter-vehicle radio networks. In *Proceedings of the IEEE International Conference on Intelligent Transportation Systems*, 2003.
[48] O. Tonguz et al. Broadcasting in VANET. In *Proceedings of IEEE INFOCOM*, 2008.

[49] A. Vahdat and D. Becker. Epidemic routing for partially connected ad hoc networks. Technical Report CS-200006, Department of Computer Science, Duke University, 2000.

[50] Z. Wang et al. A position-based clustering technique for ad hoc intervehicle communication. *IEEE Transactions on Systems, Man and Cybernetics, Part C: Application and Reviews*, 38(2), 2008.

[51] H. Wu et al. Mddv: A mobility-centric data dissemination algorithm for vehicular networks. In *Proceedings of the ACM Workshop on Vehicular Ad Hoc Networks*, 2004.

[52] C-W. Yi et al. Streetcast: An urban broadcast protocol for vehicular ad-hoc networks. In *Proceedings of IEEE VTC*, 2010.

[53] Y. Yi et al. On-demand multicast routing protocols (ODMRP) for ad hoc networks. IETF Internet Draft, draft-ieft-manet-odmrp-04.txt, 2002.

[54] D. Yu and Y-B Ko. FFRDV: Fastest-ferry routing in dtn-enabled vehicular ad hoc networks. In *Proceedings of the IEEE International Conference on Advanced Communication Technology*, 2009.

[55] J. Zhao and G. Cao. VADD: Vehicle-assisted data delivery in vehicular ad hoc networks. In *Proceedings of the IEEE INFOCOM*, 2006.

[56] W. Zhao et al. A message ferrying approach for data delivery in sparse mobile ad hoc networks. In *Proceedings of ACM MOBIHOC*, 2004.

19

Introduction to Vehicular Network Applications

Han-You Jeong[1], Dong-Hyun Je[2] and Seung-Woo Seo[2]

[1]*Institute of Logistics Information Technology, Pusan National University, Republic of Korea; e-mail: hyjeong@pusan.ac.kr*
[2] *School of Electrical Engineering and Computer Science, Seoul National University, Republic of Korea; e-mail: dhje@cnslab.snu.ac.kr, sseo@snu.ac.kr*

Abstract

This chapter gives an overview of a plethora of vehicular network applications which are expected to reduce the high number of fatalities and injuries originating from accidents in road environments. The objectives of these applications are three-fold: (1) to enhance the traffic safety through a mitigation of a vehicle collision; (2) to improve the traffic flow efficiency with cooperative information sharing; and (3) to reduce the workloads of drivers. We summarize the standard activities and the major research and development projects of vehicular network applications, and then discuss some technical challenges that must be addressed to disseminate these applications in the foreseeable future.

Keywords: vehicular network application, safety, non-safety, comfort.

19.1 Introduction

The astronomical losses due to traffic congestion and accidents, such as the high number of casualties and the corresponding social and economic expenses, have stimulated the study of the *intelligent transportation sys-*

Anand R. Prasad et al. (Eds.), Future Internet Services and Service Architectures, 413–433.

tems (ITS) that integrate transportation infrastructure with information and communication technology (ICT) to reduce the cost of the above problems. These trends have led to an increase of the consortia consisting of government agencies, the automotive industry, and academia, which have actively participated in many research and development (R&D) projects and standardization efforts.

Recent advances in automotive electronics and wireless communication technologies have encouraged the study of the vehicular networks. Vehicular networks are a new kind of wireless networks that are spontaneously organized among neighboring vehicles as well as between vehicles and road-side infrastructure. In particular, a wireless interface in a moving vehicle, called an *on-board unit* (OBU), communicates with other OBUs in different vehicles or with a fixed *road-side unit* (RSU) attached to an infrastructure network. Many vehicular network applications make use of these cooperative communication technologies, such as *vehicle-to-infrastructure* (V2I), *infrastructure-to-vehicle* (I2V), and/or *vehicle-to-vehicle* (V2V) communications. The objectives of these applications are three-fold: (1) to mitigate the chance of a vehicle collision; (2) to improve the traffic flow efficiency via cooperative information sharing; and (3) to reduce the workloads of drivers. In this chapter, we focus on the overview of vehicular network applications proposed so far.

This chapter discusses the following issues of vehicular network applications: Section 19.2 classifies state-of-the-art vehicular network applications into safety, non-safety and comfort applications. In Section 19.3 we explain the vehicular network architecture to support these applications. Then, Section 19.4 summarizes the standardization activities and the major R&D projects relevant to the vehicular network applications. Section 19.5 briefly remarks on the technical challenges to overcome the major obstacles to the proliferation of these applications. Finally, concluding remarks are presented in Section 19.6.

19.2 A Taxonomy of Vehicular Network Applications

Recent advances in global navigation and satellite systems (GNSS) and wireless communication technologies have stimulated many research works into vehicular network applications. These applications include *safety applications* which aim to prevent vehicle collisions; *non-safety applications* which enhances road traffic efficiency with cooperative information sharing; and *comfort applications* which aim at reducing the workload of the driver.

Table 19.1 Characteristics of vehicular network applications.

Application	Comm. Type	Comm. Range	Min. Freq.	Max. Delay	Priority
Safety	periodic, event-driven	100 m ∼ 1000 m	1 Hz ∼ 50 Hz	100 msec ∼ 1 sec	high
Non-safety	periodic, event-driven	100 m ∼ 500 m	1 Hz	1 sec	medium
Comfort	periodic, event-driven	10 m ∼ 1000 m	< 1 Hz	> 1 sec	low

Table 19.1 summarizes the characteristics of vehicular network applications. All safety applications have a communication range between 100 and 1000 m. Some applications periodically send a message with a minimum frequency between 1 and 50 Hz, whereas other applications send a message once an event happens. Safety applications usually have strict delay requirements with an allowable latency between 100 msec and 1 sec; as a result, safety applications are assigned to the *highest priority* among all vehicular applications. Non-safety applications have a communication range between 100 and 500 m. Some applications send messages based on an event driven transmission mode. In comparison to the safety application, most non-safety applications are relatively delay tolerant with an allowable latency within 1 sec. As a result, the non-safety applications are usually assigned to a medium priority among all vehicular applications. Comfort applications have a communication range between 10 to 1000 m. Most of the comfort applications, except for those indirectly related to safety applications, are event-driven and delay-tolerant. Consequently, they are usually assigned to the lowest priority among all vehicular applications.

In the following, we will explain the details of each category of vehicular application.

19.2.1 Safety Applications

In 2004, the World Health Organization (WHO) estimated that 1.2 million people were killed and more than 50 million were injured on the road every year [3]. This high number of casualties, as well as astronomical economic losses around the world, makes these safety applications a major concern of government agencies and automotive industry. The recent emergence of the vehicular networks has stimulated many researchers to study how such communication technologies can be utilized to enhance traffic safety. Government agencies collaborated with car manufacturers to create consortia for the design and development of safety-related vehicular applications.

In this section, we classify a plethora of vehicular safety applications in the final report of the Vehicle Safety Communications Consortium of the Crash Avoidance Metrics Partnership (CAMP) [4] into five different categories:

1. sign/signal notification and violation warning;
2. driving safety assistant;
3. visibility warning/assistant;
4. public safety; and
5. collision warning.

Each of these categories will be addressed in the following sections.

19.2.1.1 Sign/Signal Notification and Violation Warning

In these safety applications, RSUs located in the area vehicles are approaching, such as school zones, curves, intersections, and so on, use I2V communications to send a *periodic* messages to the vehicles near their zones to inform/warn them about the situation at that time. Some examples of these applications are briefly explained as follows:

1. *In-vehicle signage*: The in-vehicle signage system is designed to provide drivers with information on traffic signs along the road via *one-way I2V communications*.
2. *Curve speed warning*: The curve speed warning is designed to alert drivers, via *one-way I2V communications*, if they are going too fast.
3. *Traffic signal violation warning*: The traffic signal violation warning is designed to warn drivers via *one-way I2V communications* when the system detects that the vehicle is in danger of not complying with a traffic signal.
4. *Stop sign violation warning*: The stop sign violation warning is designed to warn drivers via *one-way I2V communications* when the system detects the vehicle is in danger of disregarding a stop sign.

19.2.1.2 Driving Safety Assistant

The driving safety assistant system collects data about all vehicles approaching certain zones, such as an intersection or a highway ramp, using sensors working at these zones or vehicular communications. In this system, the driver will be *periodically* informed and warned about vehicles, coming from different directions to the zones, which could possibly collide. Some examples of these applications are described as follows:

1. *Stop sign movement assistant*: The stop sign movement assistant system is designed to warn drivers, via *one-way V2I and I2V communications*, if they are about to cross through an intersection after having stopped at a stop sign.
2. *Left turn assistant*: The left turn assistant system is designed to inform drivers, via *one-way V2I and I2V communications*, of oncoming vehicles to help them to make a safe left turn without left turn arrow phase.
3. *Highway merge assistant*: The highway merge assistant system is designed to warn drivers on a highway inlet, via *one-way V2V communications*, if another vehicle is in its merge path.

19.2.1.3 Visibility Warning/Assistant

These applications, except for emergency electronic brake lights, receive *periodic* updates of the position, direction and speed of surrounding vehicles via V2V communications. Based on this information, drivers mitigate their limitations by broadening their awareness of road environments. These improvements can lead to a reduction of vehicle collision probability. Some examples of these applications are defined in the following:

1. *Visibility enhancement*: The visibility enhancement system is designed to enhance awareness in poor visibility situations, such as caused by snow, heavy rain, via *one-way V2V communications*.
2. *Lane change warning*: The lane change warning is designed to warn drivers, via *one-way V2V communications*, if an intended lane change may lead to a crash with a nearby vehicle.
3. *Blind spot warning*: The blind spot warning is designed to warn driver via *one-way V2V communications* when he(she) intends to make a lane change while his(her) blind spot is occupied by another vehicle.
4. *Emergency electronic brake lights*: This system is designed to alert drivers, via *one-way V2V communications*, if a vehicle in front of them is braking hard.

19.2.1.4 Public Safety

The public safety applications include *event-driven* V2I, I2V, and V2V communications. The goal of these applications is to assist the emergency team to minimize their travel time as well as to help drivers in need of aid. Some examples of public safety applications are described in the following:

1. *Approaching emergency vehicle warning*: This system is designed to warn drivers, via *one-way V2V communications*, to yield the right of way to an approaching emergency vehicle.
2. *Emergency vehicle signal preemption*: This application is designed for an emergency vehicle to request all traffic lights to be green phased in its direction via *two-way V2I and I2V communications.*
3. *Post-crash warning*: The post-crash warning is designed to help drivers approaching a disabled vehicle to avoid secondary collisions via *one-way V2I or V2V communications.*

19.2.1.5 Collision Warning

The collision warning systems *periodically* gather data about all vehicles approaching the zones, such as the intersections and vehicle presence in the vicinity, using sensors and/or vehicular communications. Based on these data, these systems check if there is a chance of a collision at these zones. If there is such a situation, they immediately send warning messages to the vehicles that might be involved in the collision. Two examples of these applications are addressed in the following:

1. *Intersection collision warning*: The intersection collision warning system is designed to warn drivers, via *one-way I2V communications*, if there is a possibility of a collision at an intersection.
2. *Cooperative collision warning*: The cooperative collision warning is designed to warn drivers, via *one-way V2V communications*, if an accident is about to occur based on the information from nearby vehicles.

19.2.2 Non-Safety Application

A rapid increase of metropolitan populations around the world causes severe traffic congestion. However, increasing the capacity of a road is not a good solution to this problem, not only because it takes at least several years to construct a highway/arterial, but also because it requires a huge amount of social overhead capital (SOC). As a result, many governments have investigated an alternative solution to enhance road traffic efficiency in their transportation systems, which is known as the ITS.

In non-safety vehicular applications, vehicles as well as the infrastructure *cooperatively* communicate with one another to better support the functions of the ITS. Non-safety application can be classified into three different categories: (1) traffic management; (2) tolling; and (3) infotainment application. Each of these categories will be addressed in the following sections.

19.2.2.1 Traffic Management

Traffic congestion has been one of the major ITS problems, due to the rapid increase of population growth, industrialization, and urbanization. Traffic congestion also leads to longer driving time, more fuel consumption and air pollution. Consequently, traffic congestion causes high inefficiency in transportation infrastructure, as well as financial loss in the economy. To alleviate this problem, the traffic management application has been introduced in the literature. Some examples of these applications are briefly described as follows:

1. *Traffic monitoring system*: The traffic monitoring system collects traffic data from the static sensing infrastructure such as RSUs or installed sensors in a specific location or moving vehicles equipped with OBU via *one-way V2I communication.*
2. *Traffic information application*: The traffic information application generates and broadcasts the traffic information to the drivers via *I2V communications*, so that the vehicles can adaptively move the fastest path to the destination.
3. *Intelligent traffic flow control*: This application utilizes facilities traffic light signal phasing based on real-time traffic flow *flow V2I communications.*

19.2.2.2 Tolling

Until now, a few countries installed tolling systems in their highways. For example, Norway firstly introduced the tolling system in Bergen in 1986. The United States has a widely spread tolling system over several states, and Austria installed the e-TAG system [17]. Also, Korean Expressway Corporation has adopted tolling system called “Hi-Pass” [18].

The tolling system is designed to alleviate the bottleneck at a toll gate by collecting tolls via V2I communications. Without the tolling system, a vehicle has to slow down speed, stop at the gate, roll down a window, and pay the fare. Afterward, the vehicle can pass the gate. But, because the tolling system enables the vehicle to pass at the gate without stopping, not only the processing time is reduced, but also the throughput is greatly increased. As a result, the tolling system can greatly reduce traffic congestion at the toll gate. According to the operating results presented in [18], the tolling system has three times the throughput of a manual toll collection system.

19.2.2.3 Infotainment Application

The goal of the infotainment application in vehicular networks is to enable the driver to access internet via *V2I* and *I2V communications* as well as communicate with the other vehicles via *V2V communications*. A straightforward approach to application provisioning via *V2I* and *I2V communications* is to make use of the existing network infrastructures, such as cellular networks, mobile WiMAX, and IEEE 802.11 WLAN technologies. To access these infrastructures, the OBU inside a vehicle can play the role of gateway to communicate with a RSU. The driver and passengers can download the road map, enjoy entertainment contents, and do their work while driving [36]. In case of *V2V communications*, the driver can exchange instant messages with his or her neighbor vehicles.

19.2.3 Comfort Application

The goal of the comfort applications is to reduce the workload of a driver. In the past few decades, many car manufacturers have adopted comfort applications to help a driver to efficiently and safely drive a car. In these applications, traffic information sensed from transportation elements such as vehicles, road, and routes is gathered in the infrastructure and then is broadcast to the vehicles. One of the simplest approaches to this application provisioning is to make use of the existing network infrastructures, such as cellular networks, mobile WiMAX [19], and IEEE 802.11 WLAN technologies [20].

Comfort applications can be classified into four categories:

1. cooperative adaptive cruise control (Cooperative ACC);
2. automotive navigation;
3. ecological drive assistance; and
4. remote diagnosis.

19.2.3.1 Cooperative Adaptive Cruise Control (Cooperative ACC)

Adaptive cruise control (ACC) is designed to prevent drivers from vehicle collision as well as to reduce the workload of the driver. Radar at the front of the hood of a vehicle collects the driving information, such as position, velocity, acceleration, and direction of the vehicle in front. If the front vehicle drives within the preset distance, the ACC automatically maintains the safety distance by controlling the acceleration and the brake based on the measured information. In the recent several years, vehicle manufacturers have adopted this system in their high grade vehicles [13, 14, 15].

Cooperative ACC is the advanced version of the conventional ACC. Different from the conventional ACC, the cooperative ACC incorporates both the measured information and the received information from the vehicle in front. In order to do that, the OBU of a vehicle exchanges driving information with its neighbor vehicles through vehicular communications.

For example, the PATH projects in [5] have shown that the freeway capacity of platoons consisting of 20 vehicles can be increased by four times that of ordinary driving, thanks to the sharing of driving information among high-speed vehicles via V2V communications. In addition, cooperative ACC can prevent the driver from breaking the driving regulations by exchanging the information on a speed limit, road state, or weather conditions.

19.2.3.2 Automotive Navigation

The automotive navigation is a system that guides a driver to the fastest/shortest route to the preset destination based on the position of a vehicle obtained from the GNSS. To do this, this system maps the current location information into the geographic information system (GIS) stored in a non-volatile memory. With vehicular communications, the automotive navigation system can obtain information on real-time traffic conditions. This information can be used to suggest a better alternative route to the destination. In addition, vehicular communication enables this system to provide driving-related information such as gas station, parking lots, and even traffic regulations or location information of a building or a street address.

19.2.3.3 Ecological Driving Assistance

Ecological driving assistance has received considerable attention as a solution to save on the energy consumption of a vehicle, as well as to protect nature. Based on the agreement of the Kyoto protocol in 2005, the car manufacturers are obliged to develop the ecological driving assistance system to reduce the emission of carbon dioxide [21]. This system is designed to satisfy the following driving rules:

1. avoid sudden acceleration and braking;
2. regularly check tire pressure;
3. avoid changing driving speed and maintain a constant speed; and
4. drive with the traffic flow. By integrating these rules, the ecological driving assistance system can reduce the gas emission, save energy, and lengthen car durability. For example, the KIA motors has shown that the

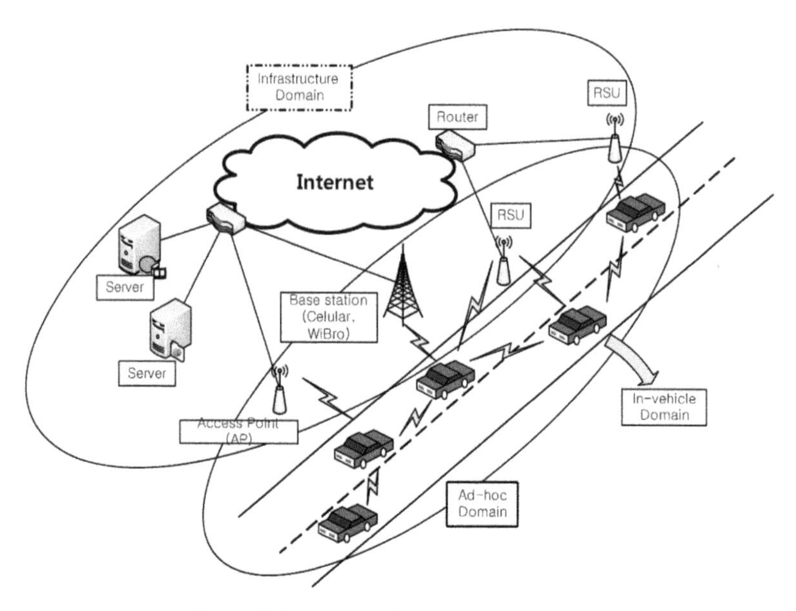

Figure 19.1 A vehicular network architecture.

ecological driving assistance system can reduce fuel consumption by 10 to 30% [16].

19.2.3.4 Remote Diagnosis

The remote diagnosis application provides diagnosis via an in-vehicle display panel as well as a remote diagnosis application. This application consists of an in-vehicle component, car server, central server, in-vehicle network, and vehicular network. The in-vehicle component, also known as an electronic control unit (ECU), is an embedded microprocessor to control engine, braking system, tires, air-conditioner, and so on. They are connected via in-vehicle networks such as controller area networks (CANs) [8] and FlexRay [10]. If the ECU detects a fault, it sends a diagnosis message to the car server which manages all components of the vehicle. Then, the car server sends this message to the central server via V2I or V2V communications. In the meantime, the car server can also display the fault information on the screen panel of the vehicle. Finally, the central server can notify the fault information to the car owner via the network infrastructure.

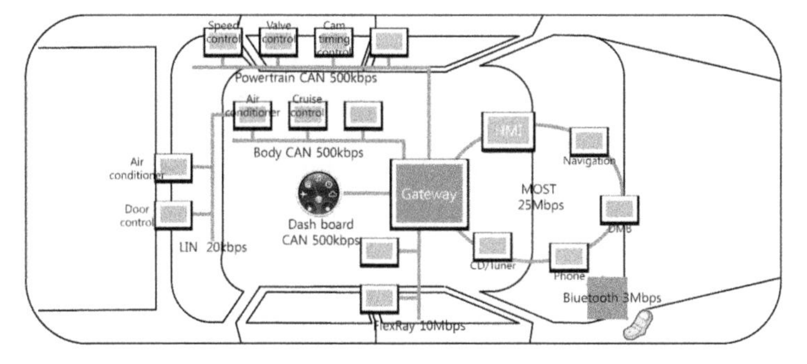

Figure 19.2 An in-vehicle architecture.

19.3 Vehicular Network Architectures

In comparison to existing wireless networks, the vehicular networks have the following unique attractive features [6]: unlimited transmission power, higher computation capability, and predictable mobility. On the other hand, they also face some challenging characteristics [7]: potentially large scale, high mobility, partitioned network, and intermittent connectivity. Taking these characteristics of vehicular networks into account, many researchers have carefully designed the architecture of vehicular networks. Figure 19.1 shows the illustration of vehicular network architecture to support various vehicular network applications. In this section, we briefly describe the vehicular network architecture consisting of in-vehicle, ad-hoc, and infrastructure domains.[1]

19.3.1 In-Vehicle Domains

The in-vehicle domain refers to a local bus network architecture within a vehicle, such as controller area networks (CANs) [8, 9] and FlexRay [10, 11], consisting of OBU and multiple ECUs as shown in Figure 19.2. The CAN is designed to efficiently communicate real-time control data with a high secrecy level. Bosch developed the CAN in 1983 and officially released it in 1986. Afterward, the CAN 2.0 specification was published in 1991. The CAN consists of the object layer for filtering messages, the transfer layer for transfer protocol, and the physical layer to actually transfer the bits data. The CAN features the messages prioritization, latency guarantee, flexible config-

[1] For the readers who are interested in the detailed architecture of the vehicular networks, we refer to the previous chapter.

uration, data consistency, automatic retransmission, etc. The CAN connects the various kinds of ECUs such as airbags, antilock, cruise control, window, doors, mirror adjustment, audio system, and navigation system as shown in Figure 19.2.

FlexRay is the communication system for the advanced automotive network communication protocol developed by the FlexRay consortium [11]. The FlexRay consortium finalizes the FlexRay development in 2009 and releases the FlexRay Communication System Specification Version 3.0. The FlexRay is designed for a robust, scalable, deterministic and fault-tolerant in-vehicle system for automotive application. The core members of the FlexRay consortium are Bayerische Motoren Werke AG (BMW), DaimlerChrysler (DC), General Motors (GM), Volkswagen (VW), Freescale, Philips, Bosch). As a convergence protocol combining time- and event-triggered communication, the FlexRay protocol provides a fault-tolerance system by supporting single or double channel, fault-tolerant clock synchronization using global time-base, collision-free bus access, and high data rates. Because of higher bandwidth requirements, the FlexRay is expected to replace the CAN. Even though the CAN protocol has the non-deterministic feature based on Carrier Sense Multiple Access (CSMA), the FlexRay protocol has deterministic bus access based on Time Division Multiple Access (TDMA) where the message delivery time on the bus is predictable.

19.3.2 Ad-hoc Domain

In the ad-hoc domain, the network consists of multiple vehicles equipped with OBUs including a wireless communication interface. Nodes, by themselves, are supposed to make a communication network, and are able to communicate with each other via intermittently connected wireless links regardless of network infrastructures. The absence of the network infrastructure, called a decentralized wireless network, leads to network configuration issues such as routing protocol, identification, and so on. During the last decade, several kinds of ad-hoc network have been introduced, and they are classified into mobile ad-hoc networks (MANETs), wireless mesh/sensor networks, and vehicular networks according to the network characteristics such as nodes mobility, processing power, power efficiency requirement, and so on. Among them, the vehicular network is a kind of mobile ad hoc network which has the following characteristics: the network has a wide range from small in rural areas to large in urban areas. The mobility is very high, but the pattern of mobility is constrained in the location topology. In addition, the vehicle can provide high computation power and storage capability as well as abundant energy for wireless communication.

The vehicular network can be configured by wireless communication technologies such as WiFi (IEEE 802.11 b/g) [20], WiMAX (IEEE 802.16e) [19], and so on. Especially, Wireless Access in Vehicular Environments (WAVE), IEEE 802.11p, has received more attention than other wireless technologies, because the WAVE is specially designed for automotive applications [22]. To communicate via the WAVE, the vehicles are equipped with the OBU including the WAVE component. By using the WAVE, the vehicles can configure the ad hoc network, communicate with neighbor vehicles and exchange vehicle information messages including the position, velocity, and acceleration for road safety applications such as the collision avoidance application.

19.3.3 Infrastructure Domain

In the infrastructure domain, the RSUs are connected to the infrastructure networks, such as the Internet. In the absence of the RSUs, a vehicle may make use of the existing public or commercial networks, such as IEEE 802.11p WLAN, IEEE 802.16e WiMAX and the cellular radio networks [19, 20]. In the transition periods, it is very important to design an application architecture that fully utilizes the existing network infrastructure.

One of the most promising vehicular network architectures is the system combination of IEEE P1609 (WAVE) [24, 25, 26], IEEE 802.11p, and dedicated short-range communications (DSRC) [12]. The WAVE architecture covering from network layer to application layer manages network resources for efficient transmission, supports security applications such as link level encryption and authentication, and provides networking applications including addressing and routing for the vehicle mobility. IEEE 802.11p and DSRC define physical and data link layers. The DSRC includes 5.9 GHz range with seven 10 MHz channels, 300 m short range radio, etc. Also, IEEE 802.11p defines MAC layer architecture which supports IP network and WAVE short Messages Protocol. In spite of the CSMA/CA wireless channel, 802.11p adopts MAC transmission priority based on 802.11e EDCA to transmit safety data as soon as possible. In addition, 802.11p includes security functions such as authentication and encrypting confidential data.

19.4 Standardization and R&D Projects

With the benefits of increased road safety and improved traffic flows, vehicular networks have recently received significant attention from academia, the automotive industry, and governments worldwide. As a result, there have been far-reaching activities not only in the standardizations but also in national and international projects. In this section, we will briefly introduce the standardization efforts and the major R&D projects.

19.4.1 Standardization Activities

DSRC are one-way or two-way short- to medium-range wireless channels designed for vehicular communications, as well as the corresponding suite of standards and protocols. In October 1999, the Federal Communication Commission (FCC) decided to allocate 75 MHz of spectrum around 5.9 GHz (5.850–5.925 GHz) primarily for use in safety applications but also for other applications, such as non-safety and comfort applications. The DSRC spectrum consists of seven 10 MHz wireless channels from Channel 172 (5.860 GHz) to Channel 184 (5.920 GHz). Here, the channel numbers are determined by the offset of the center frequency relative to 5.0 GHz in units of 5 MHz. The *control channel* (CCH) at Channel 178 (5.885–5.895 GHz) is restricted to safety applications only. In July 2006, the FCC designated two channels, 172 and 184, at the edges of DSRC spectrum, for accident avoidance and public safety applications, respectively. The remaining four channels are referred to as *service channels* (SCHs) which can be used for both safety and non-safety vehicular applications.

A WAVE system is a radio communication system intended to provide seamless and interoperable services to transportation systems. The IEEE working groups standardize a family of communication protocol stacks for the WAVE system, including IEEE 802.11p amendment [22] and a family of IEEE 1609 standards [23–27].

The popular IEEE 802.11 wireless local area network (WLAN) standards are being amended to define the PHY and MAC protocols for DSRC, which is known as *IEEE 802.11p* [22]. The original IEEE 802.11 WLAN standards are designed for a high-bandwidth data delivery service to a *nomadic* station. Instead, the IEEE 802.11p defines some amendments to better support vehicular applications, including V2V and V2I communications, to a *high-speed* station. To cope with frequent link disconnections and handoffs in vehicular environments, the IEEE 802.11p enhances the MAC functions IEEE 802.11 standard for adapting to time-varying channel condition: (1) IEEE 802.11p

MAC does not require authentication and association prior to being allowed to transmit/receive data; and (2) IEEE 802.11p MAC supports stateless channel access, caching for handoff and opportunistic frame scheduling.

A family of IEEE 1609 standards covers the remaining functionalities of the WAVE system based on IEEE 802.11p: (1) IEEE 1609.0 *Architecture* describes the WAVE architecture and services necessary for multi-channel DSRC/WAVE devices to communicate in a mobile vehicular environment [23]; (2) IEEE 1609.1 *Resource Manager* defines the services and interfaces of the WAVE Resource Manager application, including command message formats and the appropriate responses to those messages, data storage formats that must be used by applications to communicate between architecture components, and status and request message formats [24]; (3) IEEE 1609.2 *Security Services for Applications and Management Messages* defines secure message formats and processing, including the circumstances for using secure message exchanges and how those messages should be processed based upon the purpose of the exchange [25]; (4) IEEE 1609.3 *Networking Services* defines network and transport layer services, including addressing and routing, in support of secure WAVE data exchange [26]; and (5) IEEE 1609.4 *Multi-Channel Operation* provides DSRC frequency coordination and management to support WAVE operations [27].

In Europe, the European Telecommunications Standard Institute (ETSI), well known for its standards GSM and TISPAN, has created a technical committee for intelligent transportation systems (TC ITS). The TC ITS plays a key role in developing standards and specifications for the ITS services consisting of five working groups: WG 1 for user and application requirements; WG2 for architecture and cross-layer issues; WG 3 for transport and network; WG 4 for media and related issues; and WG 5 for security. In addition, the European industry standard for car-to-X communication based on WLAN technology is standardized by the car-to-car communication consortium (C2C-CC), an industry consortium initiated by six European vehicle manufacturers.

19.4.2 Major R&D Projects

Many non-profit consortia and organizations have been actively promoting many R&D projects of vehicular networks. In this section, we briefly address several R&D projects, for example, PReVENT [28], NoW [29], CVIS [30], SAFESPOT [31], COOPERS [32], SeVeCom [33], COMeSafety [34], and INTERSAFE-2 [35]:

1. *PReVENT* [28]: PReVENT is an R&D integrated project co-funded by the European Commission to contribute to road safety by developing and demonstrating preventive safety technologies and applications. The *preventive safety systems* exploit information, communication, sensor, and positioning technology to help drivers avoid or mitigate an accident. PReVENT ran from 2004 to 2008 and consisted of 13 subprojects. Two of them investigated the research issues related to vehicular networks: Wireless Local Danger Warning (WILLWARN) and Intersection Safety (INTERSAFE). The former studies a distributed system for a decentralized hazard warning application, whereas the latter focuses on the integration of the sensor and the communication technology to increase safety at intersections such as sign/signal notification and violation warning applications in Section 19.2.1.1.
2. *NoW* [29]: Network on Wheel (NoW) is a successor of the "FleeNet-Internet on the Road" project and ran from 2004 to 2008. The goal of NoW is to address technical issues on the communication protocols and data security for car-to-car communications and to submit the results to the standardization activities of the C2C-CC. NoW was founded by German car manufacturers to develop an open communication platform for supporting dual protocol stacks including the safety applications in Section 19.2.1 and the infotainment applications in Section 19.2.2.3. This platform includes the protocol stack and an open API, and provides a toolkit for application design, implementation and testing.
3. *CVIS* [30]: Cooperative Vehicle Infrastructure Systems (CVIS) is an R&D integrated project co-funded by the European Commission running from 2006 to 2010. The goal of CVIS is to design, develop and test the technologies needed to allow vehicles to communicate with each other and with the nearby roadside infrastructure. Especially, the CVIS project focuses on developing a unified network architecture for various networking technologies discussed in Section 19.3, e.g., cellular networks, WiMAX, and IEEE 802.11p. To overcome a number of non-technical obstacles, the CVIS project also creates a toolkit to address key deployment enablers, such as user acceptance, data privacy and security, system openness and interoperability, risk and liability, public policy needs, cost/benefit and business models, and roll-out plans for implementation.
4. *SAFESPOT* [31]: Cooperative vehicles and road infrastructure for road safety (SAFESPOT) is an R&D integrated project co-funded by the European Commission and is to run from 2006 to 2010. The objective

of SAFESPOT is to prevent road accidents via *safety margin assistant* to detect potentially dangerous situations in advance and extend the drivers' perception of the vehicle surroundings. To achieve this goal, SAFESPOT creates dynamic cooperative networks where the vehicles and the road infrastructure communicate to share information gathered from vehicles and roadside. Some examples of SAFESPOT applications are driving safety assistant applications as mentioned in Section 19.2.1.2 and collision warning applications such as in Section 19.2.1.5.

5. *COOPERS* [32]: Cooperative Systems for Intelligent Road Safety (COOPERS) is an R&D integrated project co-funded by the European Commission running from 2006 to 2010. The objective of the COOPERS project is the enhancement of road safety by direct communication of up to date traffic information between infrastructure and vehicles on a road. COOPERS focuses on the development of innovative telematics applications with the long term goal of a cooperative traffic management between vehicle and infrastructure, such as the traffic management applications in Section 19.2.2.1 and the automotive navigation application in Section 19.2.3.2.
6. *SeVeCom* [33]: Secure Vehicular Communication (SeVeCom) is an R&D project co-funded by the European Commission, running from 2006 to 2008. The objective of SeVeCom is to define the security architecture of such networks, as well as to propose a roadmap for integration of security functions in vehicular networks. In particular, this project focuses on providing a full definition and implementation of security requirements for the vehicular applications in Section 19.2, including both the security and privacy of V2V and V2I communication.
7. *COMeSafety* [34]: Communications for eSafety (COMeSafety) is a specific support action of the European Commission. This project, which ran from 2006 to 2009, supports the eSafety Forum with respect to all issues related to V2V and V2I communications as the basis for cooperative intelligent road transport systems. COMeSafety also provides a platform for both the exchange of information and the presentation of results from the major R&D projects, such as CVIS, SAFESPOT, PReVENT, SeVeCom, and others.
8. *INTERSAFE-2* [35]: INTERSAFE-2 is an R&D project co-funded by the European Commission, running from 2008 to 2011. This project aims to develop and demonstrate a Cooperative Intersection Safety System (CISS) that can significantly reduce injury and fatal accidents at intersections. INTERSAFE-2 goes beyond the scope of INTERSAFE

in the sense that it integrates warning and intervening functions in the vehicles. In INTERSAFE-2, different kinds of sensing technologies, such as radar, lidars, and cameras, recognize and track the objects at intersections, for example, vehicles, pedestrians and bicyclists. An example of the INTERSAFE-2 is the intersection collision warning in Section 19.2.1.5.

19.5 Technical Challenges

In this section, we briefly discuss some technical challenges to the proliferation of the vehicular network applications.

1. *PHY and MAC Protocol Design*: Similar to MANETs, vehicular networks are wireless multi-hop networks which can be potentially connected to a wired infrastructure. The PHY and MAC protocol should be designed to enable reliable multi-hop communication while adapting to the highly dynamic environment of vehicular networks. Also, these protocols should support the priority of applications to meet different quality-of-service (QoS) requirements of safety and non-safety applications. Another challenge in the PHY and MAC protocol design is that they support fast association and low handoff latency to cope with a rapid change in the wireless link condition.
2. *Dissemination and Routing*: Unlike most ad hoc networks that usually have a limited number of nodes, vehicular networks can have a diverse range of network size depending on the road traffic condition. For example, vehicular networks in urban areas are likely to form a highly dense network during the rush hour, whereas they may experience frequent network partitions in rural areas during late night hours. Furthermore, vehicular networks are expected to deal with a wide range of applications ranging from safety to infotainment. As a result, the routing and dissemination algorithms should be adaptive to these characteristics of vehicular networks and applications.
3. *Security*: Since many vehicular applications are designed to increase road safety, malicious behavior of a vehicular application may threaten the life of a driver and passengers. Besides, it is very important to design a security architecture providing secure communication between participants as well as authorized and secure service access. As a result, there is a strong research interest in the design of security mechanisms to

provide trust, authentication, access control, authorization, and privacy of the service access in vehicular network environments.

4. *IP Addressing and Mobility Management*: Non-safety and comfort applications such as vehicular Internet access have been considered to be the most promising applications in vehicular networks. However, there are two technical challenges threatening the service quality and continuity of high-speed vehicles: IP addressing and mobility management. In vehicular networks, these challenges should be addressed in an automatic and distributed manner. A number of standardization bodies, including IETF, C2C-CC, and ETSI TC ITS, currently focus on the IPv6 address auto-configuration and the network mobility solutions to efficiently support both V2I and V2V communications.

19.6 Conclusion

This chapter introduced state-of-the-art vehicular applications and then classified them into three categories: safety applications for mitigating the chance of a vehicle collision; non-safety application for increasing traffic flow on the road; and comfort applications for reducing the workload of a driver. The vehicular network architecture enabling those different vehicular network applications was also addressed by decomposing it into three domains: in-vehicle domain, ad-hoc domain and infrastructure domain. Then, the standardization activities and the major R&D projects were summarized, and then a brief remark on some technical challenges of proliferating the vehicular network applications was given in this chapter. It is our strong belief that we will witness an abundance of new vehicular network applications in the foreseeable future.

References

[1] M. Nekovee. Sensor networks on the road: the promises and challenges of vehicular ad hoc networks and vehicular grids. In *Proceedings of the Workshop on Ubiquitous Computing and e-Research,* Edinburgh, UK, May 2005.

[2] J. Blum, A. Eskandarian, and L. Hoffmman. Challenges of intervehicle ad hoc networks. *IEEE Trans. Intelligent Transportation Systems*, 5(4):347–351, December 2004.

[3] World Health Organization. World report on road traffic injury prevention. Online available at: http://www.who.int/violence_injury_prevention/publications/road_traffic/world_report/en/index.html.

[4] National Highway Traffic Safety Administration, CAMP Vehicle Safety Communications, Vehicle Safety Communications Project, Task 3 Final Report. Identify intelligent

vehicle safety application enabled by DSRC. DOT HS 809 859, National Highway Traffic Administration, Washington, DC, March 2005.

[5] U. Karaaslan, P. Varaiya, and J. Walrand. Two proposals to improve freeway traffic flow. Technical Report UCB-ITS-PRR-90-6, California Partners for Advanced Transit and Highways (PATH), Institute of Transportation Studies, UC Berkeley.

[6] M. Nekovee. Sensor networks on the road: the promises and challenges of vehicular ad hoc networks and vehicular grids. In *Proceedings of the Workshop on Ubiquitous Computing and e-Research*, Edinburgh, UK, May 2005.

[7] J. Blum, A. Eskandarian, and L. Hoffmman. Challenges of intervehicle ad hoc networks. *IEEE Trans. Intelligent Transportation Systems*, 5(4):347–351, December 2004.

[8] R. Bosch. Controller Area Network specification 2.0, 2001.

[9] International Organization for Standardization, Road vehicles.Controller area network (CAN). Part 1: Data link layer and physical signalling, ISO IS 11898-1, 2003.

[10] FlexRay. Available at: http://www.flexray-group.com.

[11] FlexRay Consortium, FlexRay Communications System. Protocol Specification, Version 2.0, June 2004.

[12] ASTM E2213-03, Standard Specification for Telecommunications and Information Exchange between Road-side and Vehicle Systems-5GHz Band Dedicated Short Range Communications (DSRC) Medium Access Control (MAC) and Physical Layer (PHY) Specifications, ASTM International, July 2003.

[13] Mercedes-Benz, http://www.mercedes-benz.com.

[14] Bayerische Motoren Werke AG, http://www.bmw.com.

[15] Hyundai Motor Company, http://worldwide.hyundai.com.

[16] H. Cho. Ecom driving System. Kia Motors Corporation, Sustainability Management Team, June 2008.

[17] e-TAG, http://www.roam.com.au.

[18] Hi-Pass, Korea Expressway Corperation, http://www.ex.co.kr/english/major_job/tra/major_job_tra_txt1.jsp.

[19] IEEE Standard for local and metropolitan area networks. Part 16: Air Interface for Broadband Wireless Access Systems, 2009.

[20] IEEE Standard for Information technology. Telecommunications and information exchange between systems. Local and metropolitan area networks. Specific requirements. Part 11: Wireless LAN Medium Access Control (MAC) and Physical Layer (PHY) Specifications, 2007.

[21] United Nations Framework Convention on Climate Change, http://unfccc.int.

[22] IEEE WG, IEEE 802.11p/D2.01, Draft amendment to part 11: wireless medium access control (MAC) and physical layer (PHY) specifications: Wireless access in vehicular environments, March 2007.

[23] IEEE WG, IEEE P1609.0/D02, Trial use standard for wireless access in vehicular environments (WAVE) – Architecture, March 2007.

[24] IEEE WG, IEEE 1609.1-2006, Trial use standard for wireless access in vehicular environments (WAVE) – Resource manager.

[25] IEEE WG, IEEE 1609.2-2006, Trial use standard for wireless access in vehicular environments (WAVE) – Security services for applications and management messages.

[26] IEEE WG, IEEE 1609.3-2007, Trial use standard for wireless access in vehicular environments (WAVE) – Networking services.

[27] IEEE WG, IEEE 1609.4-2006, Trial use standardfor wireless access in vehicular environments (WAVE) – multi-channel operations.
[28] PReVENT http://www.prevent-ip.org.
[29] NoW (Network-on-Wheels), http://www.network-on-wheels.de.
[30] CVIS (Cooperative Vehicle-Infrastructure Systems) http://www.cvisproject.org.
[31] SAFESPOT (Cooperative Vehicles and Road Infrastructure for Road Safety), http://www.safespot-eu.org.
[32] COOPERS (Cooperative Systems for Intelligent Road Safety), http://www.coopers-ip.eu.
[33] SeVeCom (Secure Vehicular Communications), http://www.sevecom.org.
[34] COMeSaftey (Communications for eSafety), http://www.comesafety.org.
[35] INTERSAFE-2, http://www.intersafe-2.eu.
[36] A. Qureshi, J.N. Carlisle, and J.V. Guttag. Tavarua: Video streaming with WWAN striping. In *ACM Multimedia*, pages 327–336. ACM, 2006.

20

A Clustering Algorithm for Vehicular Ad-Hoc Networks

Peng Fan[1], Abolfazl Mohammadian[2] and Peter C. Nelson[2]

[1]*Facebook, e-mail: pefa1@facebook.com*
[2]*Department of Computer Science, University of Illinois at Chicago, 851 S. Morgan St., Chicago, IL 60607 USA; e-mail: {kouros, nelson}@.uic.edu*

Abstract

Applications of Vehicular Ad Hoc Networks (VANETs) in the service of Intelligent Transportation Systems (ITS) have received much attention in recent years, which also has brought new challenges. Clustering is known to reduce the overhead of wireless communication in Mobile Ad Hoc Networks (MANETs) and provides an efficient hierarchical network structure. In this chapter, we propose a Directional Stability-based Clustering Algorithm (DISCA) especially for VANETs which takes direction, mobility features, and leadership duration into consideration. DISCA is distributed and enables the formation of stable clusters. An efficient cluster maintenance procedure adaptive to high mobility is also presented. Simulation experiments are conducted in a realistic vehicular network model to determine which algorithm provides optimal stability over the simulation timeline. The results show that DISCA significantly improves cluster stability. In addition, a DISCA-based broadcasting protocol is proposed to further validate the effectiveness of DISCA, and as a result it significantly reduces the number of rebroadcasts, while improving the reachability and latency in comparison to other protocols.

Anand R. Prasad et al. (Eds.), Future Internet Services and Service Architectures, 435–461.

Keywords: vehicle communication, traffic simulation, clustering algorithm, broadcasting, vehicular ad-hoc network.

20.1 Introduction

Vehicular Ad Hoc Networks (VANETs), an outgrowth of traditional Mobile Ad Hoc Networks (MANETs), provide one of the basic network communication frameworks for Intelligent Transportation Systems (ITS) applications. The U.S. Federal Communications Commission (FCC) has recently allocated the 5.85- 5.925 GHz portion of the spectrum to inter-vehicle communication (IVC) and vehicle-to-roadside communication (VRC) under the umbrella of dedicated short-range communications (DSRC) [15]. This has fuelled significant interest in applications of DSRC to driver-vehicle safety applications, infotainment, and mobile Internet services for passengers.

VANETs have significant advantages over MANETs. Vehicles can easily provide the power required for wireless communication devices and will not be seriously affected by the addition of extra weight for antennas and additional hardware. Furthermore, it can be generally expected that vehicles will have an accurate knowledge of their own geographical position, e.g., by means of Global Positioning Satellite (GPS). Thus, many of the issues making deployment and long-term use of ad hoc networks problematic in other contexts are not relevant in VANETs.

One of the main characteristics of ad-hoc networks is to use less specialized hardware for infrastructure support and leave the burden of network communication on the individual nodes within the network. After being deployed, the nodes form a chaotic unstructured flat network, which means that no efficient and reliable communication pattern is available. This flat structure is known to have a serious scalability problem [19]. To support reliable and efficient communication in large scale ad hoc networks, one frequent approach is to abstract a hierarchical clustering structure within the network. Clustering allows the formation of virtual communication backbone supporting efficient routing and broadcasting [27], and it improves the usage of scarce resources such as bandwidth. Typically, nodes are partitioned into clusters, each with a representative or clusterhead that coordinates and manages its cluster. To support the dynamic nature of MANETs and VANETs, clustering must be updated or reorganized quickly to reflect topological changes and node movements. Thus, ensuring stability is a very important challenge for clustering algorithms especially in highly dynamic networks. Therefore, a good clustering algorithm should not only focus on forming a minimal number of clusters as most of the existing algorithms, but also dynamically maintain the cluster structure without incurring a high communication overhead.

A significant amount of research [6, 11, 18] focuses on optimal methods for clustering nodes in MANETs. VANETs, however, have different characteristics with respect to clusterhead selection and network stability. VANETs must follow a more restrictive set of constraints than MANETs, and therefore can utilize specialized clustering algorithms. First, nodes or vehicles cannot randomly move within the physical space, but must instead follow constraints set in place by the existing road network topology. Second, vehicle movements follow well-understood traffic movement patterns. Each vehicle is constrained by the movements of surrounding vehicles. Third, vehicles generally travel in a single direction and are constrained to travel within a two-dimensional movement. Given these movement restrictions and the knowledge of position, velocity, and acceleration commonly available to onboard vehicle systems, it is desirable to pursue more intelligent and adaptive clustering methodologies for VANET environments. In this chapter, a directional stability-based clustering algorithm (DISCA) for VANET is proposed. DISCA is designed to stabilize the existing clusters of a desirable size by taking the vehicular mobility features into account, and also it is adaptive to node mobility. Thus, stable and robust clusters can be established and maintained. Simulation results show the improvement of both the average leadership and membership duration over other cluster algorithms. Additionally, DISCA has the minimal number of status changes per node in comparison to other algorithms.

The constrained environmental conditions of VANETs warrant a constrained simulation environment. Many simulation tools and environments have been designed for MANET implementations. These tools, however, fail to adequately model the needs of a VANET network. Compared to the random movements modeled in MANET environments, VANET simulation movements must behave according to traffic patterns in terms of car-following, directional movement, velocity, and acceleration. Current MANET simulation environments are not suitable for VANET simulations. Therefore, simulation of the network environment is best performed with traffic simulation tools. For the purpose of this study, an integrated simulation tool developed by Choffnes [13] was employed, after making the necessary modifications.

20.2 Related Work

The main goal of clustering algorithms is to establish and maintain an efficient cluster structure which provides a connected path and tends to cover the entire network. So a distributed clustering algorithm usually consists of two phases: cluster setup and cluster maintenance.

Attention on clustering in MANETs has increased considerably as wireless technologies improve and MANET theories become practice. In prior work, approaches proposed for clusterhead election are based on: (1) node ID or (2) node degree. Linked Cluster Architecture (LCA) [4] is one of the earliest ID-based clustering algorithms, which were intended to be applied for small networks of less than 100 nodes. LCA was later revised in [16] to reduce the number of clusters in the network. In the revised version, a node could only declare its leadership role if it has the lowest ID among the undecided nodes in its neighborhood, also known as Lowest-ID Clustering. In [24], the connectivity degree, instead of node ID, is used for clusterhead election. This is known as Highest-Degree Clustering in which a node with the highest connectivity degree, i.e., maximum number of neighbors is elected as a clusterhead, and the neighbors of a clusterhead become member nodes. However, this degree-based algorithm tends to have a large but very unstable cluster because the degree is more prone to topology changes than the ID.

Besides using either ID or degree to elect clusterheads, a generalized algorithm, Generalized Clustering Algorithm (GCA) is proposed in [8], which provides a generic method of preferences for choosing clusterheads. A detailed weighted clustering algorithm (WCA) is presented in [11] by taking multiple parameters such as node degree, speed and power into consideration for optimizing clusterhead election. In addition, a distributed version of GCA, namely Distributed Clustering Algorithm (DCA) is thoroughly discussed for the first time in [6], which also describes the dominance and independent properties that should be held by any ad hoc clustering algorithm. However, DCA has an unrealistic assumption that the network is static or quasi-static, and as a result it does not have a cluster maintenance procedure to deal with the topology changes caused by mobility.

In addition, many other clustering algorithms have been introduced for the purposes of solving problems under different MANET circumstances. For example, hierarchical clustering algorithms [2, 31] are focused on saving energy consumption in wireless sensor networks, and a distributed clustering algorithm in [25] aims to minimize the transmission power. However, as we discussed, energy consumption is not a concern in VANETs – therefore, it is not studied in our work.

20.3 Directional Stability-Based Clustering Algorithm (DISCA)

20.3.1 Design Philosophy

In VANETs, vehicles or mobile objects are not only highly dynamic but constrained by traffic behavior rules such as speed limits, intersections, tailgating restrictions etc., and by road network topology, such as direction, number and width of lanes, etc. Hence, VANETs are a more constrained environment and introduce new opportunities for designing more stable and efficient clustering algorithms. In this section, we describe our algorithm – DISCA. It takes multiple factors into consideration to dynamically form a stable cluster structure. Moreover, DISCA only requires 1-hop neighbor knowledge and incurs minimal overhead, and is specifically designed to ensure that the cluster structure is adaptive to mobility. Many algorithms proposed more recently require two-hop neighbor knowledge [2, 3, 10, 22] suggests that a rather simple clustering solution that can react quickly to topology changes in a mobile and distributed environment is desirable. Furthermore, the maintenance procedure of DISCA focuses on preserving the existing cluster structure as much as possible while keeping its validity.

The following four factors are utilized by the DISCA algorithm:

1. *Moving Direction* (MD): The moving direction is critical for forming a stable cluster because vehicles moving in different directions can hardly stay together for a long period of time. Each node has a mobility vector mV, and the *dot product* of $(mV1, mV2)$ gives their angle difference, so if the angle is within a predefined threshold, the two nodes are considered to have a similar moving direction.
2. *Leadership Duration* (LD): A node should join a clusterhead with longer leadership duration (the period that this node has been a leader since the last role change) due to a higher chance of staying together longer. Duration is a better indicator of the characteristics of good and stable connections among all its neighbors. LD refers to the value of duration time.
3. *Moving Pattern* (MP): A normal or steady moving pattern of a node is always a sign of good stability. For instance, a node moving on a moderate speed within a certain time period has a better chance to stay

together with its neighbors. Each node knows its mean speed mS over a certain time period, and if the condition stated below holds, this node is a "good" candidate:

$$|mS - \text{SpeedLimit} * \text{threshold}_{mS}| < \text{tolerance} \qquad (20.1)$$

The absolute difference between mS and a desirable speed indicates the degree of speed variance of a node. MP refers to the absolute value above.

4. *Projected Distance* (PD): A node with a small difference of projected distances among its neighbors over a certain time (e.g., an interval between beacons) implies how stable a node is with respect to all of its immediate neighbors. Each node is aware of its neighbors' current locations. Assume we have two neighboring nodes A and B with distance D_{AB}. Then, M_{AB} indicates whether they are moving away, moving towards or relatively stable from time step t to $t + 1$:

$$M_{AB} = D_{AB(t+1)} - D_{AB(t)}$$

Then, the local projected distance PD is calculated as the mean of D_{Ai} from node A to all its neighboring nodes 1 through i. If a node has a low PD (a predefined value, |PD| refers to the absolute value of PD), it may have a better chance to remain a stable relative distance to all its neighbors – making it a "good" candidate:

$$\text{PD} = \frac{1}{\text{degree}} \sum_{i=1}^{\text{degree}} M_{Ai} \qquad (20.2)$$

In summary, DISCA factors in all of the observations above with the goal to improve clustering perf3ormance with respect to stability and communication efficiency.

20.3.2 Algorithm Description

In this section, procedures for cluster setup and cluster maintenance in DISCA are presented in detail.

DISCA is initiated by each node and keeps running as long as the node is active. The node state transition graph is shown in Figure 20.1.

As a member node moves around, it decides which cluster it belongs to and what role it currently plays solely based on its local knowledge. Each

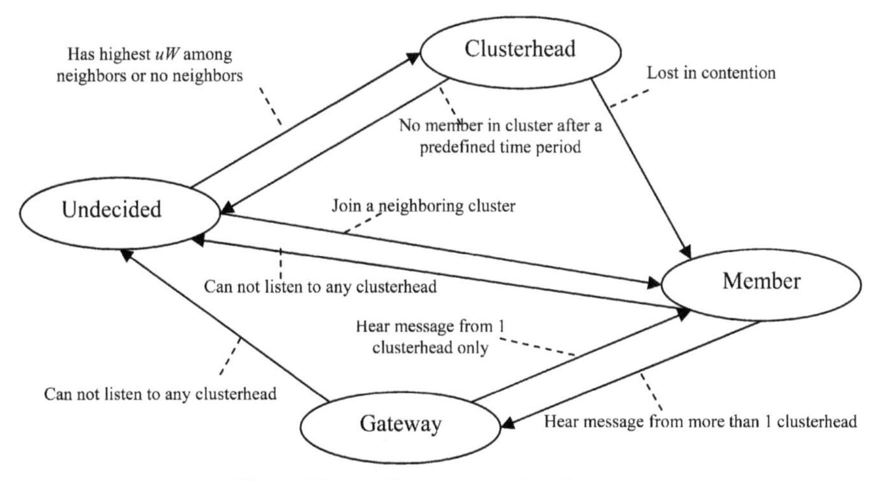

Figure 20.1 Node state transition diagram.

node responds locally to all topology changes by updating its role accordingly, if necessary. A stable and valid cluster structure will be reorganized efficiently after a certain convergence period.

In DISCA, only two types of clustering messages are used: *CLUSTER_CLAIM*(u) and *ACK_JOIN*(u, v). The *CLUSTER_CLAIM*(u) is used by node u to make its neighbors aware that it is going to be a clusterhead; the *ACK_JOIN*(u, v) is sent by node u to communicate to its neighbors that it will be a member of the clusters whose clusterhead is node v.

After exchanging the periodical beacon messages via a simple neighbor discovery protocol and some local calculation, every node a knows its own ID id(a), degree deg(a), current status Status(a), all neighboring nodes i: MD(i) (moving direction), MP(i) (mobility pattern), PD(i) (projected distance), LD(i) (leadership duration – valid for clusterhead only, 0 otherwise, and a special case is that the duration is set to 0 if a cluster has only one node). Firstly, DISCA uses MD to filter out any nodes in one's transmission range that are moving away – thereby reducing a large number of irrelevant nodes to make the algorithm run faster. The core part of DISCA is then evaluated by the following utility weight function uW:

$$\begin{aligned}\text{Utility Weight of node } i &:= uW(i)\\ &= \text{Primary}(\text{LD}(i)) + \text{Secondary}(\text{MP}(i), \text{PD}(i), \text{ID}(i))\end{aligned}$$

For clusterhead election, the LD value is taken to be the primary criterion. If it is not unique, the secondary unique criterion will break the tie (MP, PD, and ID). Precisely speaking, if a node has an LD value greater than any neighboring nodes, this node is going to initiate the claim to be a clusterhead. Otherwise, when both MP and PD threshold conditions hold, the node with the smaller PD value will claim clusterheadship. Finally, if no node has "good" MP and PD values, the one with the lowest ID value will claim clusterheadship. The reason why DISCA takes LD as the primary criterion to evaluate a cluster instead of using PD, MP, or other combinations is based on the fact that LD better reflects the characteristics of good and stable connections among all of its neighbors. In addition to LD, the utility weight of DISCA assures that nodes passing the threshold value of both MP(i) and PD(i) have higher priorities to be considered as clusterhead candidates than others. In fact, it imposes the node mobility features such as MP and PD on cluster stability. Therefore, after the above steps, DISCA keeps those stable candidates for election, and then the most stable one with the largest uW value will be elected. Detailed descriptions of all procedures in DISCA are shown in Figure 20.2.

On the Cluster_Setup Procedure

During both the initial cluster setup and when a node joins the network, the *Cluster_Setup* procedure is executed in order to determine the role of a node. Given the fact that nodes initially start this procedure asynchronously at the time they join the network, it is always true that each node's initial uW value depends on its neighboring configuration at the starting time. If a node has existing clusterheads in its neighborhood, it will join the one with the highest uW value (sends out an *ACK_JOIN* message). Otherwise it will initiate the claim to be a clusterhead (sends out a *CLUSTER_CLAIM* message). If a node becomes a clusterhead without any member node, i.e., a lone cluster, its LD value remains at 0, because the leadership duration of being a single node cluster does not have anything to do with neighbor connection and stability. When a member node joins it, its LD timing value is initiated. After a node decides its initial status, the subsequent actions will be handled by the following *Cluster_Maintenance* Procedure.

On the Cluster_Maintenance Procedure

Cluster topology changes, caused by node mobility, require the continual maintenance of node status. In DISCA, we have several subroutines that have been carefully designed to cope with this problem.

Cluster_Maintenance	

Actions on Neighbor_Loss
Node v is aware of the loss of its neighbor node u.
eH(v): The existing other clusterheads in node v's neighbors
 	
{
IF (Status(v) == clusterhead)
IF (Status(u) == member of v)
Remove u from v's member list;
 	
ELSE IF(Status(v) == member) {
IF(Status(u) == clusterhead)
IF (eH(v) !=
SendACK_Join(v,max(*uW*(i))); {i∈eH}
ELSE {
Status(v) = clusterhead;
CLUSTER_CLAIM(v);
}
}
}
	Actions on Neighbor_Establishment
	Node v is aware of the presence of a new neighbor node u.
	Fresh(v): Indicator of degree of freshness of node v's clusterhead, which depends on the value of Associativity Ticks.
	tol: The tolerance threshold for utility weight.
	{
	IF (Status(v) == clusterhead){
	IF (Status(u) == clusterhead)&&
	(*uW*(v) < *uW*(u)){
	Status(v) = member;
	my_cid = u;
	Send *ACK_JOIN*(v, u);}
	}
	ELSE IF(Status(v) == member) {
	IF((Status(u) == clusterhead)&&
	(*uW*(u) > *uW*(v) + *tol*) &&
	(!Fresh(v))){
	my_cid = u;
	Send *ACK_JOIN*(v, u);}
	}
	}

Actions on receiving CLUSTER_CLAIM
Node v receives a message from node u:
IF (on receiving *CLUSTER_CLAIM*(u)){
execute as a Neighbor_Establishment routine
} //end of *CLUSTER_CLAIM*
	Actions on receiving ACK_JOIN
	Node x receives ACK_JOIN(v, u):
	IF (on receiving *ACK_JOIN*(v, u)) {
	IF(my_id == u) // node x is clusterhead u
	Add v to u's member list.
	Otherwise execute a Neighbor_Loss routine
	}//end of *ACK_JOIN*

Figure 20.2 Cluster setup and maintenance procedures.

Neighbor_Loss: Whenever a given node v is made aware of the loss of a neighboring node u, the former checks if its own role is clusterhead and if u used to belong to its cluster. If this is the case, v removes u from its member list. If v is a member node and u was its clusterhead, then it is necessary for v to determine its new clusterhead. In order to preserve the current cluster structure as much as possible, v will first try to join to an existing clusterhead with the highest uW value (if a clusterhead c is in v's neighborhood, then c

certainly has a higher uW value in terms of LD than v's LD at this point). If it fails, v will initiate the process of claiming clusterheadship.

Neighbor_Establishment: When node v, either a member or clusterhead, is aware of the presence of a new neighbor u, it checks if u is a clusterhead. If this is the case, v should compare u with its current clusterhead, and then make a choice based on their uW values. It is clear that, depending on the degree of mobility in the network, there can be frequent cluster restructuring. However, in order to stabilize the existing cluster by minimizing the number of mobility-dependent restructurings, the maintenance procedure of DISCA operates on two additional parameters tol and fresh. A member node will consider switching to a newly arrived/established clusterhead only when the leadership duration value of the new clusterhead exceeds that of current clusterhead by a tolerance threshold of *tol*. Additionally, according to the periodic beacon mechanism, every node maintains an Associativity Tick (AT) field of its neighbors. The AT value of that entry is reset to a positive number t whenever a node receives a beacon or *CLUSTER_CLAIM* message from that node. If it is not received within the stipulated time, its AT is decremented. When AT becomes zero, the entry will be deleted. Hence, the AT value is a measurement for the degree of freshness, and a member will stick to its current clusterhead if it is still quite "fresh". After considering both parameters, a member node will decide to switch its cluster only when the newly clusterhead is significantly better than the current one. In DISCA, only nodes with good stability are considered as clusterhead candidates. Therefore the opportunities for a *Neighbor_Establishment* event happening between two clusterheads are rare. When it happens, in order to keep the independence property of a valid cluster structure, the one with a lower uW value will resign and send out a *ACK_JOIN* message, and its members will follow the procedure to either join an existing cluster or initiate the process of claiming clusterheadship.

The actions on receiving *CLUSTER_CLAIM* and *ACK_JOIN* messages are similar to what happens for *Neighbor_Establishment* and *Neighbor_Loss*. For instance, on receiving *ACK_JOIN*(v, u), a node will check its own ID, if it is node u, then it adds node v to its member list.

20.4 A DISCA-Based Broadcasting Protocol

DISCA is able to form and maintain a stable cluster structure. Related clustering work on MANETs shows that stable clusters greatly reduce the overall control overhead for many services (routing, broadcasting). However, VANETs introduce a number of features distinct from MANETs. This raises the question regarding whether clustering improves the quality and performance of similar services in VANETs.

In both MANET and VANET applications, broadcasting is a fundamental service in which essentially a mobile node sends a packet to all the nodes in the network. In the literature, it is also referred to as *flooding*. This simple broadcasting protocol has the infamous "broadcast storm problem" [21]. Much research has been done to design efficient protocols in MANETs [21, 27, 30]. However, not much emphasis is given to algorithms that are specifically designed based upon the unique features of VANETs. In this section, a broadcast protocol based on DISCA is presented for the purpose of validating and demonstrating the effectiveness of our clustering algorithm.

20.4.1 Some Existing Broadcasting Protocols

Many efficient protocols in MANETs have been proposed to alleviate the broadcast storm problem with the main goal of finding a small subset of nodes to rebroadcast the packets while trying to keep a large broadcast coverage area.

The probability-based protocol from [21] is similar to simple flooding, except that nodes only rebroadcast with a predefined probability. In a location-based protocol [21], each node uses a more accurate method to determine its location such as GPS. A node receiving a broadcast packet calculates the additional coverage area by utilizing the exact location of the sender. If this is less than a pre-defined threshold, it will drop the packet without re-broadcasting. Many other protocols, such as counter-based, and area-based have also been introduced [30] to alleviate the broadcast storm problem.

The above protocols are based on statistical and topological modeling to estimate the coverage area. The clustering concept has been adopted to derive this protocol on graph-based models. As shown in Figure 20.3, the clusterheads and gateways form the backbone infrastructure of the network. Gateways are member nodes with links to other neighboring clusterheads, so they can directly access neighboring clusters and forward information between clusters.

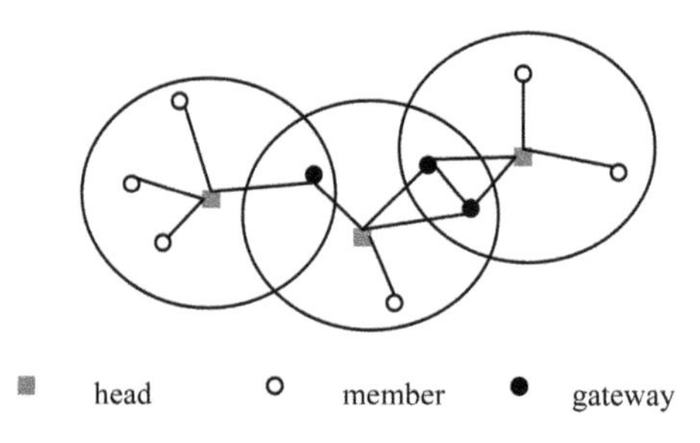

Figure 20.3 The cluster-based broadcasting.

20.4.2 Characteristics of Broadcasting in VANETs

VANETs significantly differ from MANETs, since vehicles have higher mobility than mobile nodes. Furthermore, vehicles move in specific directions on road networks, while ordinary mobile nodes move freely in all directions. Therefore, this leads to different design objectives of broadcasting algorithms.

Many new technologies have been studied for ITS to improve traffic condition and vehicle safety. Methods for broadcasting emergency information have drawn much attention recently. For instance, traffic congestion is a serious problem in many big cities. Even emergency vehicles such as ambulances, police vehicles and fire engines can be trapped in traffic jams, which could lead to loss of life or injury. Drivers either do not hear the siren of an emergency vehicle or cannot determine quickly where it comes from or the best way to let it through. Notifying drivers in advance of the approach of an emergency vehicle via IVC in VANETs would assist drivers to make a prompt reaction and correct movement on the road for an emergency vehicle. Another example is about traffic accidents. When an accident occurs, an accident warning message should be broadcasted via IVC to all approaching vehicles immediately to avoid secondary accidents. Therefore, an efficient broadcasting algorithm can provide instant information in such situations where the roadside/central communication infrastructure is unavailable or infeasible to respond immediately. In this research, a novel broadcast algorithm for VANETs is proposed, which is based on our DISCA clustering.

As discussed in the last section, with the increasing interest in VANETs, a number of related research efforts have emerged. In [28], a flooding based protocol for IVC in VANETs is proposed, in which short packets are used through broadcast communication. Furthermore, the protocol limits the channel access rate at each vehicle by defining a transmission window. In [9], an IEEE 802.11b protocol adapted for broadcast communication in IVC is introduced. It uses a distance based delay time mechanism to determine the waiting time for each vehicle before it rebroadcasts the received message. Further explanations around a concept of this mechanism will be discussed later. Also, an IEEE 802.11 based Urban Multi-hop Broadcast protocol (UMB) for IVC is proposed in [20], which consists of directional and intersection broadcast. In the directional broadcast, a source vehicle selects a furthest vehicle on a road segment to be the forwarding vehicle by sending additional signals. When vehicles arrive at a road intersection, then intersection broadcast sends packets in each road segment constitutes this intersection. Afterward, these packets are forwarded directionally through repeaters installed at the starting position of each road segment. However, UMB is an expensive broadcast protocol specifically designed for VANETs with infrastructure support because it requires additional pre-installed repeaters for packet relay. Furthermore, all these proposed protocols still suffer from problems such as hidden nodes, collisions at high packet traffic rates, reliability, and broadcast storms.

20.4.3 DISCA-Based Broadcasting Protocol

Assume that clusters have been formed in an ad-hoc network and will be controlled by our clustering procedure (DISCA). The rebroadcast from the clusterhead is going to cover all of its members if no collisions happen and gateways take the responsibility of transmitting packets to other clusters. Therefore, it is unnecessary for a non-gateway member to rebroadcast the packet.

In this section, a broadcasting protocol is proposed based on DISCA. In this protocol, the main objective is to find a small subset of re-broadcasting nodes (clusterheads and gateways) in the network. Notice that the general cluster-based broadcasting protocol may still have too many re-broadcasting nodes. Therefore, two additional pruning heuristics adaptive to VANETs are applied to further reduce the number of rebroadcasts while, simultaneously, improving broadcasting performance.

(1) *Forwarding heuristic*
A message notifying other vehicles of emergency information is transmitted by the source vehicle and relayed by others. Then a vehicle receiving this information can determine the presence, location and forwarding direction of the message. Vehicles in a particular area or direction need this information,

consisting of messages such as "emergency vehicle approaching" or "traffic accident ahead". In a conventional broadcast, this information is relayed and notified in all directions. In the direction of or in areas where vehicles do not need it, redundant notifications of this type of information can distract the drivers' attention and cause unnecessary accidents or problems. Therefore, a heuristic is proposed for relaying traffic information only toward the related direction or area. In this type of broadcasting message, the source vehicle transmits broadcast message including its own location, only in the intended forwarding direction. In case of the rise of unexpected direction changes, the moving direction of a vehicle may change drastically, and as a result this vehicle fails to deliver traffic messages because its current moving direction does not match the forwarding direction of a message. Therefore, this vehicle can use its traveling road information instead of its direction as forwarding heuristic.

A simple forwarding broadcast protocol based only on this heuristic without clustering structure is implemented to serve as a baseline for comparing the DISCA-based broadcast protocol.

(2) *Farthest gateway heuristic*

Utilizing the forwarding heuristic described above, a lot of unnecessary network traffic can be avoided. However, all gateways on the forwarding direction can still relay the same message almost simultaneously. In fact, only the farthest one is needed to relay the message. The reason for this is:

- The farthest gateway reaches the largest uncovered area in the forwarding direction.
- It is able to cover the area which is covered by other gateways in between.

Even though the additional overhead to managing clustering is a concern, simulation results show that the obtained benefits from DISCA-based protocol significantly offsets the additional overhead. Note that when no gateway exists for a cluster, the clusterhead will relay a message to its farthest node on the forwarding path, and then this node will try to send the message to the next neighboring cluster if existed.

The procedure of the protocol is presented in Figure 20.4.

```
(Each vehicle is able to generate a packet with a forwarding direction.)
Forwarding procedure

//invoked on receiving a packet
IF (isForwardingDirection & isUnseenPacket){
        IF(the node is clusterhead)
         Rebroadcast the packet immediately
         Update packet's sender ID
        ELSE IF (the node is gateway)
                IF (isFarthestGateway)
                 Rebroadcast the packet immediately
                 Update packet's sender ID
}
// The IsForwadingDirection function checks if a node is in the forwarding direction of a
packet.
// The isFarthestGateway function checks whether a gateway is the farthest gateway of a
clusterhead by comparing the relative distance of all gateways on the packet's forwarding
direction.
// The isUnseenPacket function checks the node's packet buffer to make sure that it is the
first time for a node to receive the packet.
// The function "Similar (V1, V2)" takes two vectors as parameter and apply dot product to
determine if the difference angle is within a predefined degree.
```

Figure 20.4 Broadcasting protocol.

20.5 Simulation Study

20.5.1 Traffic Simulation Model

The majority of existing MANET simulation models are developed apart from traffic network simulation and only consider random movements in fixed field areas [2, 6, 31]. It is therefore necessary to improve the awareness of the model with respect to the transportation network features. On the other hand, there are a few simulators (e.g., the NS-2 network simulator [NS-2, 2005]) that can partially address this concern by incorporating real network map data into the model [26]. However, the major limitation that remains comes from the fact that models that deal with wireless communication techniques do not account for vehicle-to-vehicle interactions in traffic flow. Therefore, the wireless communication between vehicles and the outcomes of these models can be impractical and unreliable.

In order to overcome this problem, an open-source simulator, SWANS++ [13], has been used to integrate both wireless network and traffic control models. Incorporated with TIGER/LINE data and traffic flow simulation, the model is implemented as an extension to JIST/SWANS [5], a scalable, high-performance and scalable discrete-event wireless network simulator featuring low memory consumption and fast run times.

20.5.2 Testing Scenarios and Traffic Parameters

Both clustering algorithms and broadcasting protocols have been examined on the highway network of a portion of Cook County, Illinois that includes a portion of the city of Chicago. According to vehicular mobility models of the simulator [13], a simulation map (Figure 20.5) showing the study area and demonstrating major highway/arterial network traffic has been carefully designed. The simulation network represents current conditions of the highway system where there are potentials for many VANET applications to be deployed in the near future. Therefore, all clustering algorithms are accommodated and evaluated under highway/arterial network traffic conditions in the study area.

The network simulator includes two modules that determine the travel path taken by each vehicle for the duration of the simulation: simple inter segment mobility (Simple) and mobility with origin-destination pairs (OD). In the Simple module, the next road to which a vehicle will move is determined stochastically at each intersection. In the OD pair module, the routing decision is based on the shortest path between the vehicle's specified origin and destination that is computed for each vehicle before leaving the origin point. In this study, the OD module has been implemented to improve the accuracy of the vehicular movement representation in the network. Origin and destination locations are assumed to be the centroids of Traffic Analysis Zones (TAZ) in the region and vehicles are assigned to travel paths between pairs of origin-destination zones.

20.5.3 Vehicular Traffic Data

In this study, actual counts of vehicle flow between pairs of origin-destinations were extracted from a regional travel demand model and assigned to travel routes between origin-destination zones in the regional highway network. However, zonal centroids can be off-grid and many TAZs can be external to the simulation network. Centroids of TAZs in this study are positioned adjacent to the closest highway and all trips originated from or destined to TAZs that are located outside of the simulation network are aggregated into external superzones. While input data in a simulation framework plays a pivotal point, it is important to consider the effects of a random

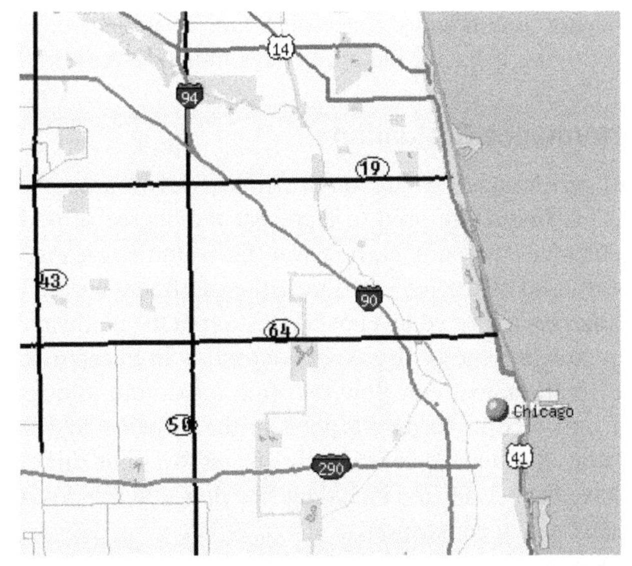

Figure 20.5 Simulation area.

traffic flow on performance of the algorithm. Hence, in order to make simulation results more reliable, realistic traffic count data (over 16,000 trips) has been imported to the simulation model. The data set was extracted from typical weekday origin-destination trip matrices in the CMAP's (Chicago Metropolitan Agency for Planning) regional travel demand model.

The Zone data file obtained from CMAP contains Zone ID, and the longitude of TAZ centroids, and latitude information while the OD trip data file contains a list of OD pairs and trip volume data between those TAZs.

The data processing procedure first extracts the related zones and their coordinates of the simulated area, and then obtains the trip data of these TAZs. Secondly, aggregations of all trip data OD pairs are performed. Finally, vehicles are assigned to the network based on the demand (rate) between each OD pairs.

It should be noted that while the algorithm is tested in a simulation model with a real highway network with actual traffic data, the simulation process used in this study has some inherent limitations including the assumption of uniform temporal distribution of travel demand in a typical weekday, restricting origin and destination points to centroids of TAZs in the network, and some simplifying assumptions concerning lane changing, gap keeping and other related driving behavior factors in the simulation process.

20.5.4 Performance Evaluation

In a standard performance evaluation of wireless networking, the term "number of nodes" is frequently used to represent the degree of node density. In a general vehicular environment, however, the volume-to-capacity ratio (v/c ratio) has been used as the preferred measure of effectiveness. This measure can characterize the expected level of congestion in the highway link which is a natural way to represent the degree of congestion in transportation network. The capacity is the maximum flow rate (the maximum allowed number of vehicles per unit of time to pass a point in the system), and the volume is the current flow rate in that location. In this study, four different levels of congestion have been evaluated including free flow, low congestion, moderate congestion, and heavy congestion.

The simulation results represent the performance of each algorithm across different v/c ratios (moderate congestion by default) and wireless transmission ranges (250 m by default). All other simulation parameters are using the default values in SWANS++ (an 802.11b MAC protocol operating at 2 Mbps, a generic path loss model with shadowing, an exponent 2.8 and standard deviation 6.0 to determine signal strength at the receiver, etc.) [13] and the simulation duration is held constant across all testing cases to reach consistency. To minimize traffic variability among simulations as well as to enable repeatable test results, the randomized features of the model are seeded with the same value at each simulation run. Finally, the results presented here are collected on average through all scenarios by multiple runs.

20.5.4.1 Clustering Performance

Five clustering algorithms (Lowest-ID, D-LID, Highest-Degree, D-HD, and DISCA) have been examined. Algorithms with prefix "D" indicate the directional heuristic proposed in DISCA is added to their original versions. For example, D-LID is the modified Lowest-ID such that all nodes in a cluster have a similar moving direction.

In a highly dynamic VANET, the stability of the cluster structure is a major concern. A desirable clustering algorithm should form a stable structure with a manageable size (it is guaranteed by DISCA's one-hop neighbor requirement). In other words, the number of node status changes from one

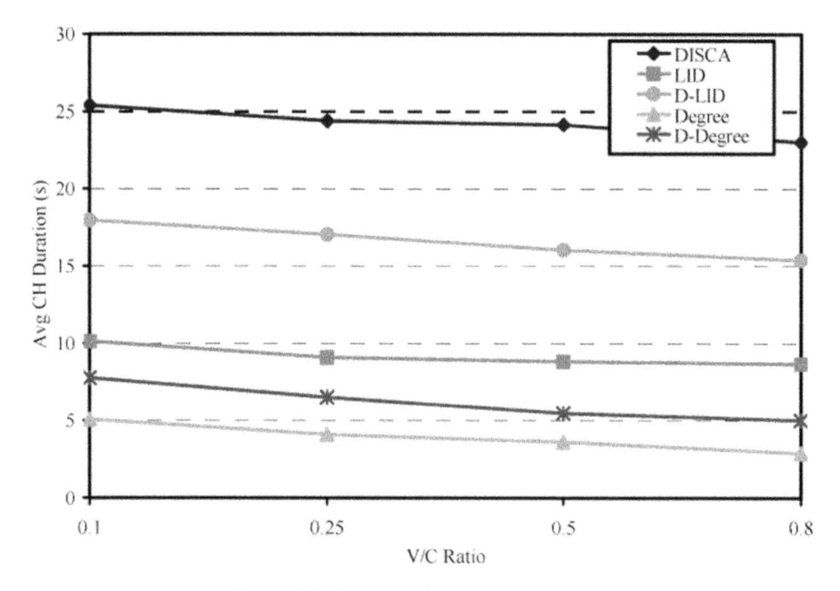

Figure 20.6 Average leadership duration.

cluster to another should be minimized. Two status changes events involved that impact cluster instability are:

1. A member or undecided node becomes a clusterhead.
2. A member node or clusterhead leaves its cluster and joins another cluster.

To evaluate the stability, three standard metrics have been considered: clusterhead leadership duration, member duration and number of status changes. Clusterhead duration is defined as the mean time period during which a clusterhead remains in its leadership role. Cluster member duration is the average time a node stays as a member of a cluster continuously, before moving to another cluster or becoming a clusterhead. The number of status changes per unit time for a node is used to reflect the frequency of cluster changes, which is the sum of these above two events.

Figures 20.6 and 20.7 summarize the variation of the average clusterhead and member duration with respect to the traffic congestion level. Notably, our DISCA algorithm improves both clusterhead and member duration at the same time.

Figure 20.8 shows the number of status changes per node per unit time. As expected, DISCA has the least number of status changes. A further increase

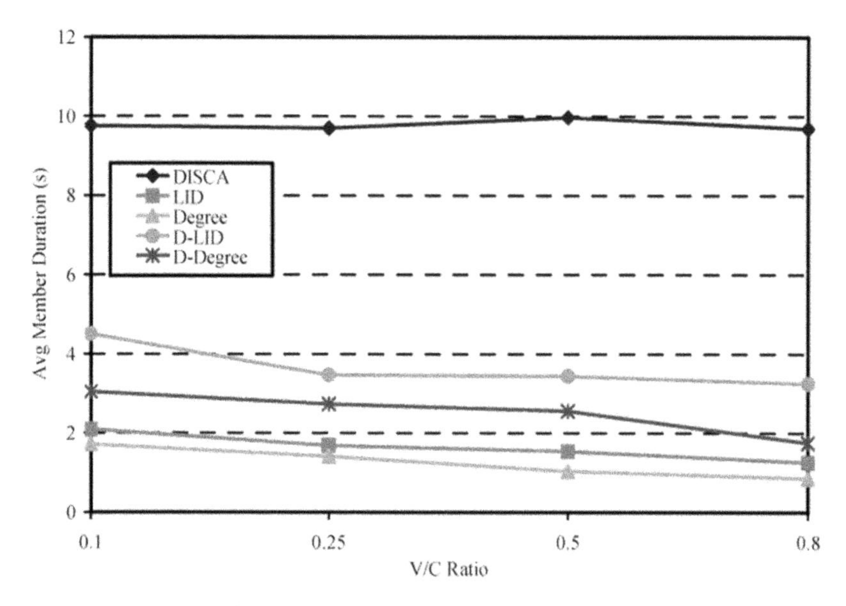

Figure 20.7 Average membership duration.

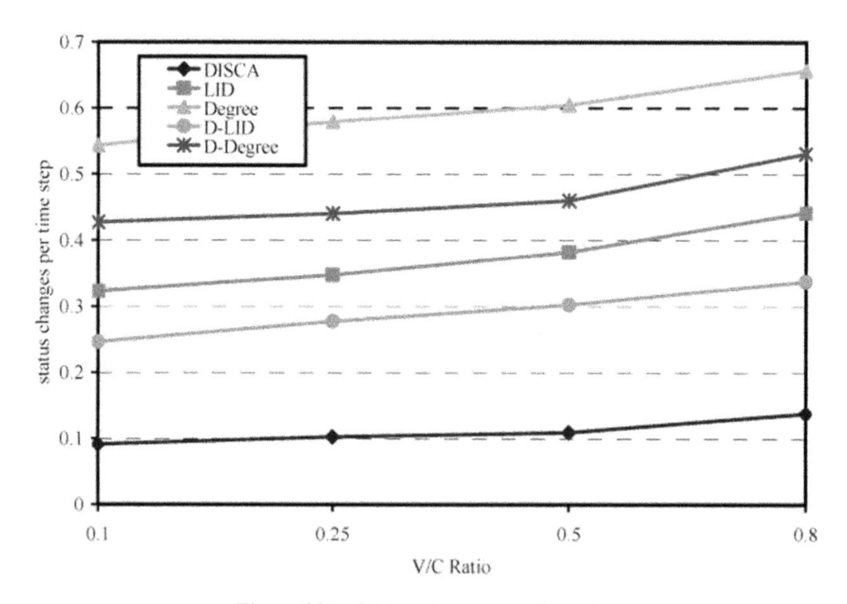

Figure 20.8 Status changes per time step.

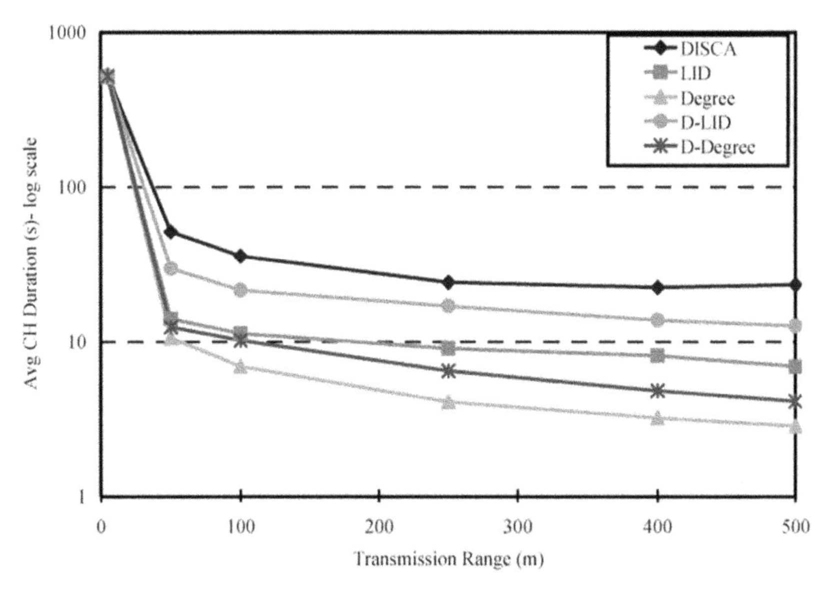

Figure 20.9 Leadership duration vs. transmission range.

in the congestion level leads to an increase in the number of changes since the nodes tend to meet "better" clusters. Figure 20.9 shows the effect of a varying transmission range on the average clusterhead duration. The performance of each algorithm is consistent with the results shown in Figure 20.6. When the range is close to 0, every node keeps its clusterhead status due to low network connectivity. As the range increases, clusters start to include more nodes, thus increasing the probability of a transition from clusterhead to member node, reducing the average clusterhead duration time.

20.5.4.2 Broadcasting Performance

Four protocol experiments (flooding, probability-based, forwarding heuristic, and our cluster-based) have been conducted. The broadcast packet size is 64 bytes and the default broadcast transmission rate is set to be 1 packet/sec. The metrics used in this chapter are reachability, rebroadcasts, end-to-end delay, normalized cost, and bandwidth utilization.

The number of nodes in the network receiving the broadcast packets divided by the total number of nodes that are reachable (directly or indirectly) is the definition of reachability. As shown in Figure 20.10, the reachability of our DISCA protocol remains high even under high traffic congestion levels.

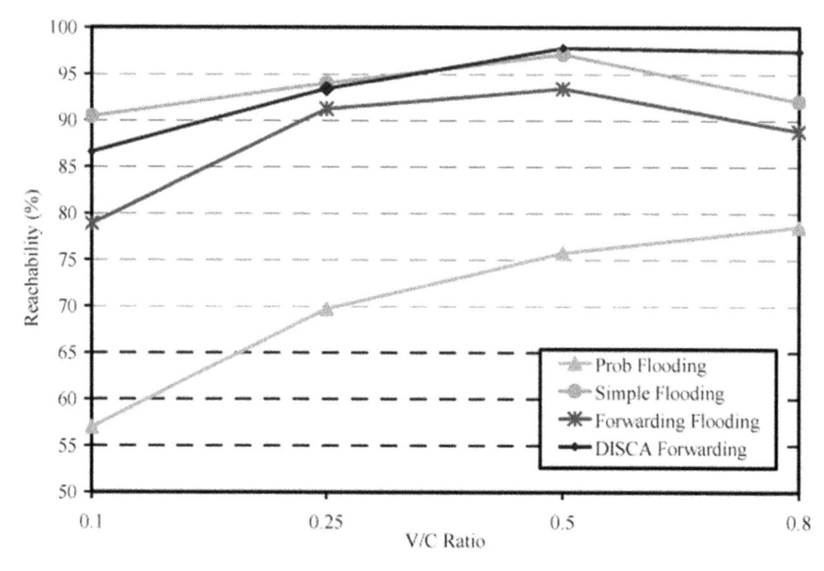

Figure 20.10 Reachability.

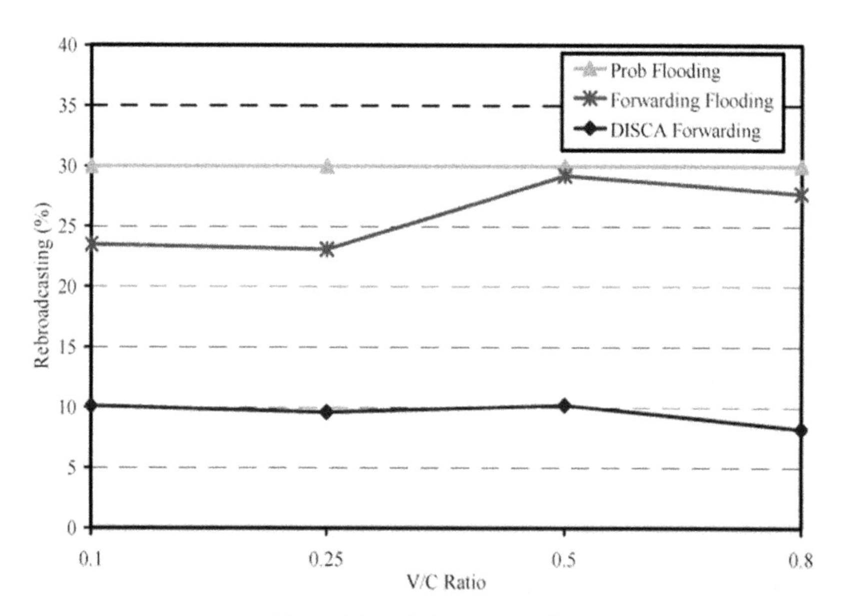

Figure 20.11 Rebroadcast ratio.

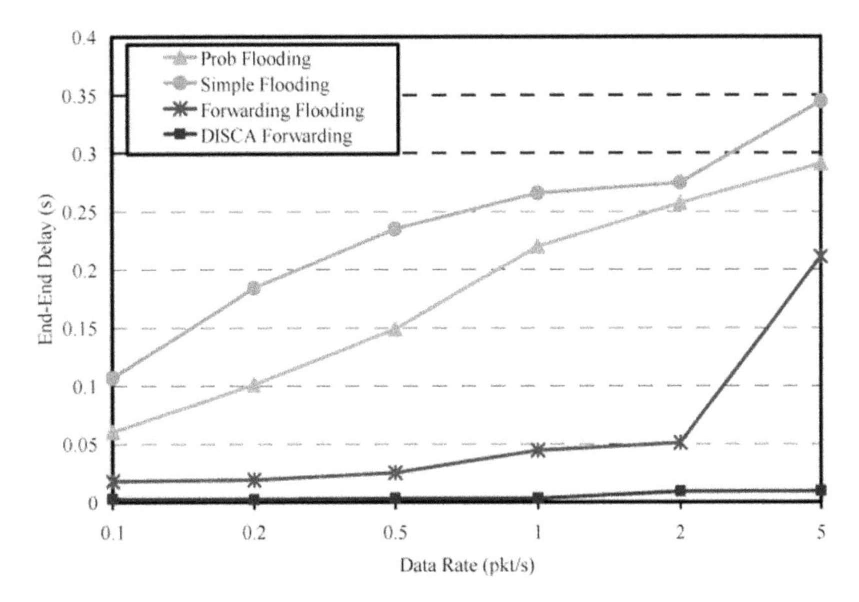

Figure 20.12 End-end delay.

Under the same conditions, however, the flooding (which has the best reachability in a low congested network) and forwarding algorithms experience the infamous broadcast storm problem.

As shown in Figure 20.10, the reachability reflects the coverage area and delivery ratio. In order to alleviate the storm problem, the number of rebroadcasting nodes is critical to evaluate a protocol. The simulation results in Figure 20.11 show our DISCA protocol outperforms others by saving the most rebroadcasts while keeping a desirable reachability.

The end-to-end delay indicates the average delay of a packet transmitted throughout the network, which is one of the most important metrics in many VANET applications, such as traffic accident notification, multimedia streaming, etc. As shown in Figure 20.12, the two forwarding methods (i.e., Forwarding Flooding and DISCA Forwarding) have shorter delays due to fewer collisions and less contention.

Unlike other non-cluster based algorithms in which broadcasting cost only consists of rebroadcasts, our algorithm requires additional control messages such as the periodical beacon and two types of clustering messages (*CLUSTER_CLAIM* and *ACK_JOIN*). The normalized cost ratio of an algorithm is defined as the additional number of messages required per node

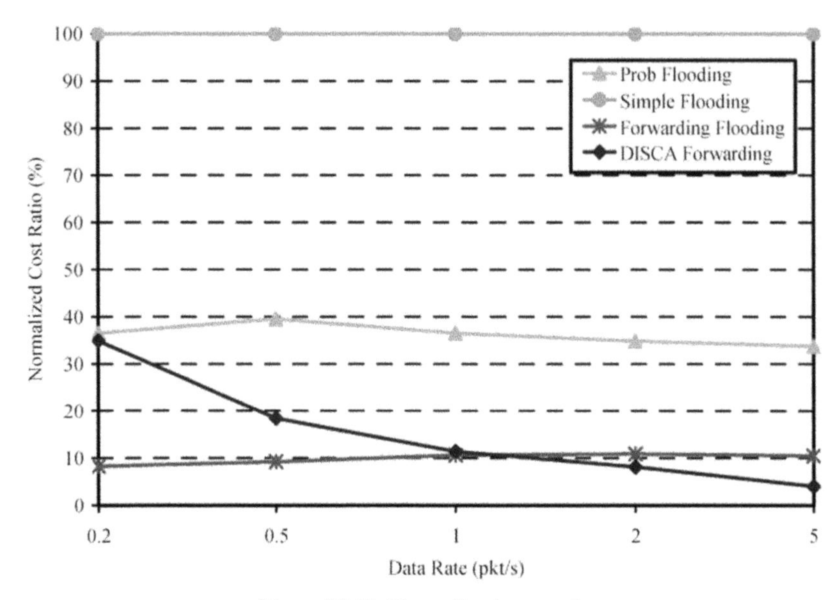

Figure 20.13 Normalized cost ratio.

per broadcast relative to the cost of the simple flooding algorithm. As shown in Figure 20.13, our algorithm outperforms others when the broadcasting rate is over 1 pkt/s because the number of rebroadcasts saved by ours offsets its clustering control cost. On the other hand, when the data rate is low, the overhead cost of our DISCA algorithm exceeds the cost of rebroadcast savings. Besides the networking overhead cost, our cluster-based DISCA algorithm also requires some additional computational costs. In this chapter, we make the assumption that on-board computation is not a limiting factor.

20.6 Conclusion

A distributed clustering algorithm, DISCA, is proposed in this chapter, which is based on the transmission of periodic beacons containing traffic-specific information about clustering. DISCA forms a stable cluster which is adaptive to node mobility. The formation of clusters in DISCA takes multiple factors into consideration, such as direction and mobility features. Simulation results show that it improves the stability of clustering in terms of both leadership and membership duration under a carefully designed realistic VANET environment.

For the purpose of demonstrating the effectiveness of DISCA, a cluster-based broadcasting protocol is presented. This broadcasting protocol utilizes a selective subset of clusterheads and gateways to rebroadcast a packet, thus significantly improving broadcast reachability and latency with a low number of rebroadcasts. More importantly, the results of the normalized cost and bandwidth utilization show that both low overhead and high efficiency are achieved by the purposed protocol when broadcast data rates are around 1 packet per second and higher. Such high broadcast rates may well occur under vehicle safety environments. Additional experiments with other types of applications and messaging traffic are also needed to classify when DISCA clustering algorithms might be advantageous.

The simulation methodology described here represented a realistic testing environment of VANETs. This work also shows how research may abstract important characteristics and reduce the overall variability and complexity of the traffic system for the purpose of modeling network and application behavior. In future research, other stability related features for clustering algorithms will be investigated, as well as the design of other VANET protocols (routing and resource discovery, etc.) on top of DISCA.

References

[1] K.M. Alzoubi, P. Wan, and O. Frieder. Distributed heuristics for connected dominating sets in wireless ad hoc networks. *Journal of Communications and Networks*, 4(1), 2002.

[2] S. Bandyopadhyay and E.J. Coyle. An energy efficient hierarchical clustering algorithm for wireless sensor networks. In *Proceedings of IEEE INFOCOM'03*, 2003.

[3] S. Banerjee and S. Khuller. A clustering scheme for hierarchical control in multi-hop wireless networks. In *Proceedings IEEE INFOCOM'01*, April 2001.

[4] D.J. Baker and A. Ephremides. The architectural organization of a mobile radio network via a distributed algorithm. *IEEE Transactions on Communications*, 29(11):1694–1701, 1981.

[5] R. Barr, Z.J. Haas, and R. Renesse. JiST: An efficient approach to simulation using virtual machines. *Software Practice & Experience*, 35(6):539–576, 2005.

[6] S. Basagni. Distributed clustering for ad hoc networks. In *Proceedings of the 1999 International Symposium on Parallel Architectures, Algorithms and Networks (I-SPAN'99)*. IEEE Computer Society, Australia, pages 310–315, 1999.

[7] S. Basagni. Finding a maximal weighted independent set in wireless networks. *Telecommunication Systems*, 18(1/3):155–168, 2001.

[8] S. Basagni, I. Chlamtac, and A. Farag. A generalized clustering algorithm for peer-to-peer networks. In *Proceedings of Workshop on Algorithmic Aspects of Communication, Satellite Workshop of ICALP'97*, Bologna, Italy, 1997.

[9] L. Briesemeister and G. Hommel. Role-based multicast in highly mobile but sparsely connected and hoc networks. In *Proceedings of IEEE/ACM Workshop on MobiHoc*, Boston, pages 45–50, August 2000.

[10] C. Bettstetter. The cluster density of a distributed clustering algorithm in ad hoc networks. In *Proceedings of IEEE International Conference on Communications (ICC)*, Paris, France, 2004.

[11] M. Chatterjee, S. Das, and D. Turgut. WCA: A weighted clustering algorithm for mobile ad hoc networks. *Journal of Cluster Computing* (Special Issue on Mobile Ad hoc Networks), 5:193–204, 2002.

[12] C.C. Chiang, H.K. Wu, W. Liu, and M. Gerla. Routing in clustered multihop, mobile wireless networks with fading channel. In *Proceedings of IEEE Singapore International Conference on Networks (SICON)*, 1997.

[13] D. Choffnes and F. Bustamante. An integrated mobility and traffic model for vehicular wireless networks. In *Proceedings of the 2nd ACM International Workshop on VANET*, 2005.

[14] B. Das, R. Sivakumar, and V. Bharghavan. Routing in ad-hoc networks using a spine. In *Proceedings of International Conference of Computers and Communications Networks*, Las Vegas, 1997.

[15] DSRC. http://wireless.fcc.gov/services/its/dsrc/, 2004.

[16] A. Ephremides, J. Wieselthier, and D. Baker, A design concept for reliable mobile radio network with frequency hopping signaling. In *Proceedings of IEEE 75*, pages 56–73, 1987.

[17] M.L. Garey and D.S. Johnson, *Computers and Intractability: A Guide to the Theory of NP-Completeness*, W.H. Freeman, San Francisco, 1979.

[18] M. Gerla and J. Tsai. Multicluster, mobile, multimedia radio network. *Wireless Networks*, 1(3):255–265, 1995.

[19] A. Helmy. Architectural framework for Large scale multicast in mobile ad hoc network. In *Proceedings of ICC 2002*, New York, USA, 2002.

[20] G.E. Korkmaz, E. Ekici, F. Ozguner, and U. Ozguner. Urban multi-hop broadcast protocols for inter-vehicle communication systems. In *Proceedings of First ACM Workshop VANET 2004*, pages 76–85, October 2004.

[21] S.Y. Ni, Y.C. Tseng, Y.S. Chen, and J.P. Sheu. The broadcast storm problem in a mobile ad hoc network. In *Proceedings of MOBICOM 1999*, pages 151–162, 1999.

[22] F.G. Nocetti, J.S. Gonzalez, and I. Stojmenovic, Connectivity based k-hop clustering in wireless networks. *Telecommunication Systems*, 22(1/4):205–220, 2003.

[23] NS-2, www.isi.edu/nsnam/ns/, accessed 2005.

[24] A.K. Parekh. Selecting routers in ad hoc wireless networks. In *ITS*, 1994.

[25] J.H. Ryu, S.H. Song, and D.H. Cho. New clustering schemes for energy conservation in two-tiered mobile ad-hoc networks. In *Proceedings of IEEE ICC*, 2001.

[26] A. Saha and D. Johnson. Modeling mobility for vehicular ad-hoc networks. In *Proceedings of the First ACM VANET*, 2004.

[27] I. Stojmenovic, M. Seddigh, and J. Zunic. Dominating sets and neighbor elimination based broadcasting algorithms in wireless networks. In *Proceedings of IEEE Hawaii International Conference on System Sciences*, 2001.

[28] K. Tokuda, M. Akiyama, and H. Fujii. Dolphin for inter-vehicle communications systems. In *Proceedings of IEEE ITS*, pages 504–509, 2000.

[29] P. Wan, K.M. Alzoubi, and O. Frieder, Distributed construction of connected dominating set in wireless ad hoc networks. In *Proceedings of IEEE INFOCOM'02*, New York, 2002.
[30] B. Williams and T. Camp. Comparison of broadcasting techniques for mobile ad hoc networks. In *MOBIHOC 2002*, pages 194–205, 2002.
[31] O. Younis and S. Fahmy. HEED: A hybrid, energy-efficient, distributed clustering approach for ad hoc sensor networks. *IEEE Trans. Mob. Comput.*, 3(4):366–379, 2004.

Author Index

Subject Index

About the Editors

Anand R. Prasad, Ph.D. & Ir., Delft University of Technology, The Netherlands, CISSP, Senior Member IEEE and Member ACM, is a NEC Certified Professional (NCP) and works as a Senior Expert at NEC Corporation, Japan, where he leads the security activity in 3GPP. Anand is Member of the Governing Council of Global ICT Standardisation Forum for India (GISFI) where he also chairs the Green ICT group and founded the Security SIG. He has several years of professional experience in all aspects of wireless networking industry. Anand has applied for over 30 patents, has co-authored 3 books and authored over 50 peer reviewed papers in international journals and conferences. He is also active in several conferences as program committee member.

John Buford is a Research Scientist with Avaya Labs Research. He is co-author of the book *P2P Networking and Applications* (Morgan Kaufman, 2008), co-editor of the *Handbook of Peer-to-Peer Networking* (Springer-Verlag, 2009) and has co-authored more than 120 refereed publications. He has been TPC chair or TPC co-Chair of more than ten conferences and workshops. He is a member of the editorial board of the *Journal of Peer-to-Peer Networking and Applications*, the *Journal of Communications*, and the *International Journal of Digital Multimedia Broadcasting*. He is an IEEE Senior Member and is co-chair of the IRTF Scalable Adaptive Multicast Research Group. Dr. Buford holds the PhD degree from Graz University of Technology, Austria, and MS and BS degrees from MIT.

Vijay K. Gurbani works for the Security Technology Research Group at Bell Laboratories, the research arm of Alcatel-Lucent. He holds a B.Sc. in Computer Science with a minor in Mathematics, a M.Sc. in Computer Science, both from Bradley University; and a Ph.D. in Computer Science from Illinois Institute of Technology. Vijay's current work focuses on security aspects of Internet multimedia session protocols and peer-to-peer (P2P) networks. He is the author of over 45 journal papers and conference proceedings, 5 books, and 11 Internet Engineering Task Force (IETF) RFCs. He is currently the co-chair of the Application Layer Traffic Optimization (ALTO) Working Group in the IETF, which is designing a protocol to enable efficient communications between peers in a peer-to-peer system. Vijay's research interests are Internet telephony services, Internet telephony signaling protocols, security of Internet telephony protocols and services, and P2P networks and their application to various domains. Vijay holds three patent and has nine applications pending with the US Patent Office. He is a senior member of the ACM and a member of the IEEE Computer Society.

RIVER PUBLISHERS SERIES IN COMMUNICATIONS

Volume 1
4G Mobile & Wireless Communications Technologies
Sofoklis Kyriazakos, Ioannis Soldatos, George Karetsos
September 2008
ISBN: 978-87-92329-02-8

Volume 2
Advances in Broadband Communication and Networks
Johnson I. Agbinya, Oya Sevimli, Sara All, Selvakennedy Selvadurai, Adel Al-Jumaily, Yonghui Li, Sam Reisenfeld
October 2008
ISBN: 978-87-92329-00-4

Volume 3
Aerospace Technologies and Applications for Dual Use A New World of Defense and Commercial in 21st Century Security
General Pietro Finocchio, Ramjee Prasad, Marina Ruggieri
November 2008
ISBN: 978-87-92329-04-2

Volume 4
Ultra Wideband Demystified Technologies, Applications, and System Design Considerations
Sunil Jogi, Manoj Choudhary
January 2009
ISBN: 978-87-92329-14-1

Volume 5
Single- and Multi-Carrier MIMO Transmission for Broadband Wireless Systems
Ramjee Prasad, Muhammad Imadur Rahman, Suvra Sekhar Das, Nicola Marchetti
April 2009
ISBN: 978-87-92329-06-6

Volume 6
Principles of Communications: A First Course in Communications
Kwang-Cheng Chen
June 2009
ISBN: 978-87-92329-10-3

Volume 7
Link Adaptation for Relay-Based Cellular Networks
Başak Can
November 2009
ISBN: 978-87-92329-30-1

Volume 8
Planning and Optimisation of 3G and 4G Wireless Networks
J.I. Agbinya
January 2010
ISBN: 978-87-92329-24-0

Volume 9
Towards Green ICT
Ramjee Prasad, Shingo Ohmori, Dina Šimunić
June 2010
ISBN: 978-87-92329-38-7

Volume 10
Adaptive PHY-MAC Design for Broadband Wireless Systems
Ramjee Prasad, Suvra Sekhar Das, Muhammad Imadur Rahman
August 2010
ISBN: 978-87-92329-08-0

Volume 11
Multihop Mobile Wireless Networks
Kannan Govindan, Deepthi Chander, Bhushan G. Jagyasi, Shabbir N. Merchant, Uday B. Desai
October 2010
ISBN: 978-87-92329-44-8

Volume 12
Telecommucations in Disaster Areas
Nicola Marchetti
November 2010
ISBN: 978-87-92329-48-6

Volume 13
Single and Cross-Layer MIMO Techniques for IMT-Advanced
Filippo Meucci
January 2011
ISBN: 978-87-92329-50-9

Volume 14
Advances in Next Generation Services and Service Architectures
Anand R. Prasad, John F. Buford, Vijay K. Gurbani
April 2011 ISBN: 978-87-92329-55-4

For Product Safety Concerns and Information please contact our EU
representative GPSR@taylorandfrancis.com
Taylor & Francis Verlag GmbH, Kaufingerstraße 24, 80331 München, Germany

www.ingramcontent.com/pod-product-compliance
Ingram Content Group UK Ltd.
Pitfield, Milton Keynes, MK11 3LW, UK
UKHW021114180425
457613UK00005B/89
* 9 7 8 8 7 9 2 3 2 9 5 9 2 *